Evolutionism in Eighteenth-Century French Thought

Currents in Comparative Romance Languages and Literatures

Tamara Alvarez-Detrell and Michael G. Paulson
General Editors

Vol. 166

PETER LANG
New York • Washington, D.C./Baltimore • Bern
Frankfurt am Main • Berlin • Brussels • Vienna • Oxford

Mary Efrosini Gregory

EVOLUTIONISM IN EIGHTEENTH-CENTURY FRENCH THOUGHT

PETER LANG
New York • Washington, D.C./Baltimore • Bern
Frankfurt am Main • Berlin • Brussels • Vienna • Oxford

Library of Congress Cataloging-in-Publication Data
Gregory, Mary Efrosini.
Evolutionism in eighteenth-century French thought / Mary Efrosini Gregory.
p. cm. — (Currents in comparative Romance languages and literatures; v. 166)
Maillet—Montesquieu—La Mettrie—Buffon—Maupertuis—Diderot—
Rousseau—Voltaire—The controversy over whether apes can be taught to speak—Race.
Includes bibliographical references and index.
1. Philosophy, French—18th century. 2. Evolution—History—18th century.
3. Evolution (Biology)—History—18th century. I. Title.
B1911.G69   146'.7094409033—dc22   2008031563
ISBN 978-1-4331-0373-5
ISSN 0893-5963

Bibliographic information published by **Die Deutsche Bibliothek**.
Die Deutsche Bibliothek lists this publication in the "Deutsche
Nationalbibliografie"; detailed bibliographic data is available
on the Internet at http://dnb.ddb.de/.

Chapter 6 on Diderot copyright © 2007 from
Diderot and the Metamorphosis of Species by Mary Efrosini Gregory.
Reprinted by permission of Routledge, Inc., a division of Informa plc.

The paper in this book meets the guidelines for permanence and durability
of the Committee on Production Guidelines for Book Longevity
of the Council of Library Resources.

© 2008 Peter Lang Publishing, Inc., New York
29 Broadway, 18th floor, New York, NY 10006
www.peterlang.com

All rights reserved.
Reprint or reproduction, even partially, in all forms such as microfilm,
xerography, microfiche, microcard, and offset strictly prohibited.

Printed in the United States of America

Contents

Introduction 1

    1. Maillet 19
    2. Montesquieu 33
    3. La Mettrie 45
    4. Buffon 69
    5. Maupertuis 93
    6. Diderot 119
    7. Rousseau 143
    8. Voltaire 167
    9. The Controversy over whether Apes Can Be Taught to Speak 195
   10. Race 217

Conclusion 247

Notes 253

Bibliography 321

Index 335

Introduction

Nature proceeds from the inanimate to the animals by such small steps that, because of continuity, we fail to see to which side the boundary and the middle between them belongs.[1]

—Aristotle, *History of Animals* (350 BC)

The *Oxford English Dictionary* defines the chain of beings as "a conception of the universe as a continuous series or gradation of types of being in order of perfection, stretching from God as the infinite down through a hierarchy of finite beings to nothingness."[2] This theory of the hierarchical arrangement of nature is a straight line or two dimensional, ranging from the most complex to the simplest creature.

Two features of the universe that are associated with the chain of beings are plenitude and continuity. The principle of plenitude states that the universe is "full," exhibiting the maximum diversity of kinds of beings; everything possible (ie: not self-contradictory) is eventuated. The principle of continuity asserts that the universe is comprised of an infinite series of forms, each of which shares with its neighbor at least one attribute.

Aristotle rejected the notion of plenitude: he did not think that the Unmoved Mover (God) was under any obligation to create every imaginable variation of living thing.[3] Aristotle refutes the principle of plenitude in *Metaphysics*: "the causes of things are not infinitely many either in a direct sequence or in kind. For the material generation of one thing from another cannot go on in an infinite progression...nor can the source of motion be moved (e.g. man be moved by air, air by the sun, the sun by Strife, with no limit to the series). In the same way neither can the Final Cause recede to infinity,"[4] "In both cases progression to infinity is impossible,"[5] and "for that which has a potentiality may not actualize it."[6]

Although Aristotle rejected the principle of plenitude, he did promote the idea of continuity. In *History of Animals*, he declared that nature passes from the inanimate to the plant kingdom and from the plant to the animal, in infinitely minute, indistinguishable gradations. Lovejoy translates Aristotle thus: "Nature passes so gradually from the inanimate to the animals that their continuity renders the boundary between them indistinguishable; and there is a middle kind that belongs to both orders. For plants come immediately after inanimate things; and plants differ from one another in the degree in which they appear to participate in life. For the class taken as a whole seems, in comparison with other bodies, to be clearly animate; but compared with animals to be inanimate. And the transition from plants to animals is continuous; for one might question whether some marine forms are animals or plants, since many of them are attached to the rock and perish if they are separated from it."[7]

In *Parts of Animals*, Aristotle continues to expound on the principle of continuity. Seals and bats are in intermediary positions on the chain: seals are land and water animals in one; bats are animals that live on the ground and also fly and so one might say that they belong to both groups or to neither; some mammals are bipeds and others are quadrupeds and so mammals, too, belong to both groups or to neither.[8]

The great chain of beings and its corollaries, plenitude and continuity, were declared and frequently reiterated by Buffon in his epic 44 volume *Natural History, General and Particular*. Buffon made a statement as to the plenitude of the chain of beings: "All that can be, is" [*tout ce qui peut être, est*]. This was accepted as true by Maupertuis, Diderot, and the materialists and it had far reaching consequences, shaping the thought of eighteenth-century biology.

In 1749 Buffon defined the chain of being as a straight line comprised of infinitely minute gradations: if man, "methodically and in succession, glances through the different objects that comprise the Universe, and places himself at the head of all created beings, he will see with astonishment that one can descend by almost imperceptible degrees, from the most perfect creature to the most unformed matter, from the most organized animal to the crudest matter..."[9] Buffon proposed a two-dimensional straight line extending from God down to inanimate matter).

That year Buffon also criticized Linnaeus' efforts to categorize living things according to their physical characteristics and declared that it is nearly impossible to confine beings to categories because characteristics frequently overlap: "But Nature works in unknown degrees, and consequently, it cannot totally lend itself to all these divisions, since it passes from one species to another species, and often from one genus to another genus, in imperceptible nuances; so that there are found a great number of species that are midway and half-way objects that one does not know where to place, and which necessarily skews the general system: this truth is too important to ignore everything that renders it clear and evident."[10]

In 1765 Buffon reiterated the gray areas between species and affirmed the fact that characteristics overlap among species. His chain of being was no longer two dimensional, but had become three dimensional or conical: "The apes make a near approach to man. The bats are the apes of birds, which they imitate in their flight. The porcupines and hedge-hogs, by the quills with which they are covered, seem to indicate that feathers are not confined to birds. The armadillos, by their scaly shells, approach the turtle and the crustaceous animals. The beavers, by the scales on their tails, resemble the fishes. The ant-eaters, by their beak or trunk without teeth, and the length of the tongue, claim an affinity to the fishes."[11]

Eighteenth-century thinkers recognized that Buffon's observation that physical characteristics overlap among species was undoubtedly true. What the polemicists did with this information was largely contingent upon their theological and biological belief systems. Viewpoints ran the gamut and were as diverse and imaginative as the *philosophes* themselves. For example, the deist Voltaire used Buffon's observation of the vast heterogeneity of forms to substantiate the beneficence of God and the magnificence of His Creation. Conversely, the atheist materialists parted company with Buffon and posited that this vast heterogeneity of living things is an iconic representation of the random motive property of atoms and the random flux that nature delivers.

The first issue was final causes (God's purpose) vs. random chance. The deist Voltaire believed that God had a purpose in mind when he created everything and that it would be sheer arrogance for man, whose intelligence is finite, to try to comprehend God's infinite wisdom. On the other hand, Diderot and the materialists rejected the existence of God and final causes and maintained that everything in the universe is a result of the random motive property of atoms. Adhering to the ancients, they believed that atoms are perpetually in motion and continually colliding; hence, every possible combination is tested for viability. Given an infinite amount of time, every possible combination will eventuate. This is the foundation of the study of games of chance (probability theory): as the number of throws increases, more combinations are realized. Given an eternal time frame, nothing is impossible and every outcome is possible. The materialists held that man should not make the mistake of reading final causes into unlikely patterns: what we interpret as organized patterns are merely events that flux delivers.

The materialists' thesis of random creation was fueled by recent discoveries of giant fossils in Siberia and the New World. There are several articles in the great *Encyclopedia*, of which Diderot and d'Alembert were the chief editors, that describe these fossils. The articles "Mammoth," "Fossil" and "Ivory Fossil" indicate that eighteenth-century France was well aware of the huge mammoth bones and ivory tusks discovered in Siberia and that it also knew that a 40 day Noachian flood was much too short to account for the massive upheavals of marine fossils and seashells that had been discovered on mountaintops and far from water. Geological and fossil evidence caused naturalists to consider that the earth was much older than 6,000 years as theologians claimed and that sequences of animal and plant populations came and disappeared one after the other.

The eighteenth century was also confronted with travelers' stories about the great apes in Africa, orangutans and pongos, that seemed much more intelligent than other animals, and that, perhaps, were a species of man. The century was forced to consider the notion that perhaps there were multiple species of man. Travelers such as Buffon, Prévost, Green, and Purchas, traveled to Africa and saw these great apes, which had an anthropomorphic form and seemed to have an intelligence midway between men and animals. Stories came back to France about how similar to man these apes were. They were reported to have a man's face, to have a body proportioned exactly like a man, to bury their dead, to kidnap people, to be attracted to campfires abandoned by men, and to put roofs on their houses.

In the article "Pongo" (1755) in the *Encyclopedia*, Jaucourt discusses the fact that pongos bury their dead and that the local populations of men regard the mounds of leaves and branches formed by the apes as "a sort of sepulcher." Recounting Andrew Battell's account of the travels of Samuel Purchas, Jaucourt says that the pongo is more than five feet high, has the height of an ordinary man, but is twice as heavy, has no hair on his face and resembles a man. Furthermore, pongos build their own shelters, live on fruits and plants, and cover their dead with leaves and branches (Africans regard the pile as a kind of sepulcher). Pongos approach campfires abandoned by men and seem to be pleased with them; however, they cannot conceive of throwing wood into them to keep the embers burning. Africans assure travelers that pongos have no language or intelligence that can cause them to be considered superior to other animals.

Hence, the notion of the chain of being became more and more problematic for eighteenth-century thinkers with every successive biological discovery. They were confronted with the task of examining the problem posed by the great apes, who looked much like men did, had the intelligence to bury their dead and put roofs on their houses, and appeared to be cheered and delighted by campfires, but who had no language and who were more like animals than man.

In the *Sequel to the Conversation* (1769) Diderot makes a joke about baptizing an orangutan. The character Bordeu asks, "Have you seen in the King's Garden, in a glass cage, an orangutan that looks like Saint John preaching in the wilderness?...Cardinal Polignac said to him one day, "Speak, and I will baptize you.""[12] Diderot raises the question of exactly how highly organized a being must be before it is endowed with a soul that requires forgiveness of sin and salvation. It also implies that animals are conscious, a notion that he ardently defends in the article "Animal."

The orangutan suggested that perhaps the insurmountable chasm between man and the animals is not so unbridgeable after all, and that perhaps 1) there are intermediary beings all over the world that need to be discovered and studied and/or 2) there may be multiple varieties of the species of man, much like there are varieties of species of flowers.

In the *Discourse on the Origin of Inequality* (1755), *Note 10*, Rousseau suggested that the way to resolve the question is 1) to have brilliant minds of the caliber of Montesquieu, Buffon, and Diderot to travel around the world and record their observations of the apes and 2) to conduct crossbreeding experiments to ascertain whether apes and humans can produce together fertile offspring. This would resolve the question of species once and for all.

This study will examine how various thinkers in eighteenth-century France approached the problem of the origin of man and specifically, the transformist hypothesis. It will be divided into ten chapters that will address

Maillet, Montesquieu, La Mettrie, Buffon, Maupertuis, Diderot, Rousseau, Voltaire, the debate over whether apes can be taught to speak, and the issue of race. Taken together, these ten chapters will show the great wealth, originality, and diversity of thought in eighteenth-century France. There were the deists (Voltaire), who defended Creationism, the atheist materialists (Maupertuis and Diderot), who posited random creation propelled by the motive and conscious properties of matter, the panspermists (Maillet and La Mettrie), who thought that preexistent seeds fertilized the land, sea, and air, and that the sea brought forth human eggs that were beached on shores, and Rousseau, who envisaged the anthropological transformism of man as he left his natural, solitary state in the woods and joined with others to form societies.

OVERVIEW

Chapter One examines the work of Benoît de Maillet, who wrote the *Telliamed* c. 1700, and who held geological and biological theses that were considered quite revolutionary in his day. His consular positions permitted him to travel extensively in the Mediterranean region and to observe, first hand, strata of fossil beds and sedimentary rock. He believed that the earth had been covered with water during a time in the distant past and that the waters had receded, exposing land surface and mountains. He also held that if we know the depth of rock and can estimate how long it takes to form, then we can calculate the age of each layer. He undertook the project of measuring the rate of sea level decline and concluded that the earth must be 2 billion years old. Although his assumptions often contained errors, his legacy was that the age of the earth can be ascertained by scientific observation, measurement, and mathematical calculations, rather than by biblical chronology.

Maillet was a contemporary of Fontenelle and so he relied on the authority of the ancients. In the *Telliamed* (although written c. 1700, it was not published until 1748), he sets forth a theory of panspermia, holding that the sea carried human eggs to land where they were beached and they hatched. He posited that there are preexistent seeds everywhere-in the air, water and land-and that they have always existed. His panspermist views placed him in the camp of Epicurus, Lucretius and La Mettrie (who would cite him in 1750).

Since he embraced the retreating ocean theory, he thought that life must have originated in the sea and that the first members of each species were aquatic. Terrestrial creatures must have metamorphosed from aquatic beings as the waters receded. His character, Telliamed, says that fish developed wings and fins that helped them to walk on the ocean floor and later on land. He also tells the tale of a Dutch cabin boy who fell overboard and reappeared years later as a merman with scales and a fish tail. Hence, Maillet's great contributions to the eighteenth century were the notions that species metamorphose from sea creatures to land creatures and vice versa, and that the physical characteristics of living beings adapt to changes in their environment.

Chapter Two demonstrates that Charles-Louis de Secondat de Montesquieu also surmised that man must have metamorphosed from animals in the distant past. In the *Persian Letters* (1721), he describes the animal-like ancestors of the savage/hunter Troglodytes, those Troglodytes of former times, who were deformed, hairy like bears, and who hissed. He derives his imagery from the ancients, most notably, Aristotle, Herodotus, Pomponius Mela, and Pliny the Elder, who wrote about man's monstrous ancestors, that could only squeak like bats and hiss. For example, Pomponius Mela described the Blemians, who did not have heads, and whose mouths and eyes were situated in their chests. This notion was also promulgated by Lucretius, who declared that in the beginning nature created many monsters and only those without significant self-contradictions survived. Classical texts held great weight during the eighteenth century and Lucretius' *De rerum natura*, which posited that when life emerged on the earth, nature created every imaginable deformed creature and only those beings without major defects survived and proliferated, provided the schema of the origin of living beings for Montesquieu, Diderot, and La Mettrie, to name a few.

In the *Spirit of Laws* (1748), Montesquieu elaborates further on the impermanence of nature and the flux that it delivers. He hypothesizes that a relationship exists between climate and human physiology and also between climate and government. Climate, geography, topography, and soil all affect man's body and temperament and contribute to an array of tendencies that a nation has. Physiology, temperament, intellect, societal customs and forms of government are influenced by, but not absolutely determined by climate. Man has the power to progress beyond the influence of climate and improve his life through legislation.

Chapter Three examines the thought of the physician Julien Offray de La Mettrie, who was a panspermist, and held views similar to those of Epicurus, Lucretius, and Maillet. He believed that the germs of all living things-plants, insects, animals, and men-came from the air and seeded the earth. He was influenced by Maillet's *Telliamed* and in *The System of Epicurus* (1750), *Chapter 32*, he refers the reader to it. La Mettrie hypothesized that at one time the ocean covered the whole earth and as the waters receded, human eggs were beached on the shores, incubated beneath the sun's warmth, and hatched human beings. He, too, was influenced by Lucretius, and thought that the flux of nature produced every imaginable monster and only those beings without serious defects survived, reproduced, and proliferated.

Even though he was a panspermist, La Mettrie made many significant contributions to Enlightenment thought. First, as a monist, he declared that physical matter is the only reality through which all phenomena can be explained. Secondly, he dispensed with the notion of the immortal soul, and declared that consciousness is solely contingent upon the functioning of the brain, central nervous system, and the five senses. Furthermore, consciousness is influenced by food, age, learning, inheritance, climate, and the environment. Thirdly, like Buffon he observed that physical characteristics overlap among species and noted the plenitude and continuity of the chain of beings: there is a seamless continuity between species as well as between kingdoms. Because he was a materialist, he felt that all life is the result of random chance and random molecular collisions: flux+time=dispersion of chaos. Hence, he held that the motive property of atoms, not God, created everything. Atoms are continually colliding and eventually form every viable organized being that exists. He rejected final causes and attributed all events to the random flux of nature.

Chapter Four discusses the great impact that George-Louis Leclerc de Buffon had on his century, even though the *philosophes* did not always agree with him. Buffon believed that all species left the Creator's hands in perfect condition at the time of Creation and that since then, they have kept their general form. He allowed for the degeneration of species, but this constituted minor changes and not a significant change in form, and he proposed it only for a limited number of species that had been domesticated by man or taken out of their native lands of origin. Voltaire and Rousseau concurred with his non-transformist point of view. Voltaire used it as propaganda in his defense of deism; Rousseau accepted it because science had not proven that man had ever been a quadruped. Diderot and Maupertuis were transformists and so they totally rejected his view of the fixity of species.

Among the vast panoply of Buffon's contributions to biology was his observation "all that can be, is {*tout ce qui peut être, est*}. He observed that nature produces every imaginable variation in each species. Furthermore, he observed that physical characteristics are shared among species. Hence, he developed the two dimensional, linear chain of beings into a three dimensional cone. The materialists Maupertuis and Diderot pondered this observation of the plenitude and continuity of the chain of beings and the fact that characteristics overlap among species. They added the fourth dimension, time, and posited the metamorphosis of species over great length of time.

Chapter Five addresses the great contributions that Pierre-Louis Moreau de Maupertuis made to the notion of the metamorphosis of species. Maupertuis' great legacy to the eighteenth century is two-fold: 1) he posited that inherited errors explain the vast diversity that the flux of nature delivers and 2) he proposed that an emergent consciousness arises when conscious particles unite. Maupertuis' study of polydactyly in a Berlinese family caused him to arrive at some landmark conclusions in biology. He identified birth anomalies as "traits" and noted that they were carried by either parent. He also noted that birth defects often skip a generation or two and reappear further down the family tree. In addition, he was able to calculate the statistical probability that polydactyly would recur in a given population. His great influence on Diderot's thought was his hypothesis that errors occur in the arrangement or pattern of parental elements. These errors are inherited and could explain how the vast variety of living beings that we see today many have developed from a single prototype.

Maupertuis' other great contribution to Diderot was an emergent consciousness: when particles unite, each particle loses its consciousness of self and acquires the consciousness of the larger body to which it belongs. It loses its individual memory and consciousness and acquires that of the whole. This explains why man is conscious of his existence and of the presence of others, but not of every molecule that constitutes his body.

In chapter Six we will see that Diderot seized upon this notion, cited Maupertuis in *Thoughts on the Interpretation of Nature* (1753), and adopted his two main contributions to biology, namely, 1) that inherited errors in the arrangement of parental elements could explain how all living beings developed from a single prototype and 2) emergent consciousness (that all matter is conscious and that when particles combine, they lose their memory and consciousness of self and acquire the consciousness of the larger body that they form).

Diderot's originality lies in the fact that he viewed species as mutable, not static, and that he posited the appearance, lifespan and extinction of species over time. He surmised that microscopic animalcules, species, star systems, and perhaps the universe itself, randomly come into existence via the motive and conscious properties of atoms, exist for a time, and then fall out of existence. Diderot's transformism rested on a fulcrum of three pivotal points: probability theory, the motive property of atoms, and the conscious property of matter.

His biology was based on William Harvey's definition of epigenesis which was the theory that the germ is brought into existence by the addition of parts that bud out of one another or by successive accretions (*partium superexorientum additamentum*) and not merely developed from a preformed seed. Because the developing embryo is formed by successive accretions and is not an exact duplicate of either parent, it is a unique new entity, different from either parent, subject to errors and accidents in the generative process. It is errors in the arrangement of parental elements supported by both parents, that are inherited, that explain how all living things arose from a single prototype. Hence, Divine agency is unnecessary. The motive and conscious properties of matter suffice to randomly create order out of chaos.

Chapter Seven will examine the thought of Jean-Jacques Rousseau, who was not a transformist. He embraced anthropological (intraspecies) metamorphosis and sociological change, but not biological (interspecies) transformism, as Diderot and the materialists did. In the *Discourse on the Origin of Inequality* (1755) he argues that there is no scientific evidence that man ever walked on all fours. In *Note 3(2)* he articulates six reasons why he rejects the notion that man metamorphosed from quadrupeds. First, possibility is not enough: probability must be shown and no one has proven the probability that man has ever been a quadruped. Hence, he cleverly takes the probability theory employed by the materialists and uses it against them. Secondly, if man had ever been a quadruped, because of the way that his head is attached to his body, his gaze would have been directed towards the ground. This would limit his perceptions, make him vulnerable to predators, and reduce his chances of survival in the world. Thirdly, quadrupeds have tails and man does not have a tail. Fourthly, the position of a woman's breast is perfect for carrying a child in her arms, but would be ill placed in a quadruped. Fifthly, man's hindquarters is high in relation to his forelegs, so that he would have to crawl about if he were on all fours. This is antithetical to survival and make's man vulnerable to prey. Sixthly, if man had ever been a quadruped, he would be able to place his feet flat on the ground the way animals do, when he crawls about on all fours. Rousseau concludes that man must have been created a biped, not a quadruped and accepts Buffon's non-transformist views on the subject.

Rousseau's legacy is that he posited anthropological (intraspecies) change. He applied Buffon's theory of the physical degeneration of species to the dissolution of man's morality. Rousseau borrowed many of Buffon's observations regarding the physical bodies of creatures, and ingeniously followed a parallel route, applying them to hypothesize a psychic and moral dissolution that occurred during man's anthropological (intraspecies) metamorphosis from his natural state to his civilized. Rousseau posited that before man joined with other men to form small groups or societies, he lived a solitary existence, roaming through the woods, living in the present moment. This "natural man" was neither good, nor evil, but a *tabula rasa*, on which his experiences would imprint. Hence, all of the vices that exist in society today, most notably war, slavery, theft, the notion of honor, pride, and greed, were unknown to natural man. It was not until natural man left his solitary state and joined civilization that every imaginable evil developed. Man consents to being unequal and to be enslaved by other men. Natural man is equal. Hence, Rousseau's *Discourse on the Origin of Inequality*, with its statement of man's journey from innocence to moral corruption, was intended to be a scathing criticism of civilized Europe and its vices, and not an essay on biological transformism.

Rousseau also noted the great similarities between man and the great apes, who also appeared to be conscious, thinking, intelligent animals. In *Note 10* he suggested widening the umbrella of the human species to include the great apes and accepted the notion that perhaps there are multiple varieties of the human species. This was a political statement: he opposed racism, slavery and subjugation of people living in extremely primitive conditions in Africa and Asia. He declared the human dignity and basic human rights of all men, even apes, who if experimentation could show had perfectibility and could be taught language, deserved the compassion and esteem accorded to all men. His proposal to include apes in the human species was not a statement as to biological transformism, which he denied, but rather, a scathing criticism of civilized Europe in which men consent to inequality and to valuing some people more than others according to the artifices of social class.

Chapter Eight will address the thought of François-Marie Arouet de Voltaire, who spent the last thirteen years of his life denouncing contemporary science. Voltaire recognized that scientific discoveries posed a threat to deism: random creationism challenged the need for a Prime Mover and fossils provided evidence that species do metamorphose over long periods of time. Furthermore, the determinism of the motive property of atoms obviated the power of God's will. Therefore, Voltaire denied scientific discoveries and their implications, the veracity of the chain of beings, the suspected antiquity of fossils, the theory of the metamorphosis of species, random creationism, microscopic animalcules, spontaneous generation, and the geological implications to the discovery of seashells on mountaintops and far from water.

Chapter Nine will explore a debate that existed among the *philosophes*: whether or not apes could ever be taught to speak. The battle lines were drawn along Cartesianism: Cartesians, embracing the dual nature of man (body and soul), held that apes can never be taught to speak because language requires intellect, intellect is a function of the soul, and only man has a soul. Conversely, the materialists Diderot and La Mettrie observed the similarities of the brains and vocal apparatus of man and apes and thought that with proper instruction, apes could learn to speak. Rousseau observed that animals that work together, such as beavers, ants, and bees, have a gestured language "that speaks only to the eyes," but because they have limited intelligence, they can never be taught to speak. He also recognized that chimpanzees belong to a different species than man, that they have limited intelligence, and that they cannot be taught to speak. However, he was open to the notion that great apes such as orangutans might be a variety of the species of man. He found their intelligent behavior to be significant (they build roofs on their shelters, exhibit pleasure at the sight of a campfire, and cover their dead with leaves and branches) and he hypothesized that if orangutans are a species of man, they may be intelligent enough to be taught language.

The question as to whether apes can be taught language is significant because it challenges man's unique place at the head of all animals on the great chain of beings. It also raises questions as to the length of the distance between man and the next most intelligent animal on the chain, whether there are multiple varieties of the human species, whether man can propagate with the great apes, and exactly how intelligent these apes are. If just one ape can be taught to speak, Cartesians have myriad problems to solve: Do apes have a soul that requires a Redeemer and forgiveness of Original Sin? Do other animals have souls, too? Can animals have everlasting life? How intelligent does an animal have to be before it can be considered to have a soul? What if the materialists are right and intellect is merely a function of physiological activity? If that is true, are they also right about the idea that all life is the result of the random motive property of matter?

Chapter Ten will examine the genealogy of the concept of race and the *philosophes'* vociferous opposition to the exploitation of dark skinned peoples in the transatlantic slave trade. Naturalists understood that by classifying similar entities, they could readily identify similarities in structures, retrieve information about relationships that exists, analyze the information, and draw conclusions. Bernier, Buffon, Diderot, and Linnaeus classified the various races of man according to distinguishing physical characteristics, perceived intelligence, personality, customs, climate, and geographical location. Buffon, Diderot, and Maupertuis were monogenecists: they held that all of the races of man could be traced back to a single ancestral pair. Voltaire, on the other hand, was a polygenecist: he believed that the different races are derived from different prototypes. Despite their biological views, the French *philosophes* defended the right of all humankind to self-determination as a basic tenet of natural law. The *philosophes* were passionate abolitionists and wrote prolifically against slavery and racial injustice. In addition, they used the virtues and vices of dark skinned people as a foil to criticize the moral turpitude of "civilized" Europe. Their lifelong struggle to make certain basic rights available to all people culminated in the Declaration of the Rights of Man and the Citizen [*La Déclaration des droits de l'homme et du citoyen*] on August 26, 1789.

WHAT CRITICS HAVE WRITTEN ON EVOLUTIONISM IN EIGHTEENTH-CENTURY FRENCH THOUGHT

To date, there exists a scarcity of material that traces the influence that probability theory, the motive property of atoms, and the hypothesis that matter is conscious had on evolutionism in eighteenth-century France. The notable exception is *Diderot and the Metamorphosis of Species*, which demonstrates how Diderot combined probability theory with the motive and conscious properties of matter to hypothesize tranformism.[13] Furthermore, there is not too much of an overview of how the eighteenth century collectively approached the problem of the chain of being when confronted with modern discoveries, or how successive thinkers modified the chain of being (ie: Buffon transformed Aristotle's two-dimensional, linear chain of being into a three dimensional cone in which physical characteristics overlap, and Diderot added the fourth dimension, time, to show that species are born, exist for a period of time, and then fall out of existence). A notable exception is Arthur O. Lovejoy's justly celebrated and highly influential work, *The Great Chain of Being: A Study in the History of an Idea*, that follows the chain of being from Plato to modern times.[14]

There are a few noteworthy overviews of eighteenth-century transformism. First, there is Jacques Roger's seminal volume, *The Life Sciences in Eighteenth-Century French Thought*.[15] Secondly, there is the anthology, *Forerunners of Darwin: 1745–1859*, edited by Bentley Glass, *et al*, which contains excellent articles on fossils, Buffon, Maupertuis, and Diderot. The articles are entitled, "Fossils and the Idea of a Process of Time in Natural History," by Francis Haber,[16] "Buffon and the Problem of Species," by Arthur O. Lovejoy,[17] Maupertuis, Pioneer of Genetics and Evolution," by Bentley Glass,[18] and "Diderot and Eighteenth-Century French Transformism," by Lester Crocker.[19] These articles, when taken together, provide a very fine overview of transformism during the century.

For example, in "Fossils and the Idea of a Process of Time in Natural History," Francis Haber provides an excellent overview of how the discovery of marine shells on mountaintops and in areas far from water caused the eighteenth century to move away from the concept of plastic forces and the Noachian flood. Haber chronicles the century beginning with Maillet's efforts to date the earth by measuring the rate of retreating sea level, and follows the *philosophes* right up to Voltaire, who spent the last thirteen years of his life exhibiting a vehement antipathy towards the new science. Haber mentions how Voltaire was opposed to creating a new system of earth history to explain marine fossils; Voltaire rejected the notion that land and sea had changed places several times and that the earth was much older than had previously been thought. Haber also shows that the geological processes that Buffon describes in his *Theory of the Earth* resembled Maillet's account in the *Telliamed*, especially the part of the formative effects of the flux and reflux of the ocean tide on mountain formations beneath the sea.

In addition, Arthur O. Lovejoy's article, "Buffon and the Problem of Species" addresses the three stages that Buffon passed through as he tried to determine whether or not species are mutable. From 1746–1756 Buffon agreed with Linnaeus that species are immutable; from 1761–1766 he explored the possibility of variability; after 1766 he decided to accept the notion of the permanence of the essential features of species and the variability of minor details. Buffon concluded that species retain their general form because of the interior molding force, but can degenerate in minor ways, due to domestication or removal from their lands of origin. Haber also discusses Buffon's treatment of the great chain of beings.

Furthermore, Bentley Glass, in his article, "Maupertuis, Pioneer of Genetics and Evolution," discusses examines Maupertuis' contributions to heredity and transformist biology. Maupertuis studied polydactyly through several generation of a Berlinese family and discovered that it was possible to predict the probability that member of a family would be born with the anomaly. Maupertuis showed that parental elements contributed to the fetus by both the father and mother, statistical probability, and birth anomalies, were intimately intertwined. Glass notes that not only did Maupertuis demonstrate that birth defects could be predicted, but that too many or conversely, missing, parental elements caused "monsters with deficiencies" [*monstres par défaut*] and 'monsters with extra parts" [*monstres par exces*]. Glass also addresses the role that random chance plays in Maupertuis' explanation of heredity. Maupertuis hypothesized that inherited errors could explain how all living things may have, over time, arisen from a single prototype.

Lastly, Lester Crocker's "Diderot and Eighteenth-Century French Transformism" addresses the development of Diderot's thoughts about transformism during the years 1746–1769. In thirty pages Crocker outlines the salient points in the contributions that Leibniz, La Mettrie, Maillet, Buffon, Maupertuis, Robinet, and Bordeu, made to Diderot's transformism.

Regarding the rarely mentioned author, Maillet, it is especially difficult to find criticism of his work or a study of his influence on Enlightenment thought. A notable exception is Jacques Roger's *The Life Sciences in Eighteenth-Century French Thought*, in which the author explains that Maillet was a panspermist, not a transformismist.[20] Otis E. Fellows and Stephen F. Milliken, in *Buffon*, mention that Maillet's far flung tales of mermen, panspermia, and transformism caused him to be the object of scorn and that Buffon's critics mentioned Maillet and Buffon together in order to ridicule the latter by association.[21]

Regarding Buffon, Jacques Roger explains that it was Buffon's concept of species that caused him to reject transformism, rather than fear of censorship. See Jacques Roger's book, *Buffon: A Life in Natural History*, and *The Life Sciences in Eighteenth-Century French Thought*. Otis E. Fellows and Stephen F. Milliken, in *Buffon*, also explain that it was Buffon's concept of an interior molding force that was the reason that he rejected the transmutation of species. For more specialized material on the subject of Buffon and species, see John H. Eddy, Jr..'s Ph.D dissertation, "Buffon, Organic Change, and the Races of Man,"[22] Eddy's article, "Buffon, Organic Alterations, and Man,"[23] Paul Lawrence Farber's Ph.D dissertation, "Buffon's Concept of Species,"[24] Farber's article, "Buffon and the Concept of Species,"[25] and Phillip R. Sloan, "The Idea of Racial Degeneracy in Buffon's *Histoire naturelle*."[26]

Excellent articles and books that address Diderot's transformism and how Buffon influenced his thought include articles by Michèle Duchet, "L'anthropologie de Diderot,"[27] Jean Ehrard, "Diderot, l'*Encyclopédie*, et l'*Histoire et théorie de la Terre*,"[28] Arthur O. Lovejoy, "Buffon and the Problem of Species,"[29] Jacques Roger's book, *Buffon: A Life in Natural History*,[30] his article, "Diderot et Buffon en 1749,"[31] and Roger's seminal volume, *The Life Sciences in Eighteenth-Century French Thought*,[32] and Aram Vartanian's article, "Buffon et Diderot."[33]

Several articles have also been written on La Mettrie's influence on Diderot: Jean E. Perkins, "Diderot and La Mettrie,"[34] Ann Thomson, "La Mettrie et Diderot"[35] and "L'unité matérielle de l'homme chez La Mettrie et Diderot,"[36] Aram Vartanian, "La Mettrie and Diderot Revisited: An Intertextual Encounter"[37] and "Trembley's Polyp, La Mettrie, and Eighteenth-Century French Materialism,"[38] and Marx W. Wartofsky, "Diderot and the Development of Materialist Monism."[39]

Critics argue that Rousseau's *Discourse on the Origin of Inequality* is not a statement of biological transformism (interspecies), which Rousseau rejected, but rather, a statement of anthropological development (intraspecies), a political statement, and a criticism of class difference in civilized Europe. The following authors cautiously point out that Rousseau was not an advocate of biological transformism: Otis Fellows, "Buffon and Rousseau: Aspects of a Relationship,"[40] and Francis Moran, "Of Pongos and Men: Orangs-Outang in Rousseau's *Discourse on Inequality*,"[41] and Leonard Sorenson, "Natural Inequality and Rousseau's Political Philosophy in his *Discourse on Inequality*."[42]

This study is significantly different from that of antecedent criticism because it is devoted to showing how eight theorists in eighteenth-century France addressed the issue of transformism. It examines the subject from the perspective of each subsequent thinker laid down additional building blocks that led to transformism and a rethinking of the chain of beings. This book will examine how the *philosophes* established a triune relationship among contemporary scientific discoveries, random creationism propelled by the motive and conscious properties of matter, and the notion of the chain of being, along with its corollaries, plenitude and continuity. Maupertuis deduced that inherited errors in the arrangement of parental elements are passed along from generation to generation and can explain how all living things arose from a single prototype. Diderot recognized Maupertuis' genius and the truth in his statement and made it the basis of his transformism. Others, like Buffon and Rousseau, were aware of the random creationist hypothesis, but rejected it-Buffon, because he embraced the notion that an interior molding force fixes the general form of species, and Rousseau, because transformism was unproven and unlikely. Then there was Voltaire, who recognized that science posed a threat to deism because it could replace God.

Chapter One

Maillet

The little Wings they had under their Belly, and which like their Fins helped them to walk in the Sea, became Feet, and served them to walk on Land.[1]
—Maillet, *Telliamed* (c. 1700)

Benoît de Maillet (1656–1738) held several consular positions that gave him the opportunity to travel extensively throughout the Mediterranean region, observe geographical phenomena, and arrive at some astute conclusions about the age of the earth. Beginning in 1692, he served as the French Consul-General to Egypt, then as Consul to Leghorn, and finally, as Inspector of the French Establishments in the Levant (Syria and Lebanon) and the Barbary Coast (Egypt to the Atlantic Ocean).[2] During his various consular tenures, his travels in Mediterranean countries permitted him to observe first hand strata of fossil beds and sedimentary rock. He surmised that these phenomena provided evidence of the gradual recession of water away from beaches that had left coastal towns at a high elevation. Because he was not only a diplomat, but also a scholar, he devised a method for calculating the age of the earth by measuring the rate at which water recedes from the beaches and extrapolating how long it must take for strata of fossils or sedimentary rock to form. He set forth his ideas in a fiction work entitled, *Telliamed: Or, Discourses between an Indian Philosopher, and a French Missionary, on the Diminution of the Sea, the Formation of the Earth, the Origin of Men and Animals, and other Curious Subjects, relating to Natural History and Philosophy*. His work was highly innovative during the period in which he lived (he was a contemporary of Fontenelle); because he anteceded Buffon, he did not have the *Natural History* (1749–1767) or the *Supplement* (1774–1789) as reference tools to gauge the accuracy of his hypotheses.

James Powell explains that Maillet used the hourglass method to calculate the age of the earth. The basic premise of the hourglass method is that if we know how much sand is in the top half of the hourglass, how much sand is in the bottom portion, and the rate at which it passes through the stricture, we can calculate how much time has elapsed since the process began.[3] When Maillet was alive, scientists began to observe layers of sedimentary rock and they concluded that perhaps the layers could reveal the age of the earth.[4] If we know the depth of the rock and can estimate how long it takes to form, then we can calculate the age of each layer.[5] Powell advises that Maillet had relied upon this method of reckoning (the rate of decline or erosion): "Instead of measuring how rapidly something builds up, one can measure how rapidly something else declines. If one knew how much erosion had lowered the land surface, and if one knew the rate, a simple calculation would reveal how long the erosion had been going on."[6]

Maillet hypothesized that the earth was once entirely covered with water and that the water gradually evaporated into outer space, causing land masses to appear. Powell says, "De Maillet assumed that a universal sea had once covered the earth but had since shrunk, stranding formerly coastal towns high above sea level. He estimated the rate of sea level decline at 3 feet in 1,000 years. At that rate, to perch a formerly seaside town at 6,000 feet would take 2 million years. But since the earth is obviously much older than its towns, de Maillet arbitrarily raised his estimate for its age to 2 billion years. Well aware that such an immense figure would incur the wrath of the Church, de Maillet presented his conclusions in the guise of a dialogue between a French missionary and as Eastern mystic named Telliamed (de Maillet spelled backwards). The manuscript remained unpublished until a decade after de Maillet's death…Though his assumptions were wrong, de Maillet did show that, starting from observation and measurement rather than from the Bible, one could calculate an age for the earth. That age might be measured not in the scores of years by which a human lifetime is counted, nor even in the thousands of years of Ussher, but in millions and billions of years."[7]

In order to avoid charges of heresy, Maillet relied on several techniques. First, he placed his speculation on the metamorphosis of the earth's topography and its creatures within the context of an oriental fantasy: the work is a dialogue between a Christian missionary and an Indian sage who is simply articulating Hindu philosophy. The transformist material is filtered through the voice of the Christian missionary, who identifies the Hindu in the third person: "he told me," "he said." This reminds the reader that it is a Hindu who is articulating the material and that the Christian missionary is simply repeating what he has heard. Secondly, the book is dedicated to Cyrano de Bergerac, who, a century earlier, had written a couple of science fiction stories entitled, *History of the States and Empires of the Moon* [*Histoire comique: contenant les états et empires de la Lune*], written in 1649, published posthumously in 1657, and *History of the States and Empires of the Sun* [*Histoire comiques: contenant les états et empires du Soleil*], written in 1650, published posthumously in 1662. Maillet hoped that his work, too would be regarded in the vein of science fiction. Thirdly, he remained an anonymous author, and cloaked his authorship by spelling his name backwards and making the word "Telliamed" both the name of the Hindu sage and the title of the book. The author was not the only one who remained anonymous: the publisher, the Abbé Jean-Baptiste Le Mascrier, did not identify himself, either. Although the book was set in print in 1735, it was not published until 1748, three years after Maillet's death. Fourthly, Maillet intersperses many mythological creatures and monsters borrowed from Pliny's *Natural History* (he mentions Pliny in *Chapter 6*) in order to make it obvious that the work is a fantasy. However, beneath the orientalism, the monsters taken from antiquity, the extraordinary mermen and mermaids (a Dutch cabin boy who falls overboard at the age of eight reappears twenty years later as a merman), there is a constant thread of science: water has retreated away from coastal towns; high hills show strata of seashells embedded in them; multiple layers of seashells of different colors indicate that they could not all have come from a single Flood; observation shows that the world cannot possibly be 6,000 years old; species have overlapping physical characteristics that suggest a common origin; life originated in the sea; fish moved from the water to the air by developing wings; birds moved to land by developing feet. The six days of Creation are mirrored in the six chapters (days) of the book.

Another innovative notion that Maillet sets forth is that the earth is two billion years old. Here he breaks with Creationism and the dating of Bishop Ussher, who had set the creation of the world at 4004 BC. Two billion years provides plenty of time for species to metamorphose from sea creatures into the present day wealth of species that populate the sea, earth, and air. This is innovative, as he arrives at his figures before Buffon comes along.

Maillet maintains that it is observation and experimentation that show us that the sea has receded and left strata of fossil beds. In the first chapter, entitled, "First Day. Proofs of the Diminution of the Sea," the Hindu sage says, "An Observation which my Grandfather made...in his Youth observ'd, that in the greatest Calm, the Sea always remained above the Rock, and cover'd it with Water: Twenty-two Years, however, before his Death, the Surface of the Rock appeared dry and began to rise."[8] The grandfather inspected the ground carefully and found "Sea-shells adhering to, and inserted in their Surfaces. He found twenty Kinds of Petrifications which had no Resemblance to each other...The Origin of this so great Variety of Soils, join'd to the *Strata* or Beds different in Substance, Thickness and Colour, of which most of these Quarries were composed, strangely perplexed his Reason..."[9] The solution to the mystery would be to measure the recession of the ocean over time.

At the end of *Chapter 3* Telliamed surmises, based on estimations that he has made, that the diminution of the sea approximates a foot every three centuries, and three feet four inches every thousand years.[10] From this data, he must necessarily conclude that this diminution has gone on for 2 billion years; the human race has existed for at least 500,000 years; mermaids are creatures that require more time to become fully human.

In the chapter entitled, "Sixth Day. Of the Origin of Man and Animals, and of the Propagation of the Species by Seeds," Maillet proffers much transformist material. He comments on how life began in the sea, the interrelatedness of all species, and how flying fish left the water and over time, were transformed into land creatures: "In a word do not Herbs, Plants, Roots, Grains, and all of this Kind, that the Earth produces and nourishes, come from the Sea? Is it not at least natural to think so, since we are certain that all our habitable Lands came originally from the Sea?"[11]; "As for the Origin of terrestrial Animals, I observe that there are none of them, whether walking, flying, or creeping, the similar species of which are not contained in the Sea; and the Passage of which from one of these Elements to another, is not only possible and probable, but even supported by a prodigious Number of Examples."[12]; "The resemblance in Figure, and even Inclination, observable between certain Fish and some Land-Animals, is highly worthy of our Attention; and it is surprising that no one has laboured to find out the Reasons of this Conformity."[13]; "There are in the Sea, Fish of almost all the Figures of Land-Animals, and even of Birds. She includes Plants, Flowers, and some Fruits; the Nettle, the Rose, the Pink, the Melon, and the Grape, are to be found there,"[14] and "...for it may happen, as it often does, that winged or flying Fish, either chasing, or being chased, in the Sea, stimulated by the Desire of Prey, or the Fear of Death, or pushed near the shore by the Billows, have fallen among Reeds or Herbage, whence it was not possible for them to resume their Flight to the Sea, by which Means they have contracted a greater Facility of flying. Then their Fins being no longer bathed in the Sea-Water, were split and became warped by their Dryness. While they found among the Reeds and Herbage among which they fell, any Aliments to support them, the Vessels of their Fins being separated were lengthened and cloathed with Beards, or to speak more justly, the Membranes which before kept them adherent to each other, were metamorphosed. The Beard formed of these warped Membranes was lengthened. The Skin of these Animals was insensibly covered with a Down of the same Colour with the Skin, and this Down gradually increased. The little Wings they had under their Belly, and which like their Fins helped them to walk in the Sea, became Feet, and served them to walk on Land. There were also other small Changes in their Figure. The Beak and Neck of some were lengthened, and those of others shortened. The Conformity, however, of the first Figure subsists in the Whole, and it will be always easy to know it."[15]

Maillet also imagines that quadrupeds have analogous counterparts in the ocean: "As for Quadrupeds, we not only find in the Sea, Species of the same Figure and Inclinations, and in the Waves living on the same Aliments by which they are nourished on Land, but we have also Examples of these Species living equally in the Air and in the Water. Have not the Sea-Apes precisely the same Figure with those of the Land? There are also several Species of them,"[16] and "The Lion, the Horse, the Ox, the Hog, the Wolf, the Camel, the Cat, the Goat, the Sheep, have also Fish in the Sea similar to them"[17]

The terrestrial counterparts metamorphosed on land from their marine counterparts: "Tis thus certainly that all terrestrial Animals have passed from the Waters to the Respiration of the Air, and have contrasted the Faculty of lowing, howling, and making themselves understood, which they had not, or which they had but very imperfectly in the Sea."[18]

Intermediary species between men and fish (mermen and mermaids) have been sighted by seafarers and the *Telliamed* recounts many such events. For example, a merman, followed by his female, was seen in one of the towns of the Delta or lower Egypt.[19] Also, a Dutch cabin boy, at the age of eight, fell into the sea and disappeared, only to have reappeared more than twenty years later as a merman, substantiating the notion that humans are transformed into fish.

Furthermore, there were a multitude of different species of men on land, some with tails and some without. When the sea receded, it beached eggs that produced every imaginable variety of human being. *Chapter 6* has subchapters on savages [*Des hommes sauvages*], men with tails [*Des hommes à queuë*], beardless men [*Des hommes sans barbe*], one-legged men and one-armed men [*Des hommes d'une jambe, & d'une seule main*], giants [*Des Géants*], and dwarves [*Des Nains*]. There were wild men of the woods that did not talk. In *Chapter 6* Maillet cites Pliny as a source of these species of wild men. Mermen are still metamorphosing into land-men and they can be seen in the polar regions where there are dwarves and giants. This happened in New Guinea: "Nothing is more common than those Savage-men; in 1702...the Dutch seized two Male Animals, which they brought to *Batavia*, and which, in the Language of the Country where they were taken, they called *Orangs-outangs*, that is, Men who live in the Woods. They had the whole of the human Form, and like us walk'd upon two Legs. Their Legs and Arms were very small, and thick-covered with Hair...These *Orang-outangs* had the Nails of their Fingers and Toes very long, and somewhat crooked."[20]

Maillet asks the ultimate question as to whether men could have metamorphosed from apes with tails: "To return to the different Species of Men. Can those who have Tails, be the Sons of them who have none? As Apes with Tails do not certainly descend from those which have none, is it not also natural to think, that Men born with Tails are of a different Species from those who have never had any?"[21]

At this juncture it is important to note that Maillet was not a transformist, but a panspermist. The *Oxford English Dictionary* defines "panspermia" as "Originally (now *hist.*) the theory that there are everywhere minute germs which develop on finding a favourable environment."[22] Maillet held that the seeds to all living things originated in the sea and species metamorphosed after that, according to their environment. He agrees with Lucretius that animals cannot have dropped down from the heavens, but must have undergone a variety of processes. In a footnote Maillet cites Lucretius' statement "For animals cannot have fallen from the sky, nor can creatures of the land have come out of the salt pools" (*De rerum natura*, 5.793–94).[23]

Now that Maillet reveals that his source is Lucretius, one needs only to read the preceding verses in *De rerum natura* to see the impact that Lucretius had on his panspermist view of creation. In 5.791–92, Lucretius says, "…so then the new-born earth put forth herbage and saplings first, and in the next place created the generations of mortal creatures, arising in many kinds and in many ways by different processes."[24] Lucretius metaphorized the earth as a giant uterus that gave birth to all living things, both vegetable and animal. Maillet modifies this and hypothesizes that the receding sea beached human eggs on its shores.

Although Maillet wrote around 1700 and was a contemporary of Fontenelle, his notions about the age of the earth, panspermia, and the metamorphosis of species were at the same time, so absurd and preposterous, yet perceptive and incisive, that he posthumously became embroiled in the polemics of the mid-eighteenth century. The panspermist La Mettrie used Maillet as propaganda in his battle to sustain random creation and deny Creationism; the deist Voltaire found his panspermist and transformist statements useful as vehicles to ridicule, by association, the contemporary science of Buffon, Needham, and Maupertuis. Jacques Roger advises that Maillet's cosmogony was antichristian and antibiblical: he believed in preexistent seeds that had existed eternally and had seeded life throughout the water, air and land of the earth, all the planets, and across the entire universe.[25] Maillet posited that matter had always existed and hence, there was no need for a Creation. As a polemical tool, he was just as useful to the panspermist La Mettrie as was Lucretius: while Lucretius believed that living matter arises from the inanimate all the time, Maillet hypothesized that eternally preexistent seeds are everywhere and so, life is no miracle.

Maillet's panspermist cosmogony was used as propaganda by La Mettrie. In the *System of Epicurus* (1750), *Chapter 32*, La Mettrie points out that Maillet's panspermia was in complete agreement with Lucretius' *On the Nature of Things*, and he defends both Maillet and Lucretius' notion of seeding: "Could all Animals, and therefore man, which no sensible Man would ever think of removing from their Category, truly be sons of the Earth, as Legend says of Giants? As the Sea perhaps originally covered the surface of our

Globe, would it not have itself been the floating cradle of all the Beings eternally enclosed in its breast? That is the system of the author of the *Telliamed*, which comes pretty near to that of Lucretius; for it is still necessary for the sea, absorbed by the pores of the Earth, consumed little by little by the warmth of the Sun and the infinite lapse of time, to have been forced, while receding, to leave the human egg, as it sometimes does the fish, high and dry on the shore. As a result of which, without any other incubation than that of the Sun, man and every other animal would come out of its shell, as certain ones still hatch today in warm countries, and also as Chickens do in warm dung through the skill of the Naturalist."[26]

La Mettrie thought that panspermia was quite plausible, since experiments in spontaneous generation indicated that animalcules are perpetually arising from nonliving matter. John Turberville Needham's *An Account of Some Microscopical Discoveries* (1745) related experiments that appeared to prove the veracity of spontaneous generation.; his book was translated into French in 1747 and then underwent a number of revisions. During his experiments, Needham saw animalcules develop after 10–20 days in an infused grain. He concluded, "It seems plain therefore, that there is a vegetative Force in every microscopical Point of Matter, and every visible Filament of which the whole animal and vegetative Texture consists: And notably this Force extends much farther; for not only in all my Observations, the whole Substance, after a certain Separation of Salts and volatile Parts, divided into Filaments, and vegetated into numberless Zoophytes, which yielded all the several Species of common microscopical Animals; but these very Animals also, after a certain time, subsided to the Bottom, became motionless, resolv'd again into a gelatinous filamentous Substance, and gave Zoophytes and Animals of a lesser Species." Animal and vegetable seeds were created in an 'exalted' matter, which is to say a matter rich in this vegetative force."[27] Because he saw animalcules present in his infusions of grain, he hypothesized that there exists a "vegetative force" that causes inanimate matter to spring to life.

Another reason that La Mettrie embraced panspermia was Trembley's polyp, whose body parts regenerated after having been cut off. In *Man a Machine* (1747), La Mettrie declares, "We do not know nature; causes hidden in her breast might have produced everything. In your turn, observe the polyp of Trembley: does it not contain in itself the causes which bring about regeneration?"[28] La Mettrie used panspermia to explain the regenerative powers of Trembley's polyp, and so he found Maillet useful because he thought that the sea produces the germs of all living things.

La Mettrie thought that germs come from the air. Jacques Roger advises, "Man came from the worm as the butterfly came from the caterpillar...plants and insects are born from germs coming from the air; likewise the germs of men had first been prepared by the air..."[29] Like Lucretius, La Mettrie asserts that in the beginning, nature produced every variety of living thing and that only those without self-contradictions survived, regenerated, and proliferated. Since La Mettrie held Lucretius' panspermist belief system, he reiterated, in *The System of Epicurus, Chapter 32*, Maillet's statement that the receding sea carried human eggs to the shoreline and beached them, where they subsequently hatched.

Voltaire, on the other hand, who was not a random creationist like La Mettrie, but a deist, had a different agenda: he satirized Maillet, and, in *The Man of Forty Ecus, Chapter 6*, used the *Telliamed* as propaganda to deride materialism, panspermia, and transformism. Maillet's book provided an excellent vehicle to ridicule the science of contemporary naturalists-Buffon, Needham, and Maupertuis-which he heaped all together. The fact that Maillet was not a transformist, but believed that the ocean carried human eggs to land where they hatched (this made him a panspermist), was irrelevant: transformism and panspermia were both dangerous and now was the time to expose modern science for the lie that it was. Voltaire mocks Telliamed, "who taught me that mountains and men are made by sea waters. First, there were handsome mermen that later became amphibians. Their beautiful forked tail transformed itself into buttocks and legs. I was full of Ovid's Metamorphoses, and a book where it was shown that the human race was the bastard of a race of baboons. I liked descending from a fish about as much as from a monkey. With time I had a few doubts about this genealogy, and even about the formation of mountains."[30]

Voltaire devotes *The Man of Forty Ecus, Chapter 6*, to a dialogue between an old hermit and Telliamed. Voltaire's chapter is a mirror of Maillet's work: while Maillet proffers a dialogue between a Hindu sage and a Christian missionary, Voltaire's chapter is a dialogue between the same Hindu sage and an old deist, who, unlike Maillet's Christian missionary, gives Telliamed a piece of his mind and argues with him about every point he makes. It is the fulfillment of any deist's wish list of arguments that Maillet's Christian missionary should have used against Telliamed's fantastic mutterings, but did not. Voltaire has a catharsis: scientists and naturalists are vain and dare to put themselves in God's place-they have created a universe with their pen as God did once with His Word. The first naturalist who presents himself to be worshipped is Telliamed. He spins yarns about handsome mermen who metamorphose into amphibians. He points out that seashells have been discovered atop high mountains and claims that this proves that the earth was once covered with water. Unlike the Christian missionary in Maillet's work, Voltaire's "Je" starts arguing and promotes the deistic point of view. Point by point, he refutes Maillet's hypotheses. When Voltaire is through refuting Maillet, he brings out the next puppet in the show, Needham with his eels, and then, after him, the Lapp, Maupertuis, who proposes digging a hole to Patagonia. It becomes obvious that not only does Voltaire not take Maillet seriously, he merely uses him as an entrée to the real dish, which is his archenemy, Maupertuis. The enmity between Voltaire and Maupertuis goes deeper than Maupertuis' random creationism, it also has to do with the feud over who was the real author of the "Principle of Least Action." Hence, in *Chapter 6*, Maillet is only a stop on the road on which Voltaire is driving: his destination is actually Maupertuis.

Otis Fellows makes the astute observation that Voltaire uses Maillet as a tool whenever he wants to ridicule someone. Since Voltaire was on a vendetta against science, one of his favorite targets was the famous Buffon. Voltaire was at odds with Buffon because of his *Theory of the Earth* (in which he hypothesizes that a comet collided with the sun and sent matter hurling through space that later cooled and became the planets), and his organic molecules (that fell to earth and seeded the planet with life). Voltaire denied both, and so, since he wanted to ridicule Buffon, one way to demean him was to associate him with Maillet and portray him as his disciple. In this instance, Maillet is simply a vehicle to insult another by association. Another way to demean Buffon was to associate him with his fellow experimenter in spontaneous generation, the eel man himself, Needham. As the saying goes, "*Noscitur a sociis*" [literally, he is known by his associates; idiomatically, a man is know by the company he keeps], and so, what better way to issue a double blow to Buffon, than to associate him with both Maillet and the eel man?

Fellows says, "In almost all of Voltaire's pieces in which Needham's connection with Buffon is more or less openly hinted at, there appears as well a second Buffonian surrogate: the much ridiculed French cosmogonist

Maillet 29

Benoît de Maillet...Voltaire, in adding a second bizarre personage to his satires, was simply expanding the device to include both of the two great areas of science in which Buffon had proposed major original theoretical systems. Without being obliged to mention Buffon by name, Voltaire was thus able to attack, within a single framework, both the Buffonian cosmogony and the Buffonian theory of organic molecules."[31]

Peter J. Bowler, in his criticism of Maillet's *Telliamed*, emphasizes the fact that during the eighteenth century, the work was considered to be highly innovative because it completely dispensed with the Flood and the book of Genesis and relied solely upon observation of layers of sedimentary rock and strata of fossils to arrive at the age of the earth: "This theory made no reference to the deluge and took it for granted that the earth was enormously old...he ignored the idea of a cooling earth and assumed that the planet originally had been covered by a great ocean. The sedimentary rocks were laid down when the surface was covered with water, and since had been exposed by a decline in sea level. This retreating-ocean theory would become popular in the eighteenth century, but seldom in so explicitly an antibiblical form. De Maillet even speculated about a natural origin for life and a process by which aquatic creatures could adapt to the emerging dry land...The planet's...history had to be read not from the Bible but from the rocks themselves. It was becoming increasingly obvious that the rocks were formed by natural processes such as sedimentation under water. Unless these deposits could all be attributed to the great flood, this would mean that natural processes-almost certainly acting over a long period of time-had shaped the earth's surface since its original formation."[32]

Bowler also discusses Maillet's panspermist view that the ocean deposited human eggs on dry land there they incubated and hatched: "If life was the product of natural forces, it need not exhibit the designing hand of the Creator...he adopted a version of the theory of preexisting germs in which the miniatures existed independently of parent bodies and were found scattered throughout nature...in the earth's early history, the great ancient ocean might have provided an environment in which germs could develop without parents. The first living things thus appeared by a natural rather than a supernatural process; once formed, they began to reproduce in the normal way."[33]

Bowler also explains Maillet's metamorphosis of species: "De Maillet also accepted the idea that change occurred within the species thus produced.

As an exponent of the retreating-ocean theory, he assumed that when life first appeared, the earth was completely covered by water. The first members of each species must have been aquatic. Each terrestrial species would have been produced by the transformation of its aquatic forebear as soon as dry land began to appear. De Maillet accepted sailors' stories of mermaids as evidence that the aquatic form of the human species still existed, and he thought that flying fish had been transformed into birds. Such ideas seemed ridiculous even to many of his contemporaries, but they show that de Maillet appreciated the need for life to adapt to changes in the earth's physical environment."[34]

Bowler notes that although Maillet embraced preexisting germs that originated in the sea, he dispensed with their divine origin: "De Maillet eliminated miracles and partly circumvented the argument from design by supposing the germs adapted to different conditions as they grew. He avoided postulating a supernatural origin for the terms by supposing that they had always existed throughout the universe."[35]

Jacques Roger maintains that Maillet cannot be regarded as a transformist for several reasons. First, he was a contemporary of Fontenelle, not of Buffon or Maupertuis.[36] His ideas were based on his time, around 1700, and it just happens that his book was published in 1748, a year before the first volume of Buffon's *Natural History* appeared. Secondly, he was a panspermist, not a transformist, and held that the sea carried human eggs to land where they were beached and they hatched. This places him in the camp of Epicurus, Lucretius and La Mettrie, rather than that of the Diderot of 1769. Roger points out that Maillet embraced the notion that preexistent seeds are everywhere-in the air, water and land, even on other planets and throughout the universe, and that they have existed through all eternity.[37] Preexistent seeds have always existed because matter has always existed, and hence, Maillet flatly denies the Creation.

Roger observes that the *Telliamed* is basically antichristian. It denies the veracity of the Bible and denies that the universe was created (matter always existed). Roger points out that the notions that life originated in the sea, that the first men were barbarians, and that there is a great diversity of the human species, are contrary to the biblical viw that God created man in His image.[38] Roger declares that Maillet was an anachronism at the time in which he wrote, totally out of step with Newton, Leibniz, Spinoza, and Locke.[39]

Nevertheless, Maillet gave the eighteenth century a lot to think about, even though, as Fellows observes, he was an object of ridicule and mentioned with Buffon in order to ridicule the latter by analogy. Maillet's legacy was that life originated in the sea. He asserted that the body parts of all living things slowly metamorphose over time, according to their environment. He provided the literary device of the dialogue between an easterner and a westerner of different religions and belief systems. He broke with tradition and suggested that the earth is at least two billion years old and that man must be at least 500,000 years old. He depicted the plenitude and continuity of the chain of being by drawing from Pliny and contemporary legends and enumerating every imaginable variation of living thing. He certainly left an impressive legacy, although he was a contemporary of Fontenelle and the science of the eighteenth century was foreign to him.

Chapter 2
Montesquieu

...those Troglodytes of former times...were more like animals than men.[1]
—Montesquieu, *Persian Letters, Letter 11* (1721)

Montesquieu's writing captures the impermanence and mutability of the physical form, its transition from animal to human, as well as the inconstancy and cyclical nature of governments. As a scientist, Montesquieu focused on the flux that nature delivers and the fact that man must accept the fact that nothing is permanent.

In the *Persian Letters, Letters 11–14*, he outlines three stages of human history: that of the savage/hunter, that of the barbarian/herdsman, and civilization. He begins with the stage of the savage/hunter, a period of development that is so primitive, men are indistinguishable from animals in appearance and behavior. *Letter 11* recounts the history of a prehistoric tribe of cave dwellers called the Troglodytes. It is to Montesquieu's credit that he draws from classical historians to hypothesize that the first men were indistinguishable from animals:

> There was in Arabia a small nation of people called Troglodytes, descended from those Troglodytes of former times who, if we are to believe the historians, were more like animals than men. Ours were not so deformed as that: they were not hairy like bears, they did not hiss, they had two eyes; but they were so wicked and ferocious that there were no principles of equity or justice among them.[2]

Troglodyte is derived from the Greek, τρωγλοδύτης, from τρώγλη, hole + δύειν, to enter, via the Latin, trōglodyta. Hence, a Troglodyte was a cave dweller or, literally, "one who enters holes." Montesquieu cedes to the authority of classical historians when he interjects, "if we are to believe the historians." There is a long list of historians who have written about these prehistoric cave dwellers: Aristotle, Herodotus, Titus-Livy, Thucydides, Pomponius Mela, Pliny the Elder, Aulus Gellius, and Gaius Julius Solinus.

Aristotle had declared that it was true that small people had once lived in caves in the land above Egypt: "Some make their moves from nearby places, but others from practically the farthest, as the cranes do: for they move from the Scythian plains to the marshes above Egypt from where the Nile flows; this is the region whereabouts the pygmies live (for they are no myth, but there truly exists a kind that is small, as reported-both the people and their horses-and they spend their lives in caves)."[3] Aristotle used the term "Troglodytes" (τρωγλοδύται) to identify this race of cave dwelling Pygmies (πυγμαι☐οι).

Herodotus also spoke of the cave dwelling Troglodytes (τρωγλοδύτας) while embellishing upon Aristotle's account: they were such fast runners, the Garamantes needed chariots to chase them and they emitted sounds that were squeaks: "These Garamantes go in their four-horse chariots chasing the cave dwelling Ethiopians: for the Ethiopian cave dwellers are swifter of foot than any men of whom tales are brought to us. They live on snakes, and lizards, and such-like creeping things. Their speech is like none other in the world; it is the squeaking of bats."[4]

Paul Vernière and Antoine Adam, who have each annotated their own editions of the *Persian Letters*, believe that Montesquieu derived his prehistoric Troglodytes from Pomponius Mela. Vernière mentions that Montesquieu had a copy of Mela's *Geography* (*De orbis situ*) at La Brède.[5] Mela wrote, in 43 or 44 AD, "To the east are:...the Troglodytes...To the interior...if it is to be believed-and with difficulty-the half wild Egyptians, the Blemians...All dwelling without walls, they have no fixed seats of habitation and possess only the land."[6] Four chapters later, Mela adds, "The Troglodytes, owning no domestic goods, can only hiss and squeak in their speaking; they occupy caverns below the ground, and live on serpents...The Blemians seem to be without heads; their faces are near the breastbone."[7] Vernière and Adam believe that this is where Montesquieu derived "they did not hiss, they had two eyes" in his description of the latter Troglodytes. Vernière notes,

"The geographer said: 'The Troglodytes, owning no domestic goods, can only hiss and squeak in their speaking'; Herodotus had them utter sharp cries like bats: 'it is the squeaking of bats.'"[8] Adam advises, "Montesquieu read Pomponius Mela. The latter wrote: *The Troglodytes, owning no domestic goods, can only hiss and squeak in their speaking*, and this word explains Montesquieu's: 'they did not hiss.' Then he adds: 'they had two eyes,' he is thinking of the same passage in which Pomponius Mela said about a people neighboring the Troglodytes: *The Blemians are without heads; their faces are in their chests*, or perhaps of Pliny the Elder: *Blemians have no head, their mouths and eyes are situated in their chests*."[9]

It is to Montesquieu's credit that he seized upon the ancients' stories about cave dwellers that resembled animals more than they did men, were hairy like bears, and that hissed. He demonstrated that in ancient times, the boundaries between men and animals were obscure, in appearance, sounds, and living quarters. A metamorphosis occurred and subsequent Troglodytes were not as deformed, they were not hairy like bears, they did not hiss, and they had two eyes. The fact that Montesquieu implies that the prehistoric Troglodytes did not have two eyes indicates that he may also have been influenced by Lucretius. In *De rerum natura* Lucretius declares that in the beginning nature created many monsters and only those that were not self-contradictory survived.[10] Montesquieu concurs with Lucretius and incorporates the notion of the extinction of self-contradictory beings into his first stage of human history, that of the savage/hunter: the first cave dwellers were deformed, they resembled animals, they did not have two eyes which indicates that they were monsters, and they became extinct; their extinction was succeeded by Troglodytes who were not as deformed.

Having provided a brief sketch of the caveman, Montesquieu proceeds to the next stage of human history, that of the barbarian/herdsman. Montesquieu specifies that the subsequent Troglodytes were farmers/herdsmen. This period of herding/barbarism lies between the hunting/savagery part of human history and that of civilized society. By Letter 12, Montesquieu tells us that the barbarian/herdsman Troglodytes perished because of their wickedness and only two virtuous families remained. This was the birth of civilization.

In the sketch of the barbarian/herdsman, Montesquieu shows that the human brain had developed to the point where the Troglodytes began to have a sense of self. The pronouns "je," "me," and "moi" indicate that a consciousness of self had arisen. Furthermore, the negative "I will not worry" (*Je ne me soucie point*), "what does it matter to me if the others are?" (*que m'importe que les autres le soient?*), and "if all the other Troglodytes are miserable" (*que tous les autres Troglodytes soient misérables*) indicate that differentiation between self and others has arisen, as well as a sense of self interest and indifference to the needs of others.

During the barbarian/herdsman era, men appointed kings to rule them. The fact that they had a foreign king who ruled them indicates that they had divided themselves into countries and they could differentiate their own country from those of foreigners. However, vestiges of their animal nature were present. They were half way between animals and civilization: they had a king and a royal family, but they killed them. They were advanced enough to hold a meeting to choose a government and elect ministers, but they killed them, too. They were semi-savage, semi-civilized-hence, they were barbarians. Vestiges of animal instincts were present: they grabbed the territories of others and were willing to fight to the death to secure them; they stole the mates of others and were willing to kill or be killed to keep them.

This period of the barbarian/herdsman ended in disaster and only two virtuous families survived. These two honest, moral families were anomalies among the wicked: "They were humane; they understood what justice was; they loved virtue."[11] This, too, is amazingly prescient: only a very few, who were radically different from all the rest, had developed a quality that permitted them to survive and proliferate; this tiny group of radically different people marked the beginning of modern civilization

Montesquieu's prescience lies in the fact that he recognized that a tiny, anomalous subset within a larger group might have qualities that permit it to survive, proliferate and reproduce. The idea is not new, as Lucretius also posited that in the beginning creatures that had no self-contradictions survived, prospered, and reproduced. However, Montesquieu is applying the principle to human behavior rather than physical characteristics. The wickedness and treachery of the evil Troglodytes were self-contradictory and led to their extinction. It is precisely the absence of this self-contradiction that led to the survival of the virtuous Troglodytes.

The virtuous Troglodytes practice a perfect form of communism-they have a desire to share with each other and this willingness to share is born of the natural virtue of the human heart and is not taught by any written code.

Montesquieu portrays these herdsman idyllically: "In the evenings, as the herds came in from the fields and the tired oxen brought in the plows, they would gather together...They described the delights of the pastoral life and the happiness of a situation that was always adorned by innocence. In this happy land, cupidity was alien. They would give each other presents, and the giver always thought that the advantage was his. The Troglodyte nation regarded themselves as a single family; the herds were almost always mixed up together, and the only task that was usually neglected was that of sorting them out."[12]

Paul Vernière believes that Montesquieu, who published the *Persian Letters* in 1721, was influenced by Fénelon's *Telemachus* (1719). Alluding to the proximate publication dates, Vernière discusses the similarities between Montesquieu's virtuous Troglodytes and Fénelon's shepherds:

> The memory of Fénelon constantly remains here. It is the Bétique of *Telemachus* (Wetstein, 1719, book VIII, p. 170), with all of the fabulous themes of the Golden Age. The Troglodytes, virtuous like the people of Bétique, are shepherds: "The innocence of morality, good faith, obedience and the horror of vice dwell in this blissful land." "All property is communal, the fruit of the trees, the milk of the flocks are of such abundant riches that such well-balanced, moderate people do not need to share them." Communism and fraternity. "They love another with a brotherly love that nothing can destroy." The same Arcadian ideal that excludes money, commerce, urban life, conquests and war. Montesquieu's same style imitates Fénelon's smooth speech.
>
>   Religion is not revealed to the Troglodytes. It is natural and arises spontaneously from virtuous hearts.[13]

Otis Fellows finds that Montesquieu had shown that a sense of justice is essential to all political organization and that without it, the survival of civilization is not possible. Fellows believes that the fable "illustrates M.'s belief that a sense of justice, the supreme political virtue, is essential to the organization and maintenance of human society. His pessimism is revealed in the last letter when he portrays the republic founded on virtue giving way to a monarchy based on the theoretically less desirable principle of honor. These ideas will be further clarified in *De l'esprit des lois* (Book III, 3–7)."[14]

J. Robert Loy points out that Montesquieu had observed a recurring, cyclical evolution from order to chaos, to order to chaos, *ad infinitum*, in governments. The Troglodyte story indicates that even these ancient men progressed through all forms of government. Loy observes:

> The Troglodytes he had invented theoretically for *The Persian Letters* now become an actual moment in history, where he can study the evolutions and mutations from chaos to republic to monarchy to despotism to chaos. To judge of the function of a particular law, the law must be comprehended in its historical context, in the particular moment of the history of a particular people.[15]

> Montesquieu sketches here for the first time his important discovery of the natural and historical cycle of governments. The parable supposes a tribe of cave-dwellers...[16]

> The cycle has come full circle and one senses that the prince will, in the absence of individual responsibility in the citizenry, become despot, then tyrant. There will be eventual revolt, disorder, small republic, large republic, monarchy; and again the cycle. The parable announces Montesquieu's theory of governments; there are three basic types: despotic, monarchic, and republican (the latter divided into oligarchy and democracy). Each government has come into existence and continues to exist because of a certain set of circumstances and therefore has a peculiar guiding principle. Virtue is the principle of republics and when it grows too burdensome to the citizens, the republic is corrupted and heading for the eternal cycle: chaos, order, chaos.[17]

> The episode of the Troglodytes is instructive. Social existence is cyclical and never constant. If Montesquieu at first thought natural virtue (contrary to Hobbes' state of war) was the regulator of social order, he later realized that in other positions along the cycle, a more sophisticated and artificial form of virtue-honor-must take over the job of regulating.[18]

In *The Spirit of Laws* (1748), Montesquieu continues to explore flux and the events that nature delivers: in his book he captures the impermanence of the forms of governments that have arisen from the moment that man had developed a sense of self and others. He begins by observing that a nexus exists between climate and human physiology and temperament.[19] He extrapolates that it is appropriate that nations situated in different climates each use forms of government suited to the physiology and temperaments of their inhabitants. In *Books 14–17*, he contends that men's physiology and temperament are influenced by the nature of the climate; in *Book 18*, he examines how the nature of the soil affects man's body and temperament. In these five books, he posits a "general spirit" or array of tendencies that a nation has because of its climate, geography, topography, and soil. By *Book 19, Chapter 14*, he calls climate "the first of all empires": the metaphor suggests the strong influence that climate has on the body and mind. This metaphor also implies that climate, like government, is a structure that has loose control in which individuals can act, but it is not absolutely determinist; legislators can pass laws to promote virtue and discourage vice.

Montesquieu describes in detail the physiology intrinsic to natives of various latitudes. J. Ehrard says, "...he arms himself with a microscope to give to this common ground of the knowledge of nations a solid scientific foundation."[20] In *Book 14, Chapter 1*, Montesquieu capsulizes his thesis thus: "...the character of the spirit and the passions of the heart are extremely different in the various climates..."[21] In *Book 14, Chapter 2*, he begins by establishing a causality between air temperature and physiology: "Cold air contracts the extremities of the body's surface fibers; this increases their spring and favors the return of blood from the extremities of the heart. It shortens these same fibers; therefore, it increases their strength in this way too. Hot air, by contrast, relaxes these extremities of the fibers and lengthens them; therefore, it decreases their strength and their spring."[22] Then he observes a relationship between the physiology of people residing in cold climates and superior physical strength: "Therefore, men are more vigorous in cold climates. The action of the heart and the reaction of the extremities of the fibers are in closer accord, the fluids are in a better equilibrium, the blood is pushed harder toward the heart and, reciprocally, the heart has more power."[23]

Having established that warm weather and cold weather have observable and measurable effects on the human body, Montesquieu proceeds to demonstrate that these physiological effects, in turn, affect intellect and temperament. Montesquieu posited that cold climates produce people who are physically superior and hence, more self-confident: "This greater strength should produce many effects: for example, more confidence in oneself, that is, more courage; better knowledge of one's superiority, that is, less desire for vengeance; a higher opinion of one's security, that is, more frankness and fewer suspicions, maneuvers, and tricks. Finally, it should make very different characters."[24]

Montesquieu observes of hot climates: "As you move toward the countries of the south, you will believe you have moved away from morality itself: the liveliest passions will increase crime; each will seek to take from others all the advantages that can favor these same passions."[25] People living in tropical climates have the tendency to be hot tempered and less in control their passions; conversely, people in cold climates tend to be more rational and less disposed to crimes of passion. Hence, a cold climate > the heart has more power > more vigorous people and superiority of strength > greater boldness and more courage, greater sense of superiority, greater self-confidence and less desire for revenge. Conversely, hot weather causes faintness and people who avoid bold enterprises.

It is significant that Montesquieu uses the metaphor "machine" (*machine*) several times to identify the human body. In *Book 14, Chapter 2*, he says: "In northern countries, a healthy and well-constituted but heavy machine finds its pleasures in all that can start the spirits in motion again: hunting, travels, war, and wine."[26]

In the eighteenth century the primary definition of *machine* was "Engine, instrument good for moving, pulling, raising, dragging, launching something. *Great machine. Admirable machine, marvelous, new machine, very ingenious machine. War machine. Ballet machine. Machine that casts bulky stone bricks, that discharges a hundred bolts at a time. Machine to draw water. Machine to raise stones to the top of a building. Hydraulic engine, or for water. To invent a machine. To set a machine in motion. This machine works well, runs well. Power transmitted by machinery. The pieces, the springs of a machine.*"[27]

The tertiary definition of *machine* was: "A certain assembly of springs whose motion and power are self-contained. *The clock is a fine machine. Robots are very ingenious machines.*"[28]

The fourth definition of *machine* was : "It is said fig. *That man is an admirable machine.* Ancient poets called the Universe, the round machine."[29]

Montesquieu is clearly using *machine* in the mechanist sense: just as the universe is a finely tuned machine whose cogs, wheels, springs, and pulleys are interdependent upon one another, so is man's body a finely tuned machine largely influenced by climate and man's temperament is contingent upon his physiology. In the man-machine metaphor, man is part of a larger structure, the earth's environment, that influences his body and mind. Visibly absent from the text is any mention of a soul. Montesquieu is a scientist here. His study of man is based on empiricism, the scientific method, experimentation, and observation. He collates observations of nations in many latitudes and arrives at hypotheses.

However, Montesquieu is very cautious. Critics call him the *libéral conservateur* (which can mean liberal conservative or conservative liberal) because of his prudence, although his caution did not prevent him from getting censored by the Church. Montesquieu, while using the term *machine*, did not go as far as La Mettrie, who, the year before, had written in the next to the last paragraph of *Man-Machine*, "Let us then conclude boldly that man is a machine, and that in the whole universe there is but a single substance differently modified."[30] Although he was a mechanist, Montesquieu did not articulate a monistic materialist cosmology. Significantly absent from the *libéral conservateur's* book is any statement that the physical world is all that there is: he presents his ethnographic analysis and the reader must extrapolate his own conclusions. Montesquieu demonstrates that man is a machine whose physiology is greatly influenced by the temperature of the environment and the food that the soil produces. This, in turn, affects his state of mind, outlook in life, and personality characteristics. The Church censored his work because if man is a machine whose temperament and state of mind are influenced by the environment, free will and therefore, sin are jeopardized. To undercut free will and sin is to negate the need for a Savior to die vicariously for man's sin.

In *Book 14, Chapter 12*, Montesquieu employs *machine* twice again. Regarding suicide among the English, Montesquieu observes that "...among the English, it is the effect of an illness; it comes from the physical state of the machine and is independent of any other cause."[31] Then again in the next sentence, Montesquieu uses the term *machine* once more as a metaphor for the body: "It is likely that there is a failure in the filtering of the nervous juice; the machine, when the forces that give it motion stay inactive, wearies of itself; it is not pain the soul feels but a certain difficulty in existence."[32] There are religious implications to explaining suicide by the effect of climate on the machine (body and mind): if man is a machine influenced by climate, then free will is gone and suicide is no longer a sin. Suicide becomes a metaphor for any action resulting from physiology and hence, the more general notion of all sin is negated.

By *Book 19, Chapter 14*, Montesquieu declares, "Climate is the first of all empires." In this sentence, he metaphorized climate as a governing body with a complex system of laws that is an iconic representation of the mechanistic laws of the universe. These laws are not deterministic, but rather, man, with the help of wise legislators, can set up laws and governments that will promote virtue and discourage vice. The fact that climate is the first of all empires indicates that others must follow.

Most critics concur that Montesquieu was not an absolute determinist. It would be incorrect to posit that he promulgated geographical and environmental determinism. Ehrard states, "...far from establishing a unilateral determinism, he believes that the legislator can and must fight the 'climate's vices' (Chap. 5 to 9)."[33] As an example, Ehrard points out to *Book 14, Chapter 5*, that is entitled, "That Bad Legislators are Those Who have Favored the Climate's Vices and Good Ones are Those Who have Opposed Them."

C.P. Courtney agrees that Montesquieu was not a determinist:

> Significantly, chapter 5 of Book XIV is entitled, "That those are bad legislators who favour the vices of the climate, and good legislators who oppose those vices," a useful reminder that, while Montesquieu believed that the various factors that make up the "general spirit" have an enormous influence on our lives, he was not an environmental determinist, if by this one means that human beings have absolutely no freedom of choice. His considered view on such factors is that environmental factors, particularly the basic ones of climate and geographical situation, are most influential on primitive and unenlightened peoples. It is in this sense that he writes in chapter 14 of Book XIX: "the empire of climate is the first of all empires."[34]

Courtney goes on to say, "Awareness of the factors that shape our life in society enables us to act in such a way that we can manipulate these factors or even oppose them."[35]

Courtney concludes that Montesquieu was opposed to slavery and he viewed slavery as contrary to natural law. Courtney points out that Montesquieu's approach to the problem, which involved taking into account environmental factors, convinced him that it was possible to find a logical explanation for the existence of slavery. It is "natural" in the sense that there is a "necessary relation" between slavery and the form of government (despotism) and climate (in hot countries). It is "unnatural," however, for intelligent beings to allow themselves to be dominated by the influence of climate or to be the instruments of the immoral power of a despot. In this sense slavery is contrary to natural law and cannot therefore be condoned even though it can be explained by "natural causes":

> But, as all men are born equal, slavery must be accounted unnatural, though, in some countries, it be founded on natural reason; and a wide difference ought to be made between such countries and those in which even natural reason rejects it, as in Europe, where it has been so happily abolished (XV, 8).[36]

Courtney contends that Montesquieu did not embrace an absolute geographic or environmental determinism, but rather, merely observed tendencies based on climate that should be negated by legislation if they are contrary to natural law. In this sense, *The Spirit of Laws* is an extension of the Troglodyte story: the universe is in continual flux and nothing stays the same forever. The earliest Troglodytes were indistinguishable from animals and they metamorphosed into less deformed creatures who eventually looked like modern man; climate is the first empire and formed the temperament of the earliest men; modern man, and legislators in particular, have the power and duty to correct the climate's vices.

Montesquieu's universe is characterized by flux: the events are species and governments. Nature continuously delivers events and the events can be improvements in physiological structures, as in the case of the Troglodytes, or in customs, manners, and laws. If climate is the first empire,[37] man can overcome the vices of climate by using his mind. Customs, manners and laws are events that change. Henri Coulet observes that during the eighteenth century, man grew accustomed to the anxiety that results from a perpetually changing universe. Coulet says that in the eighteenth century, men were ready to "tolerate contradictions" and accept the fact that "they have been immersed in the changing flux of phenomena and the inexhaustible chain of causes and effects."[38] Coulet mentions that Fontenelle had declared that the apparent permanence of the universe is illusory.[39]

Robert Shackleton is also of the mind that Montesquieu was not an absolute determinist:

> It is not a rigorous and systematic doctrine. Certain effects on men's minds are in part caused by the climate, and as men's minds influence the forms of government under which they live, so climate, vicariously, influences those forms of govern-

ment. This is a moderate and limited doctrine. It is only a part, and even a small part, of the doctrine of *L'Esprit des lois*; and the first impression one ought to retain of the theory of climate in Montesquieu's work is of the narrowness of the limits it occupies.[40]

Shackleton goes on to say:

> The imputation of determinism or of fatalism has often been made against the author of *L'Esprit des lois* on the strength of his theory of climate. A reply to this charge can be based on the instruction which Montesquieu gives to the legislator when confronted with a people climatically disposed in a certain way. The good legislator, he says, must resist the vices of the climate. The bad legislator will accept them. In hot lands, for example, in order to overcome the idleness engendered by the climate, laws should seek to remove all possibility of living without work.[41]

In summation, Montesquieu painted a universe that is in continual flux: he presciently posited that the events that flux delivers are physiological structures in living beings, as well as customs, manners, laws, and governments. He established that there were three stages of human history: that of the savage/hunter, that of the barbarian/herdsman, and civilization. Montesquieu drew from ancient historians to posit that in the far past, men were indistinguishable from animals. As time progressed, men became less deformed and approached their current form. Physiology, temperament, intellect, societal customs such as polygamy and slavery, and forms of government from democracy to despotism, are largely influenced by, but not absolutely determined by, climate, topography, and soil. Man has the power to progress beyond the "first empire," which is climate, and move on to laws and customs that are more in line with natural law. The vices of climate are no more static than anything else in the physical universe and man can improve his lot in life through legislation.

Chapter 3
La Mettrie

In those days, he did not consider himself king over the other animals, nor was he distinguished from the ape...[1]
—Julien Offray de La Mettrie, *Man a Machine* (1747)

Throughout his literary career La Mettrie observed that the chain of beings is comprised of an infinite variety of beings. However, unlike his compatriot and friend in Holland, Maupertuis, who had articulated a transformist point of view in the *Essai sur la formation des corps organisés* (1745), La Mettrie did not consider that one species may have arisen from an antecedent species. This is surprising because he and Maupertuis were friends, they both lived in Holland, and La Mettrie dedicated *The Natural History of the Soul* (1745) to his friend. La Mettrie developed his own ideas, and as Jacques Roger observed, "he followed his own path."[2] Nevertheless, La Mettrie contributed many innovative ideas to the French Enlightenment that paved the way for transformism.

For example, in *The Natural History of the Soul* (1745) he establishes that consciousness is contingent solely upon the physiology of the brain and central nervous system: all psychic phenomena (emotions, thoughts, passions, and perceptions) can be fully explained by physiology alone (the condition of the body, inheritance, the kind and amount of food that is ingested, and age), and also other environmental factors such as education and learning, and climate and the environment. Hence, he dismisses the notion of the immortal soul as fictive.

Having established that consciousness is solely contingent upon physiology, La Mettrie goes on, in *Man a Machine* (1747), to explore in great detail the physical similarities between man and "other animals." He declares that the transition from animals to man is not violent, that the apes are the animals that are closest to man in the physiology and functioning of the brain and central nervous system, and that there was a time in the distant past, before man had developed language, when he was indistinguishable from the apes.

The following year, in *Man a Plant* (1748), he examines in great detail the physical similarities and divergences between man and the vegetable kingdom.

By 1750, in *The System of Epicurus*, La Mettrie borrows Diderot's language of games of chance to show that infinite collisions of atoms, like throws of dice, will eventually yield patterns, and over an infinite amount of time, they will give rise to living beings. As Diderot had done in *Philosophic Thoughts* (1746), *Thought 21*, La Mettrie defends Epicurus' view that atoms, which have the property of motion, are continually colliding with each other: these collisions are random events, and over an infinite amount of time, they will randomly combine to form beings of increasing organizational complexity. The formula here is flux+time=the dispersion of chaos. La Mettrie derives his language and ideas pertinent to games of chance from Diderot: La Mettrie employs the phrases "random chance," "lucky combinations," "an infinite number of combinations," and "a chance arrangement."

Taken as a whole, these four books attempt to explain man by taking, as a starting point, man himself. In 1745 La Mettrie dispenses with the notion of the soul: he scrutinizes the physiology of the human body to show that consciousness can be fully explained by the functioning of the brain, nervous system and sensory organs; in 1747 he observes the great similarity between the organs of man and those of the ape; in 1748 he proceeds down the chain of being and observes similarities between animals and vegetables, and finally, in 1750, he returns to the origin of things and the creation of the universe from the random collision of atoms. However, in 1750 he was a panspermist, not a transformist. He was searching to identify the origin of man. He died in 1751; had he lived, he would have likely responded to the transformist concepts presented in Diderot's *Thoughts on the Interpretation of Nature* (1753), as he and Diderot reacted to each other's ideas in their works. This chapter will follow La Mettrie's journey, from 1745 to 1750, as he investigated the origin of man.

La Mettrie was a materialist and a monist, or one who admits that the universe is comprised of only one substance, matter. In the eighteenth century "materialist" was defined as "one who allows only for matter";[3] "materialism" was defined as "the opinion of those who do not allow for any substance other than matter."[4] *Webster's* defines materialism as "a doctrine, theory, or principle according to which physical matter is the only reality and the reality through which all being and processes and phenomena can be explained."[5]

Notably absent from the definition of materialism is any attribution of reality to spirit or soul. As a physician, La Mettrie held that consciousness is purely the result of physiological activity or more specifically, the functioning of the brain and nervous system. Ironically enough, he owes his monist view of man to Descartes, who had posited that animals, but not men, are merely machines. A century earlier, Descartes had written in a letter to the Marquis of Newcastle, "that our body is not just a self-moving machine."[6] Employing the watch metaphor again in the *Treatise on Man*, Descartes reiterated that the body is self-moving and hence, it may be compared to "clocks, artificial fountains, mills, and other such machines which, although only man-made, have the power to move of their own accord in many ways."[7] From Descartes' *The Passions of the Soul*, La Mettrie gleaned that all of man's motions, including involuntary movements associated with breathing, walking and eating, depend on "the brain, nerves and muscles. This occurs in the same way as the movement of a watch is produced merely by the strength of its spring and the configuration of its wheels."[8]

Although Descartes had held that animals are automatons and men are more than that, because they have spirit, La Mettrie extended the automaton notion to man. La Mettrie rejected the notion that a spiritual realm exists and his thesis was that everything in the universe is merely the result of the organization of matter. His starting point, then, in the *Natural History of the Soul* (1745), is to disprove the existence of the immortal soul. He does this by painstakingly examining the intricacies of the human body and demonstrating that consciousness can be satisfactorily explained solely with biology. The title itself, *Histoire naturelle de l'âme*, is a pun. Natural histories of all kinds have dated back to Pliny the Elder's *Natural History*. Instead of an *Histoire naturelle de l'homme*, La Mettrie decided to cleverly title his work, *Histoire naturelle de l'âme*. From the very first sentence in the work, however, it becomes clear that the work is decidedly not about a nonexistent entity, but rather, an essay extolling the intricacies of human physiology. In the first sentence he advises the reader, "Neither Aristotle, nor Plato, nor Descartes, nor Malebranche will teach you what your soul is. You will torture yourself in vain to learn its nature and, however much it affronts your vanity and insubordination, you will have to submit to ignorance and faith."[9]

Ernst Cassirer observes the emphasis that La Mettrie places on the study of human physiology: it is the surest way to acquire a knowledge of man:

> Therefore the conclusion which alone can assure us of the truth of nature is not deductive, logical, or mathematical; it is an inference from the part to the whole. The essence of nature as a whole can be deciphered and determined only if we take the nature of man as our starting point. Accordingly, the physiology of man becomes the point of departure and they key for the study of nature...Lamettrie begins with medical observations...Lamettrie's first book is entitled *The Natural History of the Soul*. He points out that such a history can only be written by following strictly the physical processes and taking no step not demanded and justified by exact observations. It was such observations made during an attack of fever-when he became emphatically aware that his whole emotional and intellectual life was undergoing a complete revolution-which, as he says himself, determined the nature of his studies and the tendency of his whole philosophy. Sensory, corporeal experience was to be his only guide from now on; he used to say of his senses: "Here are my philosophers' (*Voilà mes philosophes*).[10]

Therefore, what is stated to be a treatise on the soul in the work's title, is actually an essay on the physical body. La Mettrie declares, "I open my eyes and I see about me only matter or extension."[11] The *OED* defines "extension" in physics and metaphysics as "the property of being extended or of occupying space; spatial magnitude."[12] La Mettrie explains that extension is a property of all matter: all matter has length, breadth, and depth. A proponent of Lockian epistemology, he declares that all knowledge comes from our senses and the senses conceive of all matter as having three dimensions. To extension, he adds Epicurus' view that all matter has the property of motion: matter has the power to move and to be moved (kinetic and potential energy). Kinetic energy is when the movement or work is occurring; potential energy is when the movement or work is waiting to be done.

Having established that all matter has extension and motion, he goes on to establish that there exists a third property: consciousness. He bows to the authority of the ancients who acknowledged that all matter is conscious. He proves that consciousness exists all along the chain of beings. For example, animals are conscious and we know that they are because they speak the "language of feeling, such as moans, cries, caresses, flight, sighs, song, in a word all the expressions of pain, sadness, aversion, fear, daring, submission, anger..."[13] These emotions are displayed by animals as well as man. There is a "perfect resemblance" between man and animals: "for here it is only a question of the similarities between sense organs which, a few modifications apart, are completely the same and obviously indicate the same uses."[14] He criticizes Descartes for not noticing the similarity between man's sense organs and those of animals and includes a footnote praising Boerhaave who did and who had applied himself to comparative anatomy. La Mettrie's purpose is to show that the only difference between man and other animals is the complexity of organization.

In order to solidify his case that man is a highly organized animal (actually, an ape that can speak), La Mettrie goes on to examine Plato's tripartite soul-the vegetative, the animal and the rational parts; he provides definitions of each and demonstrates that both man and animals exhibit, to varying degrees, all three aspects. The vegetative soul controls generation, nutrition and growth in living beings; it has the power or faculty of growth. The animal soul or the conscious soul has the function of sensation or sense perception and it is characteristic of animals. The rational soul has the faculty of reason.

The notion that living beings have three essential functions is articulated in Plato's *Republic* (Πολιτεια), where Plato posits a tripartite soul (ψυχη) comprised of the separate vegetative (generative), animal (conscious), and rational parts. Plato was amazingly prescient and it is to La Mettrie's credit that he seized upon the three functions requisite for life: today scientists agree that the brain is comprised of a vegetative core that controls physical functions such as appetite, heartbeat and kidney functioning, an animal layer on top of that which is linked to emotions, passions and fears, and a rational layer on top of that containing the thinking and reasoning faculties. It took the genius of Plato to hypothesize this in 360 BC, and the genius of La Mettrie to reach across the ages, resurrect it, and make it the foundation of man's psychophysiology.

La Mettrie methodically proves that these three functions, which Plato explained by a tripartite soul, are purely the result of brain activity. He shows that animals, like man, exhibit the vegetative, animal and rational functions of their brains, some to a greater degree, others to a lesser degree. Both humans and animals depend on the five senses in order to feel, discern and know. It is the brain that is the center to which and from which all perceptions travel: "Many experiments have taught us that it is actually in the brain that the soul is affected by the sensations specific to animals. For when this part is seriously wounded, the animal no longer possesses feeling, discernment or knowledge."[15] Man and animals are both machines dependent on the functioning of the brain. La Mettrie remarks that many authors place the seat of the soul in the corpus callosum part of the brain, "from which, as from a throne, it governs all the parts of the body."[16]

The salient feature of La Mettrie's discussion is that he describes the process of consciousness in purely physical terms. Whenever he uses the term "soul," the reader understands that he is talking either about consciousness or the brain. Hearing carries the sensation of noise to the soul (brain); the images that come before one's eyes are carried to the soul (brain); the soul (brain) receives sensations from the senses of smell and taste; touch is universally spread over the whole of the body and heat, cold, hardness, softness, etc., are transmitted to the soul(brain). The sensations are carried via the nervous system; "since the motor nerves alone carry the idea of movements to the soul...each nerve is apt to give rise to different sensations."[17]

In chapter 10 he discusses the five senses in great detail and proves that consciousness is based solely on physiological changes that take place. He employs highly technical terms to prove that science is fully capable of explaining every aspect of consciousness that had formally been attributed to the soul. As an example he recommends, "Take an ox's eye, carefully remove the sclerotic and choroid and, in the place of the first of these membranes, put a piece of paper whose concavity fits exactly the eye's convexity. Then put any object at all in front of the hole of the pupil and you will see very clearly the image of this object at the back of the eye."[18] The sense of sight is based purely upon the relaying of images along nerves to the brain.

La Mettrie reiterates that consciousness is based purely on the physical body: "The ideas of size, hardness, etc. are determined only by our organs. If we have other senses, we would have other ideas of the same attributes, and if we had other ideas, we would think differently from the way we think about whatever is called a work of genius or feeling...In addition, feelings change with organs. In some jaundices, everything appears yellow. Change your axis of vision with your finger and you will multiply objects and vary their place and attitude at will. Chilblains, etc. remove the sense of touch. The slightest blockage in the Eustachian tube is enough to make one deaf."[19]

He also explains memory in purely physiological terms and declares that "the cause of memory is completely mechanical, like memory itself; it seems to depend on the fact that the physical impressions on the brain, which are the traces of successive ideas, are close together and on the fact that the soul is unable to discover one trace or one idea without recalling the others which habitually went with them."[20]

He demystifies the imagination, as well, by relying solely upon biology: "The imagination merges the different incomplete sensations which the memory recalls to the brain and makes images or pictures out of them, representing objects...which are different from the precise sensations previously received through the senses."[21] He also elucidates upon the passions-love, hate, fear, audacity, pity, ferocity, anger, gentleness, etc.,-long held to be contingent upon the animal soul-by engaging in purely scientific analyses.

In summation, La Mettrie's starting point in his study of man is to refute to notion of the soul. He demonstrates that all consciousness can be attributed to matter alone (the functioning of the brain, central nervous system and the five senses), and hence, there is no need for an immortal soul. He mentions that if, indeed, the soul did exist, it would have to have extension in order to have any impact on the body or perceptions. He takes exception to Descartes, who thought that extension is not necessary for a substance to have an impact on extended bodies. La Mettrie concludes that the soul must necessarily not be unextended itself (chapter 10, part 8); the reader infers that since its extension is undetected by medicine, it does not exist. Hence, La Mettrie engages in an astute observation of the physical body and dismisses unextended entities as nonexistent.

In his next work, *Man a Machine* (1747), La Mettrie begins by returning to a subject that he had taken up in *The Natural History of the Soul*, namely, the fictive nature of unextended entries, such as the immortal soul and God. He criticizes Descartes and the Cartesians for having "made the same mistake. They have taken for granted two distinct substances in man, as if they had seen them, and positively counted them."[22] Aram Vartanian remarks, "The thesis of La Mettrie's principal work springs from the persuasion that all prior efforts to clarify metaphysically the nature of mind have failed. The dualism of Descartes and Malebranche, the Leibnizian monadology, and even Locke's conjecture that God might have superadded thought to matter, all seem to him to be mere verbalizings rather than rational explanations of the mystery of mental phenomena."[23]

Vartanian also points out that La Mettrie denied the existence of God: he directly refuted the deistic stance that Diderot had taken in 1746. Vartanian observes that in 1746, while Diderot was still a deist, he had defended deism and the existence of a Prime Mover. In *Philosophic Thoughts, Thought 18*, Diderot had written that atheism receives it greatest blows from observations of the wonders of nature, such as in the evidence shown by Marcello Malpighi, Isaac Newton, Pieter van Musschenbroek, Nicholas Hartsoeker, and Bernard Nieuwentyt. The following year, La Mettrie directly refuted this statement in *L'Homme-machine*: the Fénelons, the Nieuwentyts, the Abbadies, the Derhams, the Rays, Malpighi, teach us nothing, their works are the boring repetitions of zealous writers, their wonders of nature prove nothing. Thus, a direct dialogue between the two materialists began. Hence, La Mettrie relied solely upon the physical universe and what the five senses could perceive of it, and he dispensed with metaphysical hypotheses.

As a physician, La Mettrie based his hypotheses on empiricism and the scientific method: "Experience and observation should therefore be our only guides here. Both are to be found throughout the records of the physicians who were philosophers, and not in the works of the philosophers who were not physicians. The former have traveled through and illuminated the labyrinth of man; they alone have laid bare those springs [of life] hidden under the external integument which conceals so many wonders from our eyes."[24]

As he had stated in *The Natural History of the Soul*, everything in the universe is derived from the inherent motion of matter. Therefore, nature (matter) is self-contained; nature does not require a Prime Mover to set it into motion and organize it because molecules, which have the property of motion, set themselves into motion. La Mettrie employs the watch metaphor to illustrate that the human body is self-contained: "The human body is a machine which winds its own springs. It is the living image of perpetual movement. Nourishment keeps up the movement which fever excites. Without food, the soul pines away, goes mad, and dies exhausted."[25] La Mettrie begins by establishing that man is a self-moving machine because later he will demonstrate that man has this characteristic in common with other animals. The fact that all man needs is food and heat to keep the mechanism going, is also something that he shares with the rest of the animal kingdom.

The self-moving phenomenon of the human body was a mystery inherent in nature that needed to be explained. There were two phenomena that pointed to the fact that nature is self-starting: Trembley's polyp and the irritability of muscles. Polyps were a subject that had been hotly debated in academic circles since Trembley's discovery of the freshwater hydra in 1740. Trembley had discovered a freshwater hydra on the shores of Lake Geneva that regenerated itself after being cut into pieces.[26] La Mettrie examines the polyp in detail because 1) it is a bridge between the animal and vegetable kingdoms and hence, it shows the seamless continuity between the chain of beings, 2) the polyp's ability to regenerate (as well as that of starfish and earthworms) presents problems for preformation: how many preexisting germs are needed and where are they located?, and 3) where does the polyp's consciousness go when it becomes two, four, and eight complete animals?[27]

La Mettrie discussed the polyp in 1745, 1747, 1748, and 1750 because it was an iconic representation of nature's inherent ability to start itself. In addition, there was another phenomenon that indicated that nature has a hidden, self-starting force: the irritability of muscles. La Mettrie pointed to the fact that a frog's heart continues to beat for over an hour after it has been removed from the body, and that muscles, when extracted from a body, contracted when stimulated. These two phenomena, the fact that the polyp could regenerate after having been cut to pieces and the continued activity of muscles after excision from the body, proved that there must be more to nature than simply Lucretius' atoms in motion and random chance. There must be something else in a mindless, chaotic, random universe. Lester Crocker observes that the regeneration of Trembley's polyp taught La Mettire that matter "contains within itself the power that produces its activity and-most important-its organization" and that "matter, and living organisms in particular, possess a self-organizing power or impulse."[28]

Crocker cites the passage in *L'Homme-machine* in which La Mettrie declares that there must exist "something else" in matter that causes polyps to regenerate.[29] Crocker observes that that "something else" was supplied by Leibniz: it was Leibniz who had provided La Mettrie with several significant characteristics of matter: "simple substances 'have a certain self-sufficiency which makes them the sources of their internal activities,'" "matter, too, is undergoing constant modification, perpetual change," and "'every present state of a simple substance is naturally a consequence of its preceding state, in such a way that its present is big with its future."[30] Hence, what Leibniz brought to La Mettrie's thought is the fact that "Nature had to be conceived of as a self-creating, self-patterning force, as an experimenting-and a blindly experimenting-force."[31]

Jacques Roger also observes La Mettrie's focus on an innate force in matter that animates it: "A muscle taken from a body contracted when stimulated; a frog's heart continued to beat for more than an hour; a piece of polyp reconstituted an entire polyp. Therefore, 'each little fiber or part of organized bodies is moved by a principle of its own.'"[32] He adds, "It was therefore 'clearly demonstrated...that matter is moved by itself, not just when it is organized, as in a complete heart for example, but even when this organization is destroyed.'"[33]

There is a property of nature that causes it to reorganize and we do not know what it is: nature is self-organizing. Roger cites La Mettrie's explanation: "We really do not know nature: causes hidden in her bosom may have produced everything. Take your own look at Trembley's polyp: does it not contain in itself the causes that give rise to its regeneration?...they may be something else that would be neither chance nor God; what I mean is nature..."[34]

Aram Vartanian also comments that La Mettrie's legacy is his contribution regarding the irritability concept. Vartanian says, "...the heart offered a much more dramatic proof of autonomous energy because, when resected and divided, it could contract even without the benefit of artificial stimulation-a fact that made it possible to derive the automaticity of the organism as a whole from that of the heart, which became the mainspring in the network of internal stimuli that maintained the body in a continuous state of vital activity."[35]

Vartanian goes on to say, "...La Mettrie affirms that 'each tiny fiber, or part of an organized body, moves by a principle which belongs to it, and whose activity does not depend in any way on the nerves.' With respect to the functional mode of the irritable reaction, he observes further: 'The motive principle of the whole body, and even of its parts cut into pieces, is such that it produces not irregular movements, as some have thought, but very regular ones.' The 'seat of this innate force' is placed by him in the living tissues themselves."[36]

Vartanian concludes, "In the final analysis, the chief merit of La Mettrie's discussion is that it views the phenomenon of irritability as the key to the mystery of life itself, and proposes to erect the mechanistic theory of mind on this firm biological foundation."[37]

La Mettrie also explores a gamut of factors that influence consciousness: food, age, learning, inheritance, and climate and the environment. At each juncture he shows that other animals, who are also conscious, are also influenced by the same factors. He begins by examining the need for food, which is the most basic need that man and other animals have in common. Because consciousness is contingent only upon physiology, food ingredients, the manner in which food is prepared (whether it is cooked or raw), and the quantity of food ingested, cause certain physiological changes, which in turn, affect temperament and mood). He acknowledges the impact that food has on the chemistry of the brain: "What power there is in a meal! Joy revives in a sad heart, and infects the souls of comrades, who express their delight in the friendly songs…"[38] People eating heartily at a banquet have a joyful disposition; melancholy people (he describes the melancholy as those who prefer to be alone) and studious people would be out of place at a banquet (he implies that the melancholy and studious have different temperaments because they do not regularly engage in gluttonous episodes).

As an example of the causality between food and temperament, he points to the relationship between eating raw meat and a savage disposition. In this, men and animals are similarly affected by eating raw meat. La Mettrie says, "Raw meat makes animals fierce, and it would have the same effect on man."[39] Conversely, heavy food promotes laziness and indolence. Hence, cause and effect relationships exist between food and temperament and food and aggression among nations that eat raw meat.

Temperament is also influenced by the amount of food we eat. As an example he cites a Swiss judge who, when hearing cases on an empty stomach, was indulgent and merciful towards the defendant, but when he ate a large dinner, his personality changed and "he was capable of sending the innocent like the guilty to the gallows."[40] La Mettrie concludes, "…everything depends on the way our machine is running."[41]

Because man and animals have similar body structures, the factors that influence man's behavior also influences that of animals. He hyperbolizes the commonality between man and animals by pointing out that cannibalism exists both in men and animals: "To what excesses cruel hunger can bring us! We no longer regard even our own parents and children. We tear them to pieces eagerly and make horrible banquets of them…"[42]

If eating raw meat can make man ferocious, the absence of food can cause man to viciously attack his own kind for survival. He utilizes hyperbole to paint a brutal portrait of man's animal characteristics: "to what excesses," "cruel hunger," "we tear them to shreds by the teeth," "we make horrible banquets," and "carried away with fury." These phrases hyperbolize man's similarity to the other species on the chain of beings. Among man, as among other species, "the weakest is always the prey of the strongest."[43]

He also analyzes the influence that an aging body, education, inheritance, and climate and the environment, have on the mind: "One needs only eyes to see the necessary influence of old age on reason. The soul follows the progress of the body, as it does the progress of education."[44] Here "soul" is a metaphor for consciousness. It is brain function that deteriorates with age and hence, consciousness is compromised. Similarly, education promotes a reasonable temperament. Even ethnic differences in wit are purely physiological: one nation has a heavy and stupid wit, and another, a quick, light and penetrating wit. La Mettrie asks, "Whence comes this difference, if not in part from the difference in foods, and difference in inheritance…"[45]

It is interesting that La Mettrie attributes ethnic humor to inheritance. Maupertuis had studied polydactyly and had shown that the birth anomaly is not only inherited, but it could also be predicted with mathematical precision.[46] La Mettrie goes a step further and posits that psychological phenomena, such as appreciation of a certain kind of humor, is inherited. La Mettrie also mentions that cannibalism and a penchant for criminal behavior, such as theft, are inherited. To defend this point he refers to cases from medical history such as that of the woman who used to steal when she was pregnant and whose children inherited the vice, and the daughter of a thief and cannibal who inherited her parent's proclivities even though she had been orphaned when she was a year old and raised by honest people.

He also observes the influence that climate and the environment have on physiology and the mind. When a man travels from one country to another, he thinks and acts differently: just as plants and animals improve or degenerate in a new climate, so does man. In addition, when relocated to a different country, we learn from one another and copy the gestures and accents of others. This can also be said of parrots, capable of replicating words, and monkeys, that mimic the actions of humans.

The impact of climate and learning sets the stage for a physiological comparison of man and other animals: La Mettrie compares the human brain with that of animals. He observes that the form and structure of the human brain is the same as that of other quadrupeds; it has the same shape and arrangement except that it is larger and more convoluted. He observes that of all the animals, man's brain is the largest, followed by that of the monkey, the beaver, the elephant, the dog, the fox, the cat, the bird, the fish, and the insect. Fish have no corpus callosum, and very little brain, while insects have no brain. La Mettrie demonstrates that it is the size of the brain that determines consciousness, intelligence, and placement on the chain of beings. He surmises that it is the size of man's brain that differentiates him from the rest of the animal kingdom.

La Mettrie draws conclusions about ferocity versus gentleness from the size of the brain. The less brain an animal has, the more fierce it is; the gentleness of the animal appears to increase in size and proportion to the size of the brain. Also, the more intelligence, the less instinct. Intelligence is solely contingent upon the brain; a flaw in a tiny, microscopic fiber could have made Erasmus and Fontenelle two idiots. Hence, he embraces Maupertuis' observation that birth anomalies are due to errors in the arrangement of parental elements, and expands it to include flaws in intelligence.

Having established a causality between brain size and intelligence, he is able to portray man as an ape that can speak. He compares the size of the brains of men and animals and finds that because animals have brains, they can be taught to do things: "Among animals, some learn to speak and sing; they remember tunes, and strike the notes as exactly as a musician."[47] He asks, "In a word, would it be absolutely impossible to teach the ape a language? I do not think so.'"[48] He believes that apes can be taught language. He observes, "The ape resembles us so strongly that naturalists have called it 'wild man' or 'man of the woods.'"[49] He believes that monkeys can be taught the sign language of the deaf and dumb. He asks, "Why then should the education of monkeys be impossible? Why might not the monkey, by dint of great pains, at last imitate after the manner of deaf mutes, the motions necessary for pronunciation. I do not dare decide whether the monkey's organs of speech, however trained, would be incapable of articulation. But, because of the great analogy between ape and man and because there is no known animal whose external and internal organs so strikingly resemble man's, it would surprise me if speech were absolutely impossible to the ape."[50] He believes that monkeys and apes are capable of conversation, at the very least, of the sign language of the deaf and dumb. It is to La Mettrie's credit that in 1747 he presciently foresaw the attempt to teach sign language to monkeys, a task that was undertaken by the Yerkes Institute on Washoe the monkey, in the 1970s.

La Mettrie visualizes a gray area between man and the ape that occurred in the distant past and he articulates that man distinguished himself from the ape when he learned language:

> The transition from animals to man is not violent, as true philosophers will admit. What was man before the invention of words and the knowledge of language? An animal of his own species with much less instinct than the others. In those days, he did not consider himself king over the other animals, nor was he distinguished from the ape, and from the rest, except as the ape itself differs from the other animals, i.e., by a more intelligent fact. Reduced to the bare intuitive knowledge of the Leibnizians he saw only shapes and colors, without being able to distinguish between them: the same, old as young, child at all ages, he lisped out his sensations and his needs, as a dog that is hungry or tired of sleeping, asks for something to eat or for a walk.
>
> Words, languages, laws, sciences, and the fine arts have come, and by them finally the rough diamond of our mind has been polished. Man has been trained in the same way as animals.[51]

La Mettrie notes that man has less instinct than do animals: it takes longer for a child to mature than an animal; animals can swim, but babies cannot: children do not know the foods suitable for them, that water can drown them, or that fire can reduce them to ashes.

Narrowing the divide between man and animals even more, La Mettrie maintains that animals experience remorse: a dog that bit his master seemed to repent a minute afterwards: it crouched low and appeared to be sad, ashamed, afraid to show itself, and sullen; a lion would not devour a man that he recognized as his benefactor. La Mettrie declares that it would be absurd to think that animals, who have much of the same physiology as man, do not understand and feel.

Conversely, while animals are ferocious and vicious, so is man: there are criminals, warriors and cannibals. Significantly, he ties criminal behavior to inheritance. Citing medical history as his authority, he hyperbolizes man's animal nature by reiterating his ability to engage in cannibalism: there is the case of the cruel maiden of Chalons who ate her sister; Gaston of Orleans was a compulsive kleptomaniac; there was a woman who stole while she was pregnant and whose children inherited her vice; there was a pregnant woman who ate her husband; there have been cases of women who killed their children, salted their bodies, and ate pieces of them every day; there was the daughter of a thief and cannibal who followed in his steps, although she never knew her parents and had been raised by honest people. La Mettrie further hyperbolizes the atrocities by declaring that medical records contain a thousand examples of vices and virtues that are transmitted from parents to children. Hence, man is a highly organized animal and he inherits vicious behavior from his ancestors.

It is also to La Mettrie's credit that he believes that life is the result of random chance [*hasard*]: "Perhaps he was thrown by chance on some spot on the earth's surface, nobody knows how nor why, but simply that he must live and die, like the mushrooms which appear from day to day, or like those flowers which border the ditches and cover the walls."[52] This foreshadows the panspermia that he will embrace in 1750.

La Mettrie sums up by reiterating that all of nature is self-contained and that therefore, man is a self-contained animal: "Is more needed...to prove that man is but an animal, or a collection of springs which wind each other up, without or being able to tell at what point in this human circle, nature has begun?"[53] Man, therefore, is merely a highly organized ape. He makes an analogy between man-ape relationship and that of Huygens' planetary pendulum-Julien Leroy's watch. "More instruments, more wheels and more springs were necessary to mark the movements of the planets than to mark or strike the hours."[54] Men are "these proud and vain beings" who "are at the bottom only animals and machines, which, though upright, go on all fours."[55]

La Mettrie cites many examples to illustrate that the gestation of man is similar to that of other animals, such as that of the transformation of the caterpillar into the butterfly. He describes the development of the embryo at 4, 6, 8 or 15 days: first the head alone is visible, and it is a little round egg with two black points which mark the eyes. Before that, everything is formless and one sees only a pulp, which is the brain, and in which are formed the roots of the nerves and the heart.

As William Harvey had done in 1651, he mentions the rising point (*punctum saliens*) or when the heart first begins to beat.[56] Then one sees the head lengthen from the neck. In the same way, man develops during gestation in increments, just as animals do. La Mettrie observes, "Such is the uniformity of nature, which we are beginning to realize; and the analogy of the animal with the vegetable kingdom, of man with the plant. Perhaps there even are animal plants, which in vegetating, either fight as polyps do, or perform other functions characteristic of animals."[57] Hence, the process of generation is the same between man and animals.

La Mettrie concludes, in the next to the last paragraph, "Let us then conclude boldly that man is a machine, and that in the whole universe there is but a single substance differently modified."[58]

In *Man a Plant* (1748) La Mettrie magnifies and examines the similarities between the physical structures and their functions of the animal and plant kingdoms. In *Chapter 1* he describes human organization in terms of the parts of plants: both man and plants have a main root and capillary roots; the main root in man is comprised of the lumbar region and thoracic canal, and that of plants, by the lacteal veins. Man has lungs with which to breathe, while plants breathe through their leaves. Man's lungs contain branches and plant leaves also contain tiny branches; more branches in the lungs and leaves permit more comfortable breathing. Man has a circulatory system; plants also have fluid circulating through their tubes. He goes on to expound in great detail on the similarities of the reproductive systems in man and plants.

Having examined the similarities between the animal and plant kingdoms, he devotes *Chapter 2* to addressing their differences. In this chapter he comments that nature produces organs to satisfy the needs of its creations: "First, the more needs an organism has, the more nature gives it means for satisfying them. These means are diverse degrees of sagacity known as instinct in animals and soul in man."[59] While La Mettrie wrote this in 1748, in 1769 Diderot would echo, "Organs produce needs and likewise, needs produce organs."[60] Nature has given animals instinct which they need to survive, while it has given it to man in a much lesser degree; man does not require instinct as much as other animal do, because he has intelligence or reason. He goes on to say "Second, the fewer needs an organized body has, the less difficult it is to nourish and raise, and the less its share of intelligence."[61]

He observes that in the chain of beings, after the mineral and vegetable kingdoms, come creatures which begin to be animate, namely, polyps and animal-plants that have the body parts and functions of both the vegetable and animal kingdoms. He observes that when one progresses up the chain of beings to creatures that are more highly organized than are vegetables, instinct and intelligence appear. One of the salient features of the polyp is that it shows sign of both intelligence and instinct.

In *Chapter 3* he declares that the chain of beings is comprised of a continuum of infinite gradations: "…a ladder so imperceptually graduated that nature climbs it without ever missing a step through all its diverse creations...One goes from white to black through an infinite number of nuances or degrees..."[62]

Crocker points out that La Mettrie's discussion of Trembley's polyp was an important contribution to eighteenth-century biology because he placed it in an intermediary position on the chain of beings between the vegetable and animal kingdoms; subsequent thinkers could take it from there, but at least he showed that animals that bridge kingdoms, do, indeed, exist.[63] However, Crocker concludes that it was for others to determine the concluding step: La Mettrie, himself, was not a transformist. La Mettrie observed the infinite variety in nature and that one can proceed up the chain of beings with seamless continuity, but nowhere does he articulate that one species was transformed into another species.

Jacques Roger observes that La Mettrie was not a transformist, but rather a panspermist; he agrees that La Mettrie's view of the chain of being as a continuum of infinite gradations paved the way for transformism, even thought he, himself, was not a transformist.[64] Hence, both Roger and Crocker observe a visible absence of any tranformist basis for La Mettrie's chain of beings: there is no notion that species depend on antecedents for their forms and functions; rather, La Mettrie merely observes the infinite variety of forms and the fact that all forms have organs to meet their needs.

Ascending the chain of beings past the vegetable kingdom, he reiterates, as he had done the year before, the great similarity between man and the ape. He declares, "The ape obviously resembles man in many other ways than toothwise. Comparative anatomy is proof of this, although teeth were what led Linnaeus to rank man with the quadrupeds, indeed, at their very head."[65] While the ape may be docile, man shows a much greater aptitude for education. He deems that the differences between man and the apes "obviously result from man's constitution."[66] It is man's superior intelligence that ranks him as the king of the animals and the only one apt for society. What differentiates man from the apes is that he invented language, laws, and customs. La Mettrie reiterates that while animals have a lot of instinct and little intelligence, man has a superior intelligence that distinguishes him from other animals, but very little instinct.

In *The System of Epicurus* (1750) La Mettrie embraces the philosophy of Epicurus expounded upon by Lucretius in *De rerum natura*. He contemplates what the origin of man might have been like millennia ago and, in *Chapter 10*, metaphorizes the earth as a giant uterus that gave birth to human beings: "...that it must have opened its breast to human germs, already prepared, so that this superb Animal, given certain laws, could hatch."[67] He asks, "...why would the Earth, this common Mother & nursemaid of all entities, have refused to animal seeds, what it granted to the meanest, most useless, most pernicious vegetables?"[68]

In *Chapter 11* he metaphorizes the earth, which no longer produces human beings, as an old hen that no longer lays eggs and an old woman who no longer has children. In *Chapter 13* he echoes Diderot's famous passages on Lucretian monsters in the *Letter on the Blind* (1749). Like Diderot, La Mettrie says, "The first Generations must have been very imperfect. Here the Esophagus would have been missing; there, the Stomach...the only animals which would have been able to live, survive, & perpetuate their species, would have been those that would have been found to be provided with all the necessary Parts for generation, & in which, in a word, no essential part was missing."[69] Those that lacked essential parts died after birth without leaving progeny. As Diderot had done in *Philosophic Thoughts, Thought 21* and the *Letter on the Blind*, La Mettrie observes that nature is in continual flux, and, given an infinite amount of time, randomly and haphazardly produces events that are viable combinations. Again we have the formula flux+time=dispersion of chaos. La Mettrie presciently states, "Perfection has not been accomplished in one day for Nature, any more than it has been for Art."[70] Perfection took a long time, continual flux, and after innumerable events, man's current physical form was achieved.

As Diderot had done in the *Letter on the Blind*, La Mettrie also introduces the reader to a present day monster in his narrative as an iconic representation of the mindless, random flux that nature delivers: in *Chapter 14* we meet a woman without any reproductive organs. La Mettrie puts a human face on the tragedy of birth anomalies and enumerates every organ that the woman was lacking which makes the pathos of her situation all the more striking. She was married for ten years and her husband, a naïve, ignorant, and uneducated peasant, was not capable of informing his wife of what she lacked. Plans for reconstructive surgery had to be abandoned and her marriage was annulled after ten years. She is a tragic figure, just as tragic as Diderot's Saunderson who cries, "Look at me, Mr. Holmes. I have no eyes. What have we done, you and I, to God, that one of us has this organ while the other has not?"[71] As in games of chance, nature delivers flux and given enough time, all outcomes will eventually occur. Hence, nature is mindless and there is no one to blame for birth defects.

In *Chapter 15* La Mettrie surmises that if such deformities occur today, there were many more in the past. This echoes Lucretius *De rerum natura* and Diderot's *Letter on the Blind*. In *Chapter 16* he reiterates, "Through what an infinite number of combinations must nature have passed, before arriving at the only one which could result in a perfect Animal!"[72] Again we have chaos, flux, time, and probability: given an infinite amount of time and the random motion of atoms, organized beings will eventually form. Flux+time dispel chaos. In 1750 La Mettrie echoed Diderot's *Philosophic Thoughts* (1746), *Thought 21*, in which the latter demonstrated that given an infinite number of atomic collisions, life would eventually form: the atheist professor asks whether random keystrokes on a printing press would eventually produce Voltaire's *La Henriade* or Virgil's *Æneid*. Diderot would, in turn, echo La Mettrie in 1753 when he would discuss the infinite variations in nature (*Thoughts on the Interpretation of Nature, Thought 12*).

In *Chapter 17* La Mettrie discusses lucky combinations (*combinaisons fortuites*). Again, this echoes Diderot's *Philosophic Thoughts, Thought 21*: it is purely random chance or lucky combinations that produced eyes, ears, and the five senses. In *Chapter 18* he reiterates the notions of random chance and the mindlessness of nature: "Nature no more dreamt of creating the eye to see, than water, to serve as a mirror for the simple Shepherdess."[73] Nature has no purpose: over an infinite amount of time, there is an infinite flux of events that some may interpret as purposeful, but, nevertheless, outcomes resulting in organization are merely lucky combinations.

In *Chapter 19* we have chance again: "*Chance often goes farther than Prudence.*"[74] Here a painter haphazardly throws his brush against a canvas and creates beautiful foam. The striking of the brush against the canvas becomes a metaphor for the random collision of atoms that creates life. Random collisions can form things of beauty.

In *Chapter 22* consciousness is contingent upon physiology: if there is a grain of sand in the Eustachian tubes, we cannot hear, if the optic nerve is obstructed, we see nothing. Consciousness is the result of physiology, which is the result of the random collision of atoms. As in Diderot's *Philosophic Letters*, we have the same sequence of events in La Mettrie: atomic collisions>living beings>consciousness.

In *Chapter 23* we have nature's trial and error again and a flux of events: Art's fumblings to imitate nature give us an idea of what nature's were like.

The painter struggling to create a realistic piece of art must use trial and error to see what works best. This reiterates *Chapter 19* where the painter throws the brush at the canvas. The artist uses a stream of events and selects what he likes best. Nature, however, is totally blind and creates every possible variation through trial and error. *Chapter 23* emphasizes random chance; nature fumbles by a throw of the dice; there is no Intelligent Designer.

In *Chapter 24* La Mettrie employs Diderotian verbiage,"chance arrangement," again. Nature works "by the laws of movement." Hence, it is the motive property of atoms that randomly collide and chance arrangements that have created the things we see in nature.

There are numerous instances in which La Mettrie borrows Diderotian language: infinite number of combinations (ch 16), lucky combinations (ch 17), chance (ch 19), chance arrangement (ch 24), blind cause (ch 28), and nature's blindness (ch 29).

In *Chapter 32* La Mettrie draws a conclusion as to man's origins. He suggests that perhaps all life came from the sea and refers the reader to Maillet's *Telliamed*. He surmises that once the ocean covered the whole planet and that as the waters receded, human eggs were beached on the shores, incubated beneath the sun's warmth, and hatched human beings.

Historically, "panspermia" has been defined as "the theory that there are everywhere minute germs which developed on finding a favourable environment."[75] Roger observes that La Mettrie promulgated the notion that the seeds of all living things came from the air and landed on earth.[76] Roger observes that La Mettrie, in the manner of Lucretius, embraced the notion that the first generations had been imperfect and only those without significant self-contradictions survived. Roger studies the following sentence in its context: "Through what an infinite number of combinations matter had to pass before arriving at the lone combination that might produce a perfect animal."[77] Roger's conclusion is that La Mettrie is emphasizing that nature is continually producing new combinations and not that man has metamorphosed from beings of lesser organization. Roger holds that La Mettrie adopted his panspermist ideas from Maillet's *Telliamed*: La Mettrie accepted Maillet's notion that "the human egg had been left by the sea in its retreat and hatched in the sun."[78]

Roger finds that La Mettrie has made at least three valuable contributions, based upon Lucretius, to the study of man's origins: 1) like Lucretius, he rejected the notion of final causes and observed the disorder and blindness

La Mettrie 65

of nature, 2) he posited a determinism based on the physical laws of nature, and 3) he held that it is not a divine intelligence, but rather, the motive principle of nature that created everything.[79] Roger concludes that although La Mettrie knew of Maupertuis and Diderot, "he followed his own path."[80]

Lester Crocker agrees with Aram Vartanian and Jacques Roger that La Mettrie derived his ideology from Maillet: he declared that seeds came from the air; when the oceans receded, they beached human eggs on the shores where they incubated and hatched; from Diderot and Lucretius, he gleaned that the first generations of men were imperfect and that only those without self-contradictions survived.[81]

The critics note that La Mettrie and Diderot engaged in a conversation with one another via their books. Aram Vartanian observes that La Mettrie and Diderot had a great influence on each other's thought, even though there is little evidence that they knew each other personally.[82] He cites studies done by Jean E. Perkins, Leo Spitzer, Jean-Pierre Seguin, and annotated editions of Diderot's works by Jean Mayer and Paul Vernière, which examine the points of convergence and divergence between the two materialists.[83] Even though Diderot lived in France and La Mettrie, in Holland and then in Prussia, they read each other's works and responded to one another's ideas in their books.

Vartanian points out that both men had their books shredded and burned by the public executioner in 1746. The Parlement de Paris condemned Diderot's *Philosophic Thoughts* and La Mettrie's *Natural History of the Soul*. Because the books had been written anonymously, it was believed that La Mettrie was the author of Diderot's *Philosophic Thoughts*. In 1750 La Mettrie praised the work, but denied that he had written it. He modestly, but eloquently said, "No, Sir, I am not the author of *Philosophic Thoughts*. Perhaps my grounds have never borne such beautiful fruit."[84]

In 1746, while Diderot was still a deist, he had defended deism and the existence of a Prime Mover: in *Philosophic Thoughts, Thought 18*, he wrote that atheism receives it greatest blows from observations of the wonders of nature, such as in the evidence shown by scientists. Vartanian points out that in 1747 La Mettrie directly refuted this statement in *L'Homme-machine*: scientists teach us nothing, their works are the boring repetitions of zealous writers, their wonders of nature prove nothing. Thus, a direct dialogue between the two materialists began.

Vartanian also observes that La Mettrie copied Diderot: in 1749 in the *Letter on the Blind*, Diderot surmised that in the first instants of the formation of animals, some beings had no head, some had no feet, some had no intestines, some had no lungs; only the ones that were not significantly self-contradictory survived and left progeny; in 1750 La Mettrie, in *The System of Epicurus*, speculated that the first generations must have been imperfect: some lacked an esophagus, others a stomach, others intestines. The only ones that survived were those that did not lack essential parts. Vartanian scrutinizes Lucretius' *On the Nature of Things*, 5.835–52, and extrapolates that Diderot and La Mettrie "resemble each other a good deal more than either resembles the Lucretian version."[85]

Like Vartanian, Crocker examines the influence that the Lucretian passage in Diderot's *Letter on the Blind* (1749) had on La Mettrie's *System of Epicurus* (1750). As Vartanian does, Crocker cites La Mettrie's passage in which he states that the first generation must have been imperfect and that some beings lacked an esophagus, while others lacked intestines; only the viable survived. La Mettrie explains that matter had to pass through many different combinations before the perfect generations that we have today arose.

Crocker contends that La Mettrie was not a transformist because what is significantly lacking in his writing is the notion that current forms depend on antecedent forms: "...when he speaks of animals being produced before men, 'because it requires more time to make a man' than an imperfect being, he is referring to the idea that nature reworked the same matter into various combinations. The essential elements of transformism, that present forms *depend* on earlier forms, is lacking in La Mettrie."[86]

In summation, La Mettrie contributed many valuable and innovative stepping stones to eighteenth-century transformism, even though he, himself, was not a transformist. He established that 1) physical matter is the only reality through which all phenomena can be explained, 2) consciousness is solely contingent upon the brain, nervous system and the five senses, 3) he observed the similarities in the physiology and functions among creatures in the chain of beings and that there is a seamless continuity between kingdoms as well as between species, 4) he observed that given an infinite amount of time, flux delivers all possible outcomes and organized beings are bound to arise (flux+time=dispersion of chaos), 5) he posited that the human body is a self-moving machine that does not require an outside source to set it into motion, 6) that psychophysiology is comprised of vegetative, sensory, and rational functions, 7) he held that consciousness is influenced by food, age, learning, inheritance, climate, and the environment, 8) he rejected final causes and observed the random chaos in nature, 9) he posited a determinism based on the physical laws of nature, and 10) he held that the motive property of atoms, not God, created everything.

Chapter 4

Buffon

...we should regard nothing as impossible, but believe that everything which can have existence, really exists. Ambiguous species, and irregular productions, would not then excite surprise...[1]
—Georges-Louis Leclerc de Buffon, *The Hog, the Hog of Siam, and the Wild Boar*
(1755)

In the *Natural History* (44 quarto volumes 1749–1804), Georges Louis Leclerc de Buffon escorts the reader on a vast ethnographic tour of six continents. He describes the physical characteristics and customs of various races across the globe, as well as the animals indigenous to those regions. His work is a landmark in biology not only because it is a repository of facts and it propelled Enlightenment Europe forward in the fields of anthropology and embryology, but it also provided observations that Buffon's materialist contemporaries would use to hypothesize that species metamorphose over time. Three definitive texts on Buffon's contributions to the eighteenth century are Jacques Roger's *Buffon: A Life in Natural History* (Ithaca: Cornell University Press, 1997), the chapter on Buffon in Jacques Roger's *The Life Sciences in Eighteenth-Century French Thought* (Stanford: Stanford University Press, 1997), pp. 426–74, and Otis E. Fellows and Stephen F. Milliken's *Buffon* (New York: Twayne Publishers, Inc., 1972). In a noteworthy article, Arthur O. Lovejoy examines Buffon's contribution to the definition of species and why the boundaries of species precluded him from embracing transformism in "Buffon and the Problem of Species" in *Forerunners of Darwin: 1745–*

1859, edited by Bentley Glass *et al* (Baltimore: Johns Hopkins University Press, 1959), pp. 84–113.

Buffon left a legacy of terms and concepts that gave the eighteenth century the notion that living things are not static, but that they metamorphose over time (due to their environment, climate, food, and soil). However, as innovative as he was, he believed that the changes in physical characteristics are minor, that species retain their essential character throughout time, and that they do not metamorphose into new species. Although Buffon had a fixed notion of species, the atheist materialists of his time (ie: Diderot and Maupertuis) used his ideas as building blocks to form their own hypotheses that species metamorphose into new species over time and that all living things could have originated from a single prototype.

EPIGENESIS

The first area in which Buffon moved the eighteenth century forward was in the fledgling science of embryology. By 1762 *embryologie* appeared in the French dictionary; it was defined as "Medical Term. The study of the fetus during its stay in the womb."[2] In the field of embryology, Buffon rejected preformation and embraced epigenesis, William Harvey's theory that the embryo is newly formed at the time of conception and develops by the successive accretions of parts that bud out from one another. Because the theory says that the embryo is newly formed and not a replica of either parent, it can explain birth anomalies, resemblance to either or both parents, to grandparents or distant relatives, and hybrids. Otis Fellows explains, "Only epigenesis, Buffon felt, could satisfactorily explain monstrous births, cases in which the reproductive process has gone awry, such as the stillborn calf he himself had displayed before the Académie Royale des Sciences in 1744. And only epigenesis could explain the piecemeal nature of the resemblance the offspring bears to both parents, in evidence, he noted, whenever one is led to remark that a child has 'its father's eyes and its mother's mouth' (H.N. II, p. 68), and most particularly in evidence in the cases of hybrid animals like the mule and in people of mixed race."[3] Unlike preformation, epigenesis opened the door to change: the offspring was not an exact copy of the parent, but rather, something newly formed, unique among all other individuals in its species, and different from its parents, grandparents, and all its ancestors.

Epigenesis was the eighteenth century's first step in the direction of the metamorphosis of species.

Arthur O. Lovejoy points out that Buffon was a pangenesist.[4] Pangenesis is the theory that the offspring is derived from particles originating from all parts of the bodies of its parents; those particles circulate throughout the bloodstream of the parents and go on to form the fetus. When they do form the fetus, they reproduce the body parts from which they originate. Diderot and Maupertuis also held this theory. In 1749 Buffon proposed that these particles are organic molecules that circulate through the bloodstream of the parents and eventually take on the form of the body parts from which they are derived. Buffon devotes *On the Generation of Animals, Volume 2, Chapter 4* and the *Recapitulation* to explaining how this process occurs. He summarizes it thus: "There is, therefore, in Nature, a matter common to both, which serves for the growth and nourishment of every living thing that lives or vegetates...reproduction is an effect of the same matter,when it superabounds in the body of an animal or vegetable. Every part of organized bodies sends off to proper reservoirs the organic particles which are superabundant for its nourishment: these particles are perfectly similar to the different parts from which they are detached, because they were destined for the nourishment of these parts. Hence, when the whole particles sent off from every part of the body are assembled, they must necessarily form a small body similar to the original, because every particle is similar to the part from which it was detached. It is in this manner that every species of reproduction...is effected...There are, therefore, no preexisting germs, or germs infinitely contained within each other. But there is an organic matter diffused through all animated nature, which is always active, always tending to form, to assimilate, and to produce beings similar to those which receive it. The species of animals and vegetables, therefore, can never be exhausted: As long as individuals subsist, the different species will be constantly new; they are the same now that they were three thousand years ago..."[5]

ORGANIC MOLECULES

In *Volume 2* Buffon reduces all living things to their smallest components-organic molecules. He observed that cows eat grass and then people eat cows and can subsist on plants and animals. He wondered whether there is a life force in the grass that is transferred to the cow and then to humans. Jacques Roger explains that Buffon posited that organic molecules, the seeds of all living things, fell to earth when the earth was formed; Otis Fellows clarifies what happened to Buffon's organic molecules after they fell to earth: "As the body of one living creature may be reduced to minute particles and these particles may become a part of the body of another creature, there must be, he reasoned, one single, divisible substance out of which the bodies of all living things, animal and plant alike, are formed. And, as the particles into which this substance divides, these 'organic molecules,' can sometimes be shaped into a living creature simply by bringing them together, to judge from well-known instances of spontaneous generation, some elemental form of life must actually reside in each one of the particles. Life itself, then, must be regarded as nothing other than 'a physical property of matter' (H.N. II, p. 17), probably closely also to gravitation, magnetism, electricity, and 'chemical affinity.'"[6] Buffon deduced that the number of these organic molecules must be fixed, otherwise they would take over the universe. Fellows continues, "Buffon also stated, in the earliest version of the theory, that the number of organic molecules was fixed, and that they were in the normal course of events neither created nor destroyed (H.N. II, p. 44). If this were not the case, if every act of reproduction could directly increase the numbers of organic molecules, within a little more than a century, he calculated, selecting a convenient example, a single elm seed could give rise to a quantity of wood equal in bulk to the earth itself."[7]

The organic molecule is another important building block that Diderot and Maupertuis would use in the metamorphosis of species. Buffon hypothesizes that there is a living force inherent in all matter: "...and, lastly, that animation, or the principle of life, instead of a metaphysical step in the scale of being, is a physical property common to all matter."[8] Diderot and Maupertuis would carry the idea further and posit that this living force is consciousness and that consciousness is a property common to all matter.

INTERIOR MOLD

Buffon hypothesized that living things must contain a force that permits these organic molecules to organize into life forms. Fellows explains, "Buffon concluded that each living body, every living creature, must possess the power to reshape the molecules and transform their patterns of activity to whatever extent this might be necessary in order to adapt them to its own purposes. In fact, each organ of every living body must possess such power. Confronted with the need for a new label, Buffon gave this remarkable assimilating power the deliberately equivocal name of *moule intérieur*, which may be translated as *interior mold* or, better, *interior molding force*...The *moule* of one individual creature, which was a sort of composite of the *moules* of all of its organs, was itself merely a particular exemplar of the *moule* of a species. The original exemplar of the *moule* of a species, Buffon was willing to concede, possibly for want of a better explanation, must have come into existence at the time of Creation (H.N. II, p. 426)."[9]

PROTOTYPE

In *The Horse* (1753), Buffon hypothesizes that there exists in nature a general prototype of each species on which every individual is modeled, but which undergoes minor changes due to environment. He believed that God created species perfectly at the time of Creation, and that each species left the Creator's hands in perfect condition: "There is in Nature a general prototype of every species, upon which each individual is modeled, but which seems, in its actual production, to be depraved or improved by circumstances; so that, with regard to certain qualities, there appears to be an unaccountable variation in the succession of individuals, and, at the same time, an admirable uniformity in the entire species. The first animal, the first horse, for example, has been the external and internal model, upon which all the horses that have existed, or shall exist, have been formed. But this model, of which we know only copies, has had, in communicating and multiplying its form, the power of adulterating or of improving itself. The original impression is preserved in each individual."[10] Because it was healthy, strong, and agile, it was able to protect itself from its enemies and live long enough to reproduce. Subsequent to Creation, metamorphoses began to occur within each species. Species began to degenerate when they were taken to a climate other than that of their land of origin. However, Buffon did not believe that one species can metamorphose into a new and different species. Hence, Buffon envisaged intraspecies metamorphoses, but not interspecies progressions. Diderot and Maupertuis were able to make the connection and figure out that the similarities in the physical structures among species indicate that body parts may become longer or shorter and metamorphose into different species. However, Buffon's legacy is that he made some very sharp observations about the similarities between body parts of different species.

In the same passage Buffon goes on to observe that each individual within a species has unique peculiarities that render it unique from all other individuals within that species: individuals within a species are not identical, but they are each different in some particular way: "…but although there are millions of them, not one of these individuals entirely resembles any other, nor the model from which it is born: this difference, which proves how far removed Nature is from doing anything in an absolute way, and how much it nuances its works in infinite gradations, is found in the human species, in all animals, all vegetables, in fact, in all beings that are brought forth…"[11]

Jacques Roger points out that it is precisely this notion of an interior mold of each species, that was formed at the time of Creation,, that prevented Buffon from entertaining the thought that perhaps species metamorphose into other species, and kingdoms into other kingdoms, over time. The interior mold or patterning life force is fixed and therefore, major metamorphosis is not possible. However, because he observed that species change when they are taken out of their natural climates, he allowed for minor metamorphoses, or degeneration.

LAND OF ORIGIN

In 1753 Buffon observed that each species appears to be indigenous to a certain geographical location. He called the territory where each species is found its "land of origin." Because the finest horses in the world were found in Arabia, he hypothesized that Arabia must be the land of origin of horses. Horses in other climates are not as fine as those found in Arabia, so they must have degenerated when they left their natural climate and "land of origin." Speaking of the relationship between Arabia and the Arabian horse, he surmised that Arabia, being a dry and warm country, appears to be the original climate of the horse, as it is the most conformable to its nature.

In 1761 Buffon again observed a causality between climate and the physical characteristics of animals: "The lion never inhabited the northern regions; the rein-deer was never found in the south; and perhaps no other species but that of man is generally diffused over the whole surface of the

Buffon 75

globe. Each has its peculiar country, to which it is confined by a physical necessity; each is a genuine son of the country it inhabits; and it is in this sense alone, that particular animals ought to be called natives of a particular climate."[12]

In 1764 he reiterated that the camel was perfectly adapted to Arabia, and so Arabia must be the land of origin of the camel: "It seems to be an original native of Arabia; for this is not only the country where they are most numerous, but where they thrive best. Arabia is the driest country in the world, and where water is most rare. The camel is the most sober of all animals, and can pass several days without drink. The soil is almost everywhere dry and sandy. The feet of the camel are adapted for walking on sands, and the animal cannot support itself on moist and slippery ground. This soil produces no pasture; the ox is also wanting; and the camel supplies his place. When we consider the nature and structure of these animals, we cannot be deceived with regard to their native country, which must be conformed to their frame and temperament, especially when these are not modified by the influence of other climates. In vain have attempts been made to multiply them in Spain; in vain have they been transported to America. They have neither succeeded in the one country nor in the other; and, in the East Indies, they are not found beyond Surat and Ormus."[13]

It is significant that in this passage on the camel, Buffon uses a form of *conforme* (in harmony with, adapted) three times: *conforme* (adapted), *conformité* (adaptation), and *conformée* (adapted); he also uses *se modifier* (to be transformed, to change). Buffon observed the causality that exists between climate and environment and the adaptation of the physical characteristics of animals: the camel's hump permits it to store water and its feet are perfectly made for walking on sand.

UNITY OF PLAN

Buffon observed similarities in the structures of the parts and organs of different animals. The idea was not new: Aristotle, in the *History of Animals, Book 1, Part 1*, points out that some animals resemble others in all their parts, and in other cases, their body parts are different in way of excess or defect.[14] Aristotle calls our attention to analogous parts that exist among animals of different species: bone is analogous to fishbone, nail to hoof, hand to claw, scale to feather; he observes that what the feather is for a bird, the scale is to a fish.[15]

Leonardo da Vinci also observed the similarities in the skeletal structures of various species. He laid out the leg of a man and that of a horse and pointed out how alike the two are. Likewise, in 1555, the French naturalist Pierre Belon laid out a human skeleton and a bird skeleton and showed the similarities between the two. The similarities in structures of different species were also pointed out by Marco Aurelio Severino, Claude Perrault, Jan Swammerdam, Gottfried Wilhelm Leibniz, and Pierre-Louis Moreau de Maupertuis. Pierre Daubenton also decided to lay out the skeletons of a horse and that of a man and examine the similarities between the two.

Buffon, for his part, compared the skeletons of a horse and donkey to that of a man. In *The Ass* (1753) Buffon point out the similarities between a man's hand and a horse's foot; he also compares the skeleton of the man and the horse and can turn one into the other by modifying the length of a few bones; he notes that the ribcage is present in a lot of different species- quadrupeds, birds, fish, and even in the turtle (there are furrows under its shell). Hence, Buffon extrapolates that physical characteristics are "graduated to infinity." This would be very helpful to Maupertuis and Diderot who would take this information and extrapolate the metamorphosis of all beings from a single prototype. Buffon called this general schematic that all species seem to have in common the "principle of the unity of the plan of composition" and maintained that it was devised by God Himself. Buffon concludes by asking whether "this constant uniformity of design, to be traced from men to quadrupeds, from quadrupeds to the cetaceous animals, from the cetaceous animals to birds, from birds to reptiles, from reptiles to fishes…does not indicate, that the Supreme Being, in creating animals, employed only one idea, and, at the same time, diversified it in every possible manner, to give men an opportunity of admiring equally the magnificence of the execution and the simplicity of the design?"[16]

THE TWO DIMENSIONAL LINEAR CHAIN OF BEINGS

In the *First Discourse* (1749), Buffon viewed the chain of beings as a straight line comprised of infinitely minute gradations: if man, "methodically and in succession, glances through the different objects that comprise the Universe, and places himself at the head of all created beings, he will see with astonishment that one can descend by almost imperceptible degrees, from the most perfect creature to the most unformed matter, from the most organized animal to the crudest matter..."[17] Here the chain of being is a straight line extending from God down to inanimate matter (the mineral kingdom).

BUFFON'S THREE DIMENSIONAL MATRIX

In 1749 Buffon also observed that different species share many of the same characteristics. Because physical characteristics overlap among species, any classification of living things must necessarily be arbitrary-the classifier must select the characteristics that he will use in his classification and he must ignore others. This makes all classification subjective and that is why in 1749 Buffon opposed Linnaeus' method of classifying living things according to their physical characteristics: "...one clearly sees that it is impossible to give a general system, a perfect method, not only for the entire *Natural History*, but even for one of its branches; for to make a system, an arrangement, in a word, a general method, everything must be included in it; it is necessary to divide this entirety into different classes, divide these classes into genuses, subdivide these genuses into species, and all this following an order in which it must enter into the arbitrary. But Nature works in unknown degrees, and consequently, it cannot totally lend itself to all these divisions, since it passes from one species to another species, and often from one genus to another genus, in imperceptible nuances; so that there are found a great number of species that are midway and half-way objects that one does not know where to place, and which necessarily skews the general system: this truth is too important to ignore everything that renders it clear and evident."[18]

By 1765, his chain of being was no longer two dimensional, but became three dimensional or conical: he observed that it is impossible to segregate quadrupeds into groupings by physical characteristics because traits overlap: "Let us assemble, for a moment, all the quadrupeds into one group, and let the intervals or ranks represent the proximity or distance between each species. Let us place in the centre, the most numerous genera, and on the flanks those which are the least numerous. Let us confine the whole within narrow bounds, that we may have the more distinct view of them; and we shall find, that it is impossible to round this enclosure. Though all quadrupeds are more closely connected together than to any other being; yet several of them make prominent points, and seem to fly off in order to join other classes of animated nature. The apes make a near approach to man. The bats are the apes of birds, which they imitate in their flight. The porcupines and hedge-hogs, by the quills with which they are covered, seem to indicate that feathers are not confined to birds. The armadillos, by their scaly shells, approach the turtle and the crustaceous animals. The beavers, by the scales on their tails, resemble the fishes. The ant-eaters, by their beak or trunk without teeth, and the length of the tongue, claim an affinity to the fishes. In fine, the seal, the walrus, and the manati, are a separate corps, and make a great projection, with a view to arrive at the cetaceous tribes."[19]

Buffon believed that even though species share overlapping characteristics, each species was created by God at the time of Creation and no new species have arisen since then. However, the paradigm of the three dimensional cone would be further developed by the materialists Maupertuis and Diderot. Maupertuis would hypothesize that the multitudes of species that we see came about by errors in the generative process that were passed on from generation to generation and that through these errors, perhaps all living things have arisen from a single prototype. Diderot would seize upon this notion and add the fourth dimension, time, and posit that it took millions of years for all beings to arise from a single prototype. What Buffon did was to develop the two dimensional, linear chain of beings into a three dimensional cone in which physical characteristics overlap; contemporary materialists would take it from there to new territory.

The question arises as to whether Buffon believed that one kingdom could, with time, metamorphose into another kingdom, one of a higher organization. The answer is that he did not embrace the homogeneity of kingdoms and felt that the kingdoms were created separately by God at the time of Creation. However, he did let his imagination survey this topic, he did speculate about the gray areas between kingdoms. There are implications to his statement of 1749 in which he said that if man looked down the chain of beings he would "see with astonishment that one can descend by almost imperceptible degrees, from the most perfect creature to the most unformed matter, from the most organized animal to the crudest matter." This statement erases the delineations between kingdoms and implies that there are organized stones (asbestos), lithophytes (stone-plants), and zoophytes (animal-plants such as the sponge).

Furthermore, an online word search in the text of the *Histoire naturelle* (http://www.buffon.cnrs.fr/?lang=fr) indicates that Buffon discusses asbestos (4:78–92), zoophytes (2:304, 2:322, 4:29), and polyps (2:8, 2:20–21, 2:47, 2:82, 2:261–62, 7:10). In addition, his experiments on spontaneous generation with John Turberville Needham appeared to confirm the theory that tiny animalcules are born from infusions of grain (vegetable kingdom), thus reinforcing the notion that one can go from the vegetable to the animal kingdom in the twinkling of an eye. So then, a reader would logically ask whether Buffon believed in the metamorphosis of one kingdom into another. It turns out that Buffon flatly denied the notion. Jacques Roger explains that although Buffon had declared that the chain of beings is comprised of a continuum of imperceptible gradations, the fact is that species that span kingdoms, such as lithophytes and zoophytes, do not exist. Therefore, there are definitly lines of demarcation between kingdoms.[20] Buffon had a rigid view of species: an interior molding force determines the general characteristics of a species and while climate can make minor changes, the species does not greatly change over time.

Otis Fellows agrees with Jacques Roger that Buffon believed that God created species and kingdoms as separate entities. Fellow shows that Buffon's hypothesis that organic molecules seeded the earth with life at the time of Creation led to the notion of the interior mold, which in turn, necessitates that each species is a distinct entity with an interior molding force that forms it. Organic molecules fell to earth and grouped themselves into species, each species having its own interior molding force.[21]

Although Buffon accepted the separation of kingdoms, his speculations left quite a legacy to the eighteenth century. He was forced to part company with Maupertuis and Diderot on this issue, as they did erase all boundaries between kingdoms and argued that all livings things developed from a single prototype. Diderot mentions Buffon's denial that intermediaries exist that bridge kingdoms and adds that more experimentation must be done. In *Thoughts on the Interpretation of Nature* (1753), *Thought 12*, Diderot says, "When we observe the successive outward metamorphoses which take place in this prototype, whatever it may be, pushing one realm of life closer to another by imperceptible stages, and populating the regions where these two realms border on each other (if they can be referred to as 'borders' in the absence of any true divisions); and, populating, as I said, the border regions of the two realms with vague, unidentifiable beings, largely devoid of the forms, qualities and functions of one region and assuming the forms, qualities and functions of the other; who, then, would not be persuaded that there had never been more than one single prototype for every being? But, whether this philosophic conjecture is admitted as true with Doctor Bauman [Maupertuis] or rejected as false with M. de Buffon, we will not deny that it must be adopted as a hypothesis that is essential to the progress of experimental physical science..."[22]

TO DEGENERATE [DEGENERER] AND TO PERFECT [PERFECTIONNER]

In *The Lion* (1761), Buffon observes a causality between climate and the physical characteristics of animals: each animal has a natural homeland in which it prospers and proliferates; each one is truly the "son of the earth" that it inhabits, to which it is totally adapted: "Among the other animals...the influence of climate is stronger, and marked by sensible characters; because they differed in species, and their nature is less perfect, and less diffused than that of man. The varieties of each species are not only more numerous, and more strongly marked, but even the differences of species themselves seem to depend upon the differences of climate. Some are unable to propagate but in warm, and others cannot subsist but in cold countries. The lion never inhabited the northern regions; the rein-deer was never found in the south; and perhaps no other species but that of man is generally diffused over the whole surface of the globe. Each has its peculiar country, to which it is confined by a physical necessity; each is a genuine son of the country it inhabits; and it is in this sense alone, that particular animals ought to be called natives of a particular climate."[23]

Conversely, when animals are domesticated or moved from their natural homelands, they suffer degeneration-they become smaller, paler, sickly, weaker, less active. As an example, Buffon uses the breeding of the French horse. The finest horses in the world are in Arabia and hence, one must necessarily extrapolate that Arabia is their land of origin. When horses are shipped from Arabia to France for breeding, they suffer degeneration. Climate, environment, food, and soil are crucial factors that contribute to degeneration.

For Buffon, this opened the door to the concept of the minor modification of species: he believed that species can degenerate, but not to the extent that they can metamorphose into a new and different species. However, the atheist materialists with whom Buffon differed would carry the notion to its extreme, namely, the mutability of species, and then kingdoms.

Buffon also points out that the solution to the problem of degeneration is to mate varieties of the same species from various climates. By doing this, the best qualities of animals of various regions can be expressed: "In order, therefore, to obtain good grain, beautiful flowers, &c. the seeds must be changed, and never sown in the same soil that produced them...Without this precaution, all grain, flowers, and animals degenerate, or rather receive an impression from the climate so strong as to deform and adulterate the species...by mixing races, on the contrary, or by crossing the breed of different climates, beauty of form, and every other useful quality, are brought to perfection; Nature recovers her spring, and exhibits her best productions....We know by experience, that animals or vegetables, transported from distant climates, often degenerate, and sometimes come to perfection, in a few generations. This effect, it is obvious, is produced by the difference of climate and of food...The operation of these two causes must, in process of time, render such animals exempt from, or susceptible...certain diseases. Their temperament must suffer a gradual change. Of course, their form, which partly depends on food...must also, in the course of generations, suffer an alteration."[24]

In 1755 Buffon considered the possibility that perhaps species can be perfected through man's care and breeding (an idea that he would later abandon). He entertained the notion that perhaps sheep might be perfected goats and that horses might be perfected asses. By 1764 he understood that species cannot be perfected through domestication: domestication can only destroy what nature has made perfect. He examined various kinds of sheep and concluded that all sheep have been degenerated at the hands of man. He observed that no sheep are strong enough, agile enough or lively enough to escape predators; they all need to be protected by man. Therefore, somewhere, there must exist a strong, wild, carnivorous sheep which is their prototype, and which has not yet been discovered.[25] By 1764 Buffon concluded that perfection can occur only in the wild. Domestication can bring only degeneration. The camel's callosities and hump are "imprints of servitude and stigmata of pain."[26] Animals degenerate at the hands of man because men treat them badly, feed them poorly, mistreat them, and force them to work very hard.

CLIMATE, GEOGRAPHY, FOOD, AND WAY OF LIFE

Buffon believed that minor intraspecies changes can occur due to climate, geography, food, and way of life. As far as interspecies changes are concerned, he did not believe that one species can metamorphose into another species over time: he held that God created all species perfectly at the time of Creation and that no new species have arisen since then. Furthermore, all of his unsuccessful attempts at crossbreeding failed to create any new species. He concluded that species must be static, although minor changes (degeneration) can and do occur due to climate.

Buffon points out that certain species are found only in certain climates and that in those climates they are the healthiest and the most numerous; therefore, those regions must be their land of origin. The healthiest and most numerous horses are found in Arabia; therefore, one must necessary extrapolate that Arabia is the horse's land of origin. The healthiest and most numerous camels are found in Arabia; their physical characteristics are adapted to their environment (he uses "adapted" three times: *conforme, conformité, conformée*); therefore, the desert must be the land of origin of camels.

Buffon maintained that climate, food and way of life are the causes of human diversity, as well. Jacques Roger explains, "Food 'does a lot for the [physical shape]': 'all peoples who live miserably are ugly and badly built,' and that was true in the French countryside as well as elsewhere. Air and soil also played a large role. On 'the hillsides and hilltops,' the countrymen were 'agile, full of energy, well built, spirited, and...the women there are generally pretty; while in the plains, where the earth is crude, the air thick, and the water less pure, the countrymen are coarse, heavy, badly built, stupid, and the countrywomen are almost all ugly.' This was valid for animals and plants as well as for men. As for the manner of life, it was enough to compare the Tartars with the Chinese in order to see its importance. The Tartars 'are always exposed to air' and 'live in a hard and wild way.' The Chinese were whiter 'because they lived in cities, because they are civilized, because they have all the means to protect themselves from the injuries of the air and earth.'"[27]

However, climate does not cause degeneration in humans as it does in animals. Humans are much more resilient than are animals and more immune to the influences of climate: "The influence of climate, in the human species, is only marked by slight varieties; because this species is single, and extremely distinct from every other. Man, white in Europe, black in Africa, yellow is Asia, and red in America, is the same animal, tinctured with the colour peculiar to the climate. As he is formed to exercise dominion over the earth, and, as he has the whole globe for his habitation, his nature seems to be accommodated to every situation. Under the fervors of the south, or the frozen regions of the north, he lives, multiplies, and is so universally and so anciently diffused over every country, that he appears to have no peculiar climate. Among the other animals, on the contrary, the influence of climate is stronger..."[28]

THE UNITY OF THE HUMAN RACE

Buffon takes the reader on a tour of the world, arranging his itinerary by latitudes and climate zones. As we tour the world, he points out that the skin color of people is a continuum of infinitely minute gradations and not either black or white. Roger explains that Buffon had shown that men are "more or less tanned" than one another and that skin color is a continuum comprised of gradations; there are white people who have naturally frizzy hair; there are no sharp distinctions among the races.[29]

Hence, skin color was a continuum, comprised of infinite gradations, like the chain of being itself. The differences in body build, strength, and health were determined by climate: in big cities, life is easier and people have a certain comfort. The inhabitants of cities are sheltered from misery and for that reason they are stronger and better looking than those who are forced to fight for survival every day.[30] Hence, climate, food, soil, and way of life influence physical characteristics and health, but do not cause degeneration of the human species. Buffon believed that if a white European were to move to Africa, after seven or eight generations, his progeny would turn black. Similarly, if an African moved to Denmark, his progeny would eventually turn white, but it would take several centuries.

Buffon concludes *On the Varieties of the Human Species* (1749) by affirming the unity of the human race. He declares that there is only one human race that originally multiplied and spread throughout the earth: "Upon the whole, every circumstance concurs in proving, that mankind is not composed of species essentially different from each other; that, on the contrary, there was originally but one species, who, after multiplying and spreading over the whole surface of the earth, have undergone various changes by the influence of climate, food, mode of living, epidemic diseases, and the mixture of dissimilar individuals; that, at first, these changes were not so conspicuous, and produced only individual varieties; that these varieties became afterwards specific, because they were rendered more general, more strongly marked, and more permanent, by the continual action of the same causes; that they are transmitted from generation to generation, as deformities or diseases pass from parents to children; and that, lastly, as they were originally produced by a train of external and accidental causes, and have only been perpetuated by time and the constant operation of these causes, it is probable that they will gradually disappear, or, at least, that they will differ from what they are at present, if the causes which produce them should cease, or if their operation should be varied by other circumstances and combinations."[31]

In 1765 Diderot copied Buffon's declaration of the unity of the human race when he cited Buffon's text in the *Encyclopedia*. In the article "Human Species (*Nat. Hist.*)" ["*Humaine espece (Hist. nat.)*"], Diderot retraces Buffon's itinerary as he escorts the reader on a tour of the world. Relying upon the authority of Buffon's *On the Varieties of the Human Species*, Diderot distinguishes among the different physical characteristics, customs, and belief systems of people on different continents. It is significant that after Diderot enumerates the differences among nations, he concludes that these differences are minor in view of the fact that there is only one human race that consists of people that are more or less tanned: "From the preceding material it follows that in the entire new continent that we have just traversed, there is only one and the same human race, more or less tanned. Americans come from the same source. Europeans come from the same source. From north to south we see the same varieties in one hemisphere and the other. Therefore, everything goes to prove that humankind is not comprised of essentially different species. The difference between whites and browns arises from food, morals, customs, climate; the difference between browns and blacks has the same cause. Therefore, originally, there was only

Buffon 85

one race of humans, which, having multiplied and spread out over the surface of the earth, resulted in all the varieties that we have just mentioned. Varieties that would disappear in time, if we suppose that people would suddenly move and find themselves subjected, either by necessity or voluntarily, to the same causes that will act on them in their newly occupied regions."[32]

CROSSBREEDING

In 1749 Buffon defined species thus: two animals are of the same species if they can produce fertile offspring together.[33] In 1753 he reiterated this definition and added that all human beings are of the same species because they can all, regardless of race, produce fertile offspring together.[34] The horse and the ass are of different species because their offspring, the mule, is infertile.[35]

Buffon experimented with crossbreeding because he wanted to know the extent of its limitations. In 1755 he humbly admits that more experimentation needs to be done: "We know not whether the zebra can produce with the horse or ass, or the broad-tailed Barbary ram with the common ewe; whether the chamois goat be only the common goat in a wild state, and whether an intermediate race might not be formed by their mixture; whether the monkeys really differ in species, or whether they form but one species, diversified, like that of the dog, by a great number of different races; whether the dog can produce with the fox and the wolf, the stag with the cow, &c. Our ignorance of all these facts is almost invincible; for the experiments necessary to ascertain them would require more time, attention, and expense, than the life or fortune of most men can permit. I employed several years in making trials of this kind, of which an account shall be given when I treat of mules. But, in the mean time, I acknowledge, that they afforded me very little information, and that most of my experiments were abortive."[36]

Buffon relentlessly and unceasingly conducted numerous experiments trying to mate animals of different species because he wanted to ascertain whether transformism is a scientific fact. If he could succeed in creating just one new species, then perhaps there was some validity to the assertion of his materialist contemporaries that all species metamorphosed from a single prototype. After repeated attempts over years, Buffon was unsuccessful in creating a new species and this failure confirmed 1) his definition of species, 2) his belief that transformism is impossible. In 1765 Buffon reiterated his definition of species. Arthur O. Lovejoy summarizes Buffon's conclusions thus: "In the infertility of hybrids he imagined that he had found a proof that species are objective and fundamental realities-are, indeed, '*les seuls êtres de la Nature*, as ancient and as permanent as Nature herself,' while 'an individual, of whatever species, is nothing in the universe.' A species is 'a whole independent of number, independent of time; a whole always living, always the same; a whole which was counted as one among the works of the creation, and therefore constitutes a single unit in the creation.'"[37]

It was Buffon's inability to create a new species through crossbreeding, after repeated attempts, that caused him to conclude that species are immutable: it appeared that a species is truly an entity unto itself and that it could not be created by engendering individuals that can yield only infertile offspring.

BUFFON TAKES A STAND AGAINST TRANSFORMISM

In 1753 Maupertuis asserted that errors occur in the generative process and that over time, these errors could explain how all living things arose from a single prototype. He asked, "Could we not explain from that how from two single individuals, the multiplication of the most dissimilar species could have resulted?"[38] This notion was echoed by Diderot in *Thoughts on the Interpretation of Nature, Thought 12*, that same year. In *The Ass* (1753), Buffon responded to the transformist hypothesis by flatly denying that one species can metamorphose into another; he reiterated this belief in 1755, 1758, and 1765.

Arthur O. Lovejoy, in his article, "Buffon and the Problem of Species," examines how Buffon methodically refutes the transformism of the atheist materialists, who, in 1753 had posited a common origin of all living things based on the observation of homologous parts. Lovejoy carefully follows Buffon's reasoning as the latter begins by admitting that homologies do exist in nature. In *The Ass* Buffon observes, "If...we select a single animal, or even the human body, as a standard, and compare all other organized beings with it, we shall find that...there exists, at the same time, a primitive and general design...When, for example, the parts constituting the body of a horse, which seems to differ so widely from that of man, are compared in detail with the human frame, instead of being struck with the difference, we are astonished at the singular and almost perfect resemblance...Let us next consider, that the foot of a horse, so seemingly different from the hand of a man, is, however, composed of the same bones, and that, at the extremity of each finger, we have the same small bone, resembling a horse-shoe, which bounds the foot of that animal."[39]

Because he is a naturalist in search of the truth, Buffon concedes that homologies do exist among different species. He then asks, as if speaking directly to the atheist materialists, whether these homologies prove that apes are degenerate men or that men and apes have a common origin: "And, if it be once admitted...that the ass belongs to the family of the horse, and differs from him only by degeneration; with equal propriety may it be concluded, that the monkey belongs to the family of man; that the monkey is a man degenerated; that man and the monkey have sprung from a common stock, like the horse and ass; that each family, either among animals or vegetables, has been derived from the same origin; and even that all animated beings have proceeded from a single species, which, in the course of ages, has produced, by improving and degenerating, all the different races that now exist."[40]

Having articulated the question as to whether homologies provide evidence that all beings have originated from a single prototype, Buffon then denies the mutability of species. He answers the question from two different angles. First, he answers it according to the revelation that God gives in the Bible. He begins by emphatically and vehemently declaring, "But no..." ["*Mais non...*"]. Then he continues, "We are assured by the authority of revelation, that all animals have participated equally of the favours of creation; that the two first of each species were formed by the hands of the Almighty; and we ought to believe that they were then nearly what their descendants are at present."[41]

In the pages that follow this declaration, Buffon enumerates several reasons, based on observation and scientific experimentation, why he does not believe the materialists' hypothesis that all living things metamorphosed from a single prototype. Lovejoy calls the reader's attention to the fact his biblical repudiation is followed by a series of formidable scientific arguments. Lovejoy points out that if Buffon had offered only a cursory biblical injunction against transformism, one could argue that he was being ironic. However, that was not the case. Lovejoy reiterates the three arguments that Buffon provides as a response to the materialists:

1. "...since the days of Aristotle to those of our own, no new species have appeared, notwithstanding the rapid movements which break down and dissipate the parts of matter, notwithstanding the infinite variety of combinations which must have taken place during these twenty centuries, notwithstanding those fortuitous or forced commixtures between animals of different species, from which nothing is produced but barren and vitiated individuals, totally incapable of transmitting their monstrous kinds to posterity."[42] Buffon recognizes the materialists' argument that flux+time=dispersion of chaos. He articulates the notion that nature is in perpetual flux ("the infinite variety of combinations")[*le nombre infini de combinaisons*], but points out that despite the infinite number of combinations (that connotes a continuous throws of the dice, the perpetual movement of atoms), no new species have appeared in twenty centuries.
2. After years of trying to create a new species, Buffon observed that hybrids are sterile and that infertile individuals can never engender a new species. Hence, species must be fixed, rather than mutable. Buffon comments that it is extremely improbable that the requisite factors would come together so that degenerated animals could reproduce: "But, what an amazing number of combinations are included in the supposition, that two animals, a male and a female, of a particular species, should degenerate so much as to form a new species, and to lose the faculty of producing with any other of the kind but themselves?"[43]
3. Buffon observes that there are no missing links: "For, if one species could be produced by the degeneration of another, if the ass actually originated from the horse, this metamorphosis could only have been effected by a long succession of almost imperceptible degrees. Between the horse and ass, there must have been many intermediate animals…What is become of these intermediate beings? Why are their representatives and descendants now extinguished? Why should the two extremes alone exist?"[44]

Lovejoy calls the reader's attention to the fact that one of Buffon's arguments for rejecting transformism was based on mathematical probability: it is highly improbable that animals degenerate in increments because there are no intermediaries between the horse and the ass that exist in nature. Buffon concluded, "Though, therefore, we cannot demonstrate, that the formation of a new species, by means of degeneration, exceeds the powers of Nature; yet the number of improbabilities attending such a supposition, renders it totally incredible..."[45] Lovejoy extrapolates that Buffon knew about the hypothesis of transformism, "and recognized that there was some probable evidence in its favor, he then seriously believed that the preponderance of the probability was enormously against it. It is certain that contemporary readers must have understood this to be his position."[46]

In several instances after 1753 Buffon reiterated the permanent and immutable boundaries separating species from one another. In 1755 Buffon declared, "Though the species of animals are separated from each other by an interval, which Nature cannot overleap; yet some species approach so near to others, and their mutual relations are so numerous, that space is only left for a bare line of distinction."[47] Buffon maintains that there is "an intervale that Nature cannot broach" [*intervalle que la Nature ne peut franchir*] and there also exists the "space that is necessary to draw the line of separation" between species [*l'espace nécessaire pour tirer la ligne de séparation*].

In 1758 Buffon wrote that nature is attentive to conserving each species and preserving the distinct lines of demarcation between them: "Nature, by descending gradually from great to small, from strong to weak, counterbalances every part of her works. Attentive solely to the preservation of each species, she creates a profusion of individuals, and supports by numbers the small and the feeble, whom she hath left unprovided with arms or wit courage. She has not only put those inferior animals in a condition to perpetuate and to resist by their own numbers, but she seems, at the same time, to have afforded a supply to each by multiplying the neighbouring species. The rat, the mouse...form so many distinct and separate species..."[48] Here, nature has a propensity for preserving each species by providing an ample supply of prey for each species; there are numerous distinct and separate species. Buffon held firm to his theory that the essential character of each species is constant and inalterable.

In 1765 Buffon again reiterated the immutability of species: "Individuals, whatever their kind or number may be, are of no value in the universe. Species are the only existences in Nature; for they are equally ancient and permanent with herself. To form a distinct idea of this subject, we shall not consider species as a collection or succession of similar individuals, but as a whole, independent of number and of time, always active and always the same; a whole, which has been reckoned on in the works of creation, and, therefore, constitutes only a unit in Nature…A day, a year, an age, or any given portion of time, constitutes no part of her duration. Time itself relates only to individuals, to beings whole existence is fugitive. But the existence of species is constant; their permanence produces duration, and their differences give rise to number. Let us consider species in this light; let us give to each an equal right to the indulgence and support of Nature. To her they are all equally dear; for, on each of them, she has bestowed the means of subsisting, and of lasting as long as herself."[49]

Jacques Roger agrees with Lovejoy that Buffon discounted the metamorphosis of species. Roger explains that Buffon rejected the notion that the interior molding force is plastic; the opposite is true, the interior mold defines species forever; furthermore, no new species have appeared since Aristotle.[50] Roger cites the arguments presented in *The Ass*, namely, that a mule and a female of a species must degenerate to the exact same degree that they can create a fertile offspring and also, that there are no intermediary species between the ass and the horse.[51]

In *The Life Sciences in Eighteenth-Century French Thought*, Roger again examines Buffon's view that the fixity of species is evident in the sterility of hybrids. The ass and the horse are two different species because their offspring, the mule, is sterile.[52] Roger cites Buffon's unsuccessful attempts at trying to cross a dog with a wolf and a fox and numerous other failed experiments with hybridization as the experiences that reinforced his belief that species are static.[53]

In summation, Buffon's lifelong exploration of nature opened the door to the metamorphosis of species, even though he, himself, held that species are immutable. First, Buffon embraced epigenesis, which allowed for the creation of an offspring that is uniquely different from either parent and also from any other individual in its species. Secondly, he posited that organic molecules, which rained down from Heaven at the time of Creation, are the substance of all living things; after the cooling of the earth, they came together to form all species. Since they have the ability to form life as soon as they come together, as evidenced by Buffon and Needham's flawed experiments with spontaneous generation, there must be life in them. Hence, life is a property inherent in all matter. The materialists, Diderot and Maupertuis, would take this a step further: all molecules are conscious-they have desire, aversion, memory, and intelligence.[54]

Buffon also held that there is a molding force inherent in every living thing that shapes the organic molecules and causes them to organize their activity to the purpose of the life form. Furthermore, each individual organ has a molding force, as well. The materialists would take this notion of an interior molding force that is inherent in each body part as well as in the entire body and extrapolate that an emergent consciousness arises when matter combines: consciousness exists at every level of organization and when particles combine to form a more complex structure, the latter has a consciousness of its own that is greater than that of the sum of its parts.[55]

Buffon's astute observations regarding homologies existing among the body parts of various species opened the door to the metamorphosis of species. He commented on the infinite number of gradations in the chain of beings and the great variety that exists in nature-"All that can be is." The materialists would use these homologies as evidence that all living things metamorphosed from a single prototype.

Buffon and Needham's experiments with spontaneous generation appeared to confirm the theory. The materialists would use this as evidence that the animal kingdom (animalcules) can arise from the vegetable kingdom (infusions of grain) or from nonliving matter in the twinkling of the eye, and therefore, there is no need for God.

Buffon's theory of the gradual degeneration of species also opened the door to the metamorphosis of species. While he staunchly defended the immutability of species, he conceded that minor changes do occur due to environmental factors (ie: climate, soil, food, and way of life). He copiously documented the adaptation that various species have to their environment in their land of origin (ie: the camel has the ability to store water and easily walk on sand) and what he thought was degeneration when they are taken to less favorable climates (in which they are treated poorly, fed poorly, and overworked). The materialists would not limit metamorphosis to individuals within a species: because of homologies, they would extend it and declare that species and kingdoms change over time and that all things in the chain of beings arose from a single prototype. Thus Buffon, a conservative naturalist who consistently denied the mutability of species, who tried and failed to create a new species via crossbreeding and concluded that species are forever fixed, unwittingly supplied the building blocks for the transformism of his atheist materialist contemporaries. His fairness, attention to detail, and tireless observation of nature, propelled the Enlightenment forward into hitherto uncharted territory, despite his adherence to his conception of species and an unchanging interior molding force.

Chapter 5

Maupertuis

...each degree of error would have made a new species; and by virtue of repeated digressions there would have risen the infinite diversity of animals that we see today.[1]
—Pierre-Louis Moreau de Maupertuis, *Essay on the Formation of Organized Bodies*
(1754)

Maupertuis conducted extensive research on the inheritance of polydactylism and concluded that birth anomalies are due to errors that occur in the parental elements supplied by either parent. Over time, these errors lead to the appearance of new species and can thus explain how all living things, in their great diversity, may have arisen from a single prototype. Maupertuis reiterated this hypothesis throughout his literary career: in the *Physical Dissertation on the Origin of the White Negro* (1744, reprinted as part of *Physical Venus*, 1745), the *Essay on Cosmology* (1750), the *Inaugural Dissertation on Metaphysics* (in Latin, 1751, translated into French as the *System of Nature*, 1751, and again as the *Essay on the Formation of Organized Bodies*, 1754), and his *Letters* (1752).

Physical Venus (1745) is comprised of two parts: *Part 1* is entitled, *Dissertation on the Origin of Men and Animals*, and *Part 2*, *Dissertation on the Origin of Blacks*. The second part had already been published the previous year under the title, *Physical Dissertation on the Origin of the White Negro* (Leyden, 1744).

In *Physical Venus*, Maupertuis sets out to prove two hypotheses. First, he engages in a detailed defense of epigenesis at a time when Europe was embroiled in the preformation vs. epigenesis debate (*Part 1*). Having established that both the father and mother equally contribute to the formation of a new product that is brought into existence by successive accretions, he observes that errors in the generative process randomly occur; he then extrapolates that these errors could explain how all races, all species and the three kingdoms, may have been derived from a single prototype (*Part 2*).

In his preface to *Physical Venus*, Maupertuis declares that the objective of his work is to explain the origin of white Negroes (*Negre-Blancs*), as well as that of the variety of different races that exist around the world. He entices the reader to continue the investigation by asking why the inhabitants of the Torrid Zone are black, why the most numerous populations are found in the Temperate Zones, and why the glacial zones are inhabited only by deformed nations. He asks how such a variety of different races could have arisen from two original parents through preformation.[2]

Maupertuis devotes *Part 1* to an in-depth examination of epigenesis. He discusses the work of William Harvey, who, in the *Anatomical Exercitations Concerning the Generation of Animals* (1651), had demonstrated that all living things are derived from the egg. In 1651 Harvey had declared the famous phrase, "everything comes from the egg" [*ex ovo omnia*], which appeared on the title page of various editions of the *Exercitations*. Harvey was uncertain as to how fertilization was accomplished, but he concurred with Aristotle's theory of the gradual formation of the embryo, part by part, as opposed to the preformed homunculus. Harvey had used the term "epigenesis," which he defined as "partium superexorientium additamentum," holding that the germ is brought into existence by the addition of parts that bud out of one another or by successive accretions.[3] Maupertuis points out that successive accretions allow room for the introduction of errors in the arrangement of parental elements, and that therefore, it is epigenesis, and not preformation, that can fully explain heredity, teratisms, and hybrids.

Chapter 13 is entitled, "Reasons that Prove Why the Foetus Proceeds Equally from the Father and Mother." Maupertuis cites the offspring of interracial marriages as proof of epigenesis: "When a black man marries a white woman, it appears that the two colors are mixed together; the child is born olive, and is equally divided between its mother's traits and those of its father."[4] The child is not a perfect copy of either parent (as preformation would hold), but rather, a composite of traits supplied by each parent.

Maupertuis observes that dissimilarities between the offspring and parents are even more pronounced in the union of two different species. The ass and the mare produce an animal that is neither a horse, nor a donkey, but that has traits of both. The dissimilarity between elements contributed by each parent is so great, the offspring is sterile. Hence, mulattos and sterile hybrids constitute proof that the embryo is formed part by part, by succession accretions: Maupertuis concludes that offspring so dissimilar from their parents could not possibly have been encased, as homunculi, within their ancestors going back to Adam and Eve.

To further strengthen his case, he devotes *Chapter 14* to a discussion on monsters. He distinguishes between "monsters by default" [*monstres par défaut*], or offspring that are missing body parts, and "monsters by excess" [*monstres par excès*], or offspring that have superfluous body parts. These teratisms are evidence of epigenesis, not preformation, since most of them do not resemble either of their parents. Maupertuis hypothesizes that at some time during gestation, some parts are destroyed in the egg by some error and that this creates a "monster by default," or a mutilated child. Conversely, the union or jumbling of two eggs, or two germs in the same egg, produces a "monster by excess," or a child having superfluous parts. In the case of Siamese twins, which adhere to one another, no principle part of the egg has been destroyed. Some superficial parts of the foetus are torn away in some particular place, and mend together, and cause the adherence of two bodies. Monsters having two heads on one body, or those with two bodies with one head, are like Siamese twins, except that more parts in one of the eggs have been destroyed: in one, those that formed one of the bodies, in the other, those that formed one of the heads. Finally, a child that has an extra finger or toe is a monster that has been produced by two eggs, and in one of the eggs, all of the parts, with the exception of the extra digit, have been destroyed. In this way, Maupertuis demonstrates that epigenesis succeeds in explaining a variety of phenomena, a feat that preformation has been unable to accomplish.

In *Chapter 15* Maupertuis ridicules the notion that teratisms are caused by the mother's imagination, fear, admiration or desire; neither does a sudden fright experienced by the mother cause a birth defect in the offspring. Physical processes alone, taking place in the fetus, account for birth anomalies.

Maupertuis devotes *Part 2* of *Physical Venus* to explaining the origin of the diverse races in the world. He takes the reader on an ethnographic tour of the world, emphasizing that people on different continents have different skin color, facial characteristics, language, and customs. Maupertuis highlights the variations that exist among different races: he enumerates the physical differences between blacks and whites; from the Tropic of Cancer to the Tropic of Capricorn, Africa has only the black race; as we leave the Equator and approach Antarctica, the inhabitants' skin lightens; as we draw near to the Orient, facial characteristics become less pronounced, but skin color remains dark; in America the natives have red skin; on the tip of South America there are giants; the people at the poles and the Lapps are very small; in Panama the whitest people in the world live; their hair resembles the whitest wool, their eyes are too feeble to bear the light of day, and they open them only in the obscurity of night.

In his passage describing the albino Indians in Panama, Maupertuis provides a footnote in which he refers the reader to "Wafer's voyage, description of the isthmus of America."[5] An investigation of this little footnote yields surprising and significant results: in his work entitled, *A New Voyage and Description of the Isthmus of America*, Lionel Wafer's description of the anomaly of the albino Indian would one day provide information that Maupertuis would use to posit that inherited errors could explain how all life metamorphosed from a single prototype. Because Wafer's contribution to Maupertuis is significant, it would be appropriate to examine it more fully.

LIONEL WAFER'S LEGACY TO MAUPERTUIS

In 1680 the British explorer and surgeon Lionel Wafer crossed Panama, then called the Isthmus of Darien. His journey back across the isthmus from the Pacific side to the Caribbean was interrupted by an accident that he suffered: one of the men in his exploratory party was drying gunpowder and mismanaged his task. The gunpowder exploded and the flesh on Wafer's leg was completely torn away, leaving the bare bone exposed. To make matters worse, the slave that was assigned to him escaped with his medical bag containing all of his salves and ointments. Fortunately for Wafer, he still had some small medical instruments, such as his surgical lancet, that he had been carrying in his pocket. The exploratory party was forced to return to the Caribbean without him, leaving him behind with the resident natives, the Cunas or San Blas people, to recover from his wounds.

Wafer befriended the Cuna Indians and they undertook the task of curing him of his ailment. They selected certain herbs, chewed them to the consistency of a paste, put the substance on plantain leaf, and laid it on his wounds. Each day they applied a fresh poultice; much to his delight, in twenty days he was well. To return the favor that they had done for him, Wafer shared his knowledge of surgery with them. During this time he mingled among both the tribes that dwelt in the hills as well as with the seafaring branch of the population on the Archipelago de San Blas. Because he had surgical experience aboard vessels on the high seas as well as in his practice in Port-Royal, Jamaica, the natives welcomed his medical expertise.

What surprised Wafer the most during his stay with the Cunas was his discovery of blond haired children among the dark skinned natives. This intrigued him greatly and he carefully recorded his observations thus:

> There is one complexion so singular...they are white...'tis a milk-white lighter than the colour of any Europeans...Their seeing so clear as they do on a moon shiny night, we us'd to call them moon-ey'd. For they see not very well in the sun...their eyes being but weak and running with water if the sun shines towards them...when the moon shiny nights come, they are all life and activity...neither is the child of...these white Indians white...but copper-coloured as their parents were.[6]

Today we know that this narrative describes oculocutaneous albinism. The three salient points in Wafer's description are albinism, photophobia, and the fact that the offspring are normal (recessive inheritance). It is fortunate indeed for posterity that Wafer was a seafaring surgeon with a keen interest in human physiology and all kinds of ailments. The astute observations that he made in this simple narrative would ultimately provide vital new information to Maupertuis, who would use it to posit that errors in the arrangement of parental traits could explain how all life may have arisen from a single prototype.

Maupertuis seized upon this passage in Wafer's book. He gave a great deal of thought to the interrelatedness of albinism, photophobia, and the skipping of a generation. He used the connection among these three characteristics to buttress two arguments: 1) epigenesis or the hypothesis that both the father and mother equally contribute to the formation of a new product, that, as Harvey had posited, is brought into existence by successive accretions and 2) the notion that errors in the arrangement of parental elements occur, that these errors are random, and that they are passed on from one generation to the next. These errors in the generative process could explain how, over time, all races, all species, and all three kingdoms, may have originated from a single prototype. He thought that this hypothesis might also explain why another anomaly, polydactyly (a birth with many fingers or toes), also skips generations.

After Maupertuis refers the reader to "Wafer's Voyage, description of the American isthmus" in a footnote, he devotes three pages to expanding upon the albinism that exists in Panama. Maupertuis' passages describing Wafer's discovery are not only a scientific treatise, but also constitute a lovely poetic essay. Note that the North Sea here is the Caribbean:

> On this isthmus that separates the North Sea from the Pacific Ocean, it is said that men, whiter than any we know of, are found: their hair resembles the whitest wool; their eyes, too weak for the light of day, open only in the obscurity of night. They are, among men, what bats and owls are among birds. When the star of daylight has vanished, and leaves nature in gloom and silence, when all the other inhabitants of the earth, overcome by their work, or exhausted by their pleasures, surrender to sleep, the Darien awakens, extols his Gods, rejoices in the absence of unbearable light, and comes to fill nature's void. He listens to the screech-owl's cries with as much pleasure as the shepherd of our regions listens to the lark's song at the first sight of dawn: out of view from the sparrow-hawk, it seems to go search in the cloud the daylight that is not yet on earth; it marks by the flapping of its wings, the rhythm of its warbling; it ascends and is lost in the cloud, it is no longer seen, yet it is still heard: its sounds are no longer distinct, inspire tenderness and reverie; this moment reunites the tranquility of night with the pleasures of the day. The Sun appears: it is coming to bring motion and life back to the earth, to mark the hours, and determine men's different labors. The Dariens have not waited for this moment: they have already retired.[7]

Maupertuis added this freakish people to his toolbox of assorted anomalies-the albino Negro, the dwarfs of the polar regions, Siamese twins, polydactylous individuals, and the interesting breeds of eighteenth-century canines that Pierre Lyonnet concocted, the harlequin (a small, dappled Dane) and the mopse (a pug-dog or miniature bull-dog)-to buttress two hypotheses that he held, epigenesis and the inheritance of random errors that occur in the generative process.

First, the moon-ey'd children of the Central American isthmus provided excellent proof of epigenesis. At a time when a war was raging between the preformationists and the epigenesists, Maupertuis held up the Cunas, who did not resemble their parents at all, as evidence that offspring are not little Russian dolls contained one inside the other going all the way back to Adam and Eve. Cunas are decidedly unlike their parents-as are the albino Negro, Siamese twins, those blind from birth, and other anomalies. The existence of photophobic Panamanian Indians indicates that William Harvey must have been right about the germ developing by successive accretions.

The notion that the embryo grows by the addition of parts that bud out of one another was revolutionary because it allowed for the introduction of errors in the birth process. Maupertuis pointed out that this gradual process allows the time and opportunity for mistakes to occur and that unlike preformation, these errors can fully explain heredity, teratisms, and hybrids.

In *Part 2, Chapter 2*, Maupertuis undertakes the task of explaining how the multitude of varied races in the world could have arisen from a single prototype. He declares his belief that all of humanity, despite its great diversity, came from a single being. His task will be to prove how, from a single individual, so many different races could have arisen.

Maupertuis hypothesizes that it is random chance [*le hasard*] that causes new races to arise: "Nature contains the basis of all these varieties: but chance or art sets them in motion. It is in this way that those whose industry applies itself to satisfying the taste of the curious, are, so to speak, the creators of new species. We see the appearance of species of dogs, pigeons, canaries, which did not exist before in nature. In the beginning they were just fortuitous individuals; art and repeated generations made them species. Every year the famous Lyonnet creates some new species, & then destroys those that are no longer in style. He corrects forms, & varies colors: he has invented the *Harlequin*, the *Mopse* species, etc."[8] Pierre (Pieter) Lyonnet, a Dutch naturalist and engraver best known for his dissections and illustrations of insect anatomy, experimented with the hybridization of dogs. Diderot's *Encyclopédie* defines the harlequin as "a variety of small Dane; but unlike Danes, that are nearly all in one color, harlequins are dappled, some white and black, others white and cinnamon, others of another color."[9] The mopsy was a pug-dog (a dwarf breed resembling a bull-dog in miniature).

Maupertuis asks why the art of hybridization should be limited to animals. He suggests that sultans, who keep women of every known race in their seraglios, should experiment and see whether they can create a new race of woman.[10] He points out that Frederick William of Prussia had bred tall, good looking men in order to have an army of giant soldiers and puns, "A King of the north managed to elevate his nation in stature and to beautify it."[11]

Maupertuis devotes *Part 2, Chapters 4 and 5,* to hypothesizing how the origin of the white Negro may have come about. The albino Negro was a current topic of conversation ever since a child of 4 or 5, a white Negro, had been paraded around the salons of Paris. Maupertuis recalls that the child had the facial characteristics of blacks, white skin, reddish white, woolly hair, and large, awkward hands. His clear blue eyes seemed to be hurt by the light of day. Maupertuis attributes this phenomenon to an accident or error in the arrangement of parental elements (*la production des variétés accidentelles*).[12] It is pure chance or a scarcity of family traits that sometimes causes a white child to be born of black parents or a black child, of white parents.[13] These productions are at first accidental. Throughout his work, Maupertuis reiterates that all variations are the products of random chance: he uses the words and phrases, *combinaisons fortuites des parties des semences, le hasard, accidentel,* and *erreur*. In addition to random errors in the arrangement of parental elements, he also believes that climate and food influence physical traits.

In *Part 2, Chapter 6,* Maupertuis hypothesizes that white was the original color of man and that black was a variety that occurred later and became hereditary after many centuries. His theory that one race is older than another race and that the younger race is derived from errors in the arrangement of parental elements that were passed on from generation to generation over centuries or millennia, is, indeed, amazingly prescient. Today science acknowledges that the oldest race is the black race, followed by the white race, and that they youngest race on earth is the yellow race.

Maupertuis devotes *Chapter 7* to explaining why blacks are found only in the Torrid Zone, and dwarves and giants, towards the poles. He believes that as soon as dwarves, giants, and blacks began to randomly appear, they were chased to the poles and to the Equator by others because of fear or pride. Hence, dwarves populated the Arctic pole, giants migrated to the Straits of Magellan, and blacks moved to the Torrid Zone.

In summation, *Physical Venus* is an extraordinary work in that by using epigenesis as a template for the introduction of new physical traits via errors in the generative process, Maupertuis was able to hypothesize that all life has been derived from a single prototype.

Another important biological work that Maupertuis wrote was the *Essay on Cosmology* (1750). Although this book was published in 1750, there is evidence that it was written many years earlier. On August 10, 1741, Voltaire wrote to Maupertuis, "If you were to be generous enough to send me your *Cosmology*, I would truly swear, by Newton and by you, not to make a copy of it, and to send it back to you after I have read it."[14] Voltaire asked for a copy "so that he could unstitch himself" from the cosmology of Johann Christian von Wolf.[15] Jacques Roger points out that Maupertuis placed the *Essay on Cosmology* at the beginning of his 1756 *Works*, ahead of the *Discourse on the Different Shapes of Heavenly Bodies*, that had been published in 1732, "proving at least the fundamental importance of the 'Essai' in Maupertuis's own eyes."[16]

In the *Essay on Cosmology*, Maupertuis assumes a mechanist stance and argues against teleology at a time when the mechanism vs. teleology debate was raging in Europe. *Webster's* defines "teleology" as "1a: the philosophical study of evidence of design in nature b: the doctrine or belief that ends are immanent in nature c: the metaphysical doctrine explaining phenomena and events by final causes 2: the fact or the character of being directed toward an end or shaped by a purpose-used of natural processes or of nature as a whole conceived as determined by final causes or by the design of a divine Providence and opposed to purely mechanical determinism or causation exclusively by what is temporarily antecedent 3: the use of design, purpose, or utility as an explanation of any natural phenomenon."[17] Leibniz provides an example of teleological reasoning in the following statement: "In order to explain a machine, the best way would be to state what it is intended to do, and to show how all its parts serve this intention."[18] Another example that Leibniz offers is: "If God is the Author of things, and if he is sovereignly wise, it is not possible to reason effectively on the structure of the Universe, without bringing in the intentions of his wisdom, just as it is not possible to effectively reason on the structure of a building, without bringing in the intentions of the Architect."[19]

Teleology may be contrasted to mechanism, which *Webster's* defines at "3a: nature or a natural process conceived as like a machine or as functioning purely in accordance with mechanical laws b: a philosophical doctrine that holds that natural processes and esp. the process of life are mechanically determined and capable of complete explanation by the laws of physics and chemistry."[20]

In the *Essay on Cosmology*, Maupertuis maintains that arguments that prove the existence of God by the wonders of nature can be easily refuted. He attacks Newton's thesis that Divine Intelligence is evidenced by the fact that the six known planets all travel in the same direction around the sun and in more or less concentric orbits. Maupertuis counters that this fact does not prove choice in the choice vs. chance debate. Maupertuis observes that Newton might have also noticed that the planets move in almost the same axis. He points out that there is no wonder in the fact that the planets move around the sun in *nearly* the same plane: if the planets had moved around the sun in *exactly* the same plane, that might be striking, but *nearly* the same plane is not convincing.

Maupertuis also finds that the near uniformity in the course of the planets is not impossible to achieve by chance, and since it is not impossible, one cannot say that it is the result of choice. Maupertuis argues that the only reason that choice vs. chance is an issue is because Newton was unable to explain the reason for the uniformity. Some thinkers posit that the planets reside in a fluid that moves them along, and so for them, they do not have to rely on the explanation of Divine Intelligence. Jacques Roger advises that Maupertuis concluded that the motion of matter does not prove the existence of God: "In other words, an explanation based upon chance was not necessarily absurd-here one sees the emergence of the calculus of probabilities whose role was to be so important in later developments…"[21]

Having shown that the uniformity of the planets could well be the product of random chance, Maupertuis went on to address biology and to argue that the great diversity in nature and the way that organs conform to the need for survival are also the results of random chance. Maupertuis reiterates Lucretius' argument (in *De rerum natura*) that the random collision of molecules created everything in the universe and that the first beings were imperfect and lacked various organs; only those that were not self-contradictory survived. However, it is significant that Maupertuis ingeniously uses the term "conformity" [*convenance*] to identify the traits that surviving beings had that permitted them to live, reproduce, and proliferate. Use of the term "conformity" distinguishes Maupertuis from his contemporaries, Diderot and La Mettrie, who also relied upon Lucretius, but who did not use the term in their works.

Maupertuis speculates, "Could we not say that in the random combination of Nature's productions, since only those in whom were found a certain affinity for conformity could survive, it is not marvelous that this conformity is found in all species that actually exist? Chance, one would say, had produced a multitude of countless individuals; a small number were found to be made in a way that the animal's parts could satisfy its needs; in another, infinitely greater number, there was neither conformity, nor order: all these latter ones perished; animals without a mouth could not live, others that lacked reproductive organs could not perpetuate themselves: the only ones left were those in which order and conformity were found; and these species, that we see today, are only the smallest portion of what blind destiny had produced."[22]

In this passage, Maupertuis employs the terms "conformity" [*convenance*] (4x), "fortuitous combination" [*la combinaison fortuite*] (1x), "chance" [*le hasard*] (1x), and "blind destiny" [*un destin aveugle*] (1x).

In the eighteenth century, the primary definition of *convenance* was "Relation, conformity. *Those things have no relation to one another. What is the relationship between such different things? In order to conduct a discourse on things well, it is necessary to observe the similarities and differences among them.*"[23]

The *Dictionary of the French Academy* (1694) begins its definition of *convenance* by providing two synonyms for the term: *rapport* and *conformité*. Rapport was defined as "Relation, likeness, conformity. *The Italian language was a great likeness to the Latin language. There is a great likeness in the temperaments of these two men. This man's face has a great likeness to the other man's. What you are saying has no connection to what you said yesterday.*"[24] Therefore, connotations include affinity, analogy, resemblance, conformity, correspondence, harmony, agreement, relation, connection.

Conformité was defined as, "The relationship that exists between similar things. *Similar inclinations. Similar feelings. Similar temperaments. Similar minds. Similar court decisions, discourses.*"[25] Connotations include likeness, agreement, consistency, conformity, analogy, and compliance.

In Maupertuis' passage, surviving species were found to have a certain affinity for conformity [*certaines rapports de convenance*]; "this conformity is found in all species that actually exist" [*cette convenance se trouve dans toutes les espèces qui actuellement existent*]; "there was neither conformity, nor order" [*il n'y avait ni convenance, ni ordre*]; "in which order and conformity were found" [*où se trouvaient l'ordre et la convenance*]. From the context of these four phrases, one must extrapolate that *convenance* means "conformity" or "compliance."

Bentley Glass points out that Maupertuis' passage, "Could we not say...that it is not marvelous that this conformity is found in all species that actually exist?...the only ones left were those in which order and conformity were found," is a statement that biology has no need of teleological explanations. Hence, he casts doubt on the belief that there is an ultimate purpose to life, and in fact, on the existence of God.

Maupertuis also points out the absurdity in seeing God's design in the creation of the fly: he asks what the purpose of creating a fly could be: flies are troublesome to humans, they are devoured by the first bird that comes by, and they fall into spiders' webs. Maupertuis argues that the cleverness of execution is not enough, the motive must be reasonable. He fails to see the motive in creating the fly: since there is no evidence of a motive, there is no evidence of choice.

Maupertuis' stance against teleology was one of many reasons that he had a falling out with Voltaire. Voltaire, who was a fervent deist, was offended by the *Essay on Cosmology* and engaged in diatribes against Maupertuis. Thus, the two men, who had formerly enjoyed a close friendship, became enemies. Jacques Roger advises, "In a long review published in the *Bibliothèque raisonnée*, he hammered at Maupertuis, while maintaining an even tone. He reproached Maupertuis's "Essai de cosmologie" with having tried to destroy the proof of the existence of God through final causes, by use of the argument that spiders ate flies and the earth was covered with seas or uninhabitable mountains. But flies were made *in order to be* eaten, the seas and the mountains were made to make water circulate and to fertilize the earth."[26]

Voltaire believed that God "created and arranged everything freely."[27] God bestowed gravitation and motion to matter, and consciousness to certain beings. Voltaire declared, "If the planets turn in one direction rather than in another, in unresisting space, the hand of the Creator therefore directed their course in this direction with absolute freedom."[28]

Another reason that Voltaire opposed Maupertuis was because Voltaire embraced preformation and the notion that all species left the hands of the Creator perfectly at the time of Creation. Conversely, Maupertuis not only promulgated epigenesis, he used it to hypothesize that all beings arose from a single prototype. Hence, when Voltaire wrote a review of Charles Bonnet's *Considerations on Organized Bodies* in 1764, he used it as an opportunity to attack the epigenesis set forth in *Physical Venus* and declared that it was necessary to return "to the ancient opinion that all germs were formed at the same time by the hand that arranged the universe."[29] Maupertuis had posited that parental elements from the mother and father are attracted to each other and arrange themselves in specific patterns. He did not know exactly why they are attracted to each other, but the laws of Newtonian attraction suggested that particles may have an affinity for one another or relationships of union. Voltaire made fun of this and quipped that children are formed by attraction in their mothers' bellies and that the left eye attracts the right leg.[30]

Aside from the differences that Voltaire and Maupertuis had regarding final causes vs. mechanism and preformation vs. epigenesis, there was another reason for the enmity between the two men. In 1744 Maupertuis had articulated what he called "the principle of least action," which was published in 1750 in the *Essay on Cosmology*. The "principle of least action" states, When some change occurs in nature, the quantity of action used for this change is always the smallest possible."[31] The "principle of least action" also posits that "in all the changes that take place in the universe, the sum of the products of each body multiplied by the distance it moves, and by the speed with which it moves, is the least that is possible."

Maupertuis observed that the universe obeys certain laws of motion and one of them is the principle of least action. He declared that all things have been so arranged that a blind and necessary mathematics executes all the activity in the universe, including the movement of animals, the vegetation of plants, and the revolutions of heavenly bodies; the laws of motion are beautiful and simple and bring about all the phenomena of the visible world. Voltaire clearly understood Maupertuis' point: the universe operates according to the laws of motion that obviate the need for God's will. If all of biology, chemistry, and physics are dictated by the principle of least action, then it is the motive property of matter that has set the universe into motion and continues to drive it. What infuriated Voltaire was that the determinism of the principle of least action negates the free will of God: it demonstrates that God is not free to act, that He is limited by the laws of motion. This is why when Voltaire refuted the principle of least action, he said, "What is necessary excludes a choice. It is in the choice of means that the great geometrician Newton found one of the most striking points of conviction for the existence of the creative and governing Being."[32]

Voltaire used his satirical pen to attack Maupertuis relentlessly. Jacques Roger describes Voltaire's satire of Maupertuis thus: "Maupertuis, the 'native of Saint-Malo,' would henceforth be the leading character in Voltaire's puppet show. He would always reappear escorted by his Patagonians and Laplanders."[33] Voltaire's mention of Laplanders was a reference to the fact that in 1736 Maupertuis had led an expedition to Lapland to measure the length of a degree along the meridian. His measurements verified Newton's theory that the earth is an oblate sphere (a sphere flattened at the poles). Maupertuis brought back with him two native girls from Finland and was painted in a famous portrait, wearing a fur hat and fur collar, leaning on a globe of the earth, flattening it with his hand. Voltaire called him "the earth flattener" and made fun of him for having had himself painted in a thick, furry cap.

In the *History of Doctor Akakia and the Native of Saint-Malo* (1752–1753), Voltaire satirizes Maupertuis as the native of Saint-Malo, who, suffering from a chronic case of philotimy (the love of ambition or honor) and acute philocracy (love of power), wrote against doctors and against the proofs of the existence of God; he acquired revelations about the soul while dissecting monkeys; he imagined himself to be as great as the giant of the past century, Leibniz, even though he was not even five feet high; this admirable philosopher discovered that nature always acts according to the simplest laws, and so wisely adds that nature always moves towards economy, should have certainly spared the small number of readers capable of reading his works the trouble of reading the same thing twice in his *Works* and *Letters*: one third of one volume is copied word for word in the other; he dissected two toads; he had himself painted in a thick, furry cap; he wrote that there are stars made like millstones; he declared that children are formed by attraction in their mothers' bellies and that the left eye attracts the right leg; he imagined the nature of the soul by means of opium; he dissected the heads of giants.[34]

In the article "Atheism" in the *Philosophical Dictionary* (1764) Voltaire constructs a dialogue between worshippers of God and modern atheists. He attacks Maupertuis directly, not just by mentioning him by name, but by making him one of the interlocutors of the dialogue; he subtitles sections of his work "New Objections of a Modern Atheist" and "Maupertuis' Objections." Voltaire begins the article by articulating the classic deistic argument: when we see a beautiful machine, we say that there is a good engineer and that this engineer uses excellent judgment; by analogy, the functioning of all living bodies and the movement of heavenly bodies also demonstrate that there is an intelligent engineer or an "Eternal Geometer." Voltaire refutes the atheist's argument that given an eternal time frame, random molecular movement will eventually form the world. Voltaire, on the contrary, contends that it is the existence of intelligent beings in the universe that refutes the atheist's arguments: atheist scientists are unable to prove how random molecular motion alone can bring about consciousness and intelligence. Because they do not know how this happens, they cannot claim that their hypothesis is correct. For the purpose of satire, as is his style, Voltaire focuses on a detail in an issue, magnifies it, and ridicules its proponent. He does not address the fact that in the *Essay on the Formation of Organized Bodies* Maupertuis states that it is precisely because random molecular motion alone is insufficient to explain consciousness and intelligence, that it is necessary to hypothesize that in addition to motion, molecules also have the property of consciousness (aversion, desire, memory, and intelligence).

Voltaire directly responds to Maupertuis' argument that the reason that parts of animals conform to their needs is because all organisms whose organs did not conform to their needs have perished. Voltaire's reply is, "This objection, timeworn since Lucretius, is sufficiently refuted by consciousness given to animals, and intelligence given to man. How could combinations, *that chance has produced*, produce this consciousness and this intelligence..."[35] Voltaire also argues that the limbs of animals are made for their needs with incomprehensible art, and so, the disposition of a fly's wing, a snail's organs "brings you to the ground" ("you" is Maupertuis).[36]

In the dialogue the interlocutor Maupertuis contends that if deists have found God in the folds of the skin of the rhinoceros, one could with equal reason, deny His existence because of the tortoise's shell. Voltaire's reply to this statement is that the "tortoise, the rhinoceros, and all the different species, prove equally, in their infinite variety, the same cause, the same design, the same goal, which are preservation, generation and death. There is unity in this infinite variety; the shell and the skin bear witness equally."[37]

Voltaire excoriates Maupertuis in the following: "And you be silent too, since you cannot conceive its utility any more than I can,"[38] "Some are venomous, you have been so yourself,"[39] "You ask why the snake does harm? And you, why have you done harm so many times? Why have you been a persecutor, which is the greatest of all crimes for a philosopher?"[40] and "But frauds! What are they? frauds."[41] Voltaire uses the term "fripon," which, in the eighteenth century was defined as "Deceiver, who has neither honor, nor faith, nor integrity."[42] An adjective for "fripon" was "fourbe," which meant deceiver.[43] Therefore, "fripon" connoted a fraud, deceiver or swindler. Hence, Maupertuis drew a violent reaction from Voltaire because he epitomized modern science's position that random chance brought about the universe and because the principle of least action obviates God's will.

In *Letter 14* of his *Letters* (1752), Maupertuis recounts his case study of the inheritance of polydactyly in the Ruhe family in Berlin. Jacob Ruhe, a surgeon in Berlin, had been born with six fingers on each hand and six toes on each foot. He inherited this anomaly from his mother, Elisabeth Ruhe, who had inherited it from her mother, Elisabeth Horstmann, of Rostock. Jacob's parents, Elisabeth Ruhe (polydactylous mother) and Jean Christian Ruhe (normal father) had eight children: of these eight, four were polydactylous and four were normal. Jacob (polydactylous) married Sophie-Louise de Thüngen, a normal woman, and they had six children: of these six, two were polydactylous and four were normal. One of them, Jacob Ernest, had six toes on the left foot and five on the right; he had six fingers on his right hand, one of which was amputated, and on his left hand he had only a wart instead of a sixth finger.[44]

Maupertuis concludes that errors in the generative process such as those in polydactyly may explain how all species arose from a single prototype: "I truly want to believe that these supernumerary digits were originally merely accidental varieties, whose production I have tried to show in *Physical Venus*: but these varieties, once confirmed by a sufficient number of generations in which both sexes had them, create species; and perhaps it is thus that all species have arisen."[45]

Maupertuis extrapolates that polydactyly is equally transmitted by both the father and the mother. He posited that eventually, after repeated matings, the trait will disappear. Conversely, it will be perpetuated by marriages in which both parents have the trait. Maupertuis concluded that continual matings with normal people will cause the trait to disappear from the family.

Maupertuis applied mathematics to predict the statistical probability that polydactyly would occur in a given population: "But if we wanted to regard the continuation of sexdigitism as an effect of pure chance, we would have to see what the probability is that this accidental variety in a first parent would not be repeated in his descendants. After a search that I made in a city with 100,000 inhabitants, I found two men who had this anomaly. Let us suppose, which is hard to do, that three others escaped me; and that for every 20,000 men one could reckon one sexdigitary person: the probability that his son or daughter would not be born with sexdigitism is 20,000 to 1: and that his son and his grandson not be sexdigitary is 20,000 x 20,000, or 400,000,000 to 1: finally, the probability that this anomaly would not last three consecutive generations would be 8,000,000,000,000 to 1; numbers so great that the certainty of best proven things in physics do not approach these probabilities."[46]

David Beeson explains that the issue before Maupertuis was to decide whether the appearance of polydactyly was due to chance or heredity. Beeson says that Maupertuis decided to settle the issue by setting up what is today called a "null hypothesis." *Webster's* defines "null hypothesis" as "a statistical hypothesis to be tested and accepted or rejected in favor of an alternative; *specif*: the hypothesis that an observed difference...is due to chance alone and not due to a systematic cause."[47] Beeson advises that Maupertuis' null hypothesis was that polydactyly is not hereditary. Beeson says, "If the null hypothesis were true, the probability against the trait appearing so often in the Ruhe family would be massive. He therefore concluded that he could, with what we would now call a high degree of confidence, reject the null hypothesis as false. His statistical analysis was sound."[48] Hence, Maupertuis concluded that it is unlikely that the anomaly would appear in the third generation of the family by chance alone, and therefore, because it did, it was hereditary.

Bentley Glass credits Maupertuis for having accomplished a long list of impressive achievements that distinguished him from his peers. Glass enumerates Maupertuis' legacy to biology thus: 1) he "recorded and interpreted the inheritance of a human trait through several generations," 2) he "applied the laws of probability to the study of heredity, 3) he "was led by the facts he had uncovered to develop a theory...that heredity must be due to particles derived both from the mother and from the father, that similar particles have an affinity for each other that makes them pair, and that for each pair either the particle from the mother or the one from the father may dominate over the other, so that a trait may seemingly be inherited from distant ancestors by passing through parents who are unaffected. From an accidental deficiency of certain particles there might arise embryos with certain parts missing, and from an excess of certain particles could come embryos with extra parts, like the six-fingered persons or the giant with an extra lumbar vertebra whom Maupertuis studied. There might even be complete alterations of particles...and these fortuitous changes might be the beginning of new species..."[49]

Glass also points out that Maupertuis was able to distinguish between strong traits and weak traits. Glass remarks, "Maupertuis was quite struck by this apparent weakening of the trait with time, and it led him to the conclusion that through repeated matings with normal individuals the trait might in time disappear."[50]

Glass observes that Maupertuis had acknowledged that polydactyly occurs frequently from affected parents, whereas albinism occurs sporadically among blacks: "Maupertuis also arrived though vaguely, at the idea of dominance...Maupertuis was therefore aware that whereas polydactyly descends regularly from affected persons, married to normals, to some but not all of the offspring, albinism, on the other hand, seemed to appear sporadically among negroes, albino negroes being born of parents both of whom were black...Maupertuis...concluded: 'There could be, on the other hand, arrangements so tenacious that from the first generation they dominate (*l'emporte*) over all the previous arrangements, and efface the habitude of these."[51]

Maupertuis' *Essay on the Formation of Organized Bodies* (1754) was reputedly his doctoral dissertation. He had written it in Latin and published it in 1751 under the pseudonym "Dr. Baumann," under the title, *Inaugural Dissertation on Metaphysics*.[52] The *Dissertation* was translated into French and published under the titles, *System of Nature* (1751) and *Essay on the Formation of Organized Bodies* (1754). Diderot recognized Maupertuis' genius and praised his work profusely in *Thoughts on the Interpretation of Nature* (1753). Maupertuis, in turn, was delighted to have the acknowledgement of the great Diderot, and printed portions of Diderot's kudos in the preface of his 1754 edition of the *Essay*.

Maupertuis' *Essay* is a also landmark document in that it presents two innovative ideas that were to influence contemporaries such as Diderot. First, Maupertuis reiterated that heredity is based on the arrangement or order of parental elements. When the memory of the patterns or arrangement of parental elements is preserved and there are no errors, offspring resemble their parents. When errors occur in these arrangements, the errors are transmitted from generation to generation; they can result in birth defects (such as six fingers on one hand) and, ultimately, the appearance of new species can be explained by these errors. Maupertuis carries this logic to the limit: these errors in the arrangement of parental elements can explain how all living things arose from a single prototype. Hence, he posited that the animal kingdom arose from the vegetable kingdom, and the vegetable kingdom, from the mineral kingdom. Secondly, Maupertuis hypothesized an emergent consciousness: molecules have the property of consciousness, and when they combine, the newly organized entity has its own consciousness, which is better and superior to that of the sum of its constituent parts.

Maupertuis begins the *Essay* by affirming that all biological hypotheses must be solidly grounded in physical, verifiable phenomena. He extols empiricism and the scientific method, as well as Lockian epistemology that all knowledge is acquired through the five senses. His first step, then, is to debunk the myths of his time, such as preformation and unextended entities (ie: plastic natures and intelligent substances). He dismisses plastic natures, which, "without intelligence & without matter, are imagined to exert all of the influence on the universe that matter and intelligence can exert." He equally dismisses the power of intelligent substances to "move stars and oversee the production of animals, plants & all organized bodies." He ridicules philosophers who, not wanting to resort to plastic natures and intelligent natures to explain the formation of organized bodies, have resorted to the preformation myth. He humorously refers to homunculi as "inexhaustible warehouses of individuals." Preformation posed problems for its adherents: "They were at a loss to know where to place these inexhaustible warehouses of individuals..."[53] Some held that the preformed homunculus resides in the mother (ovists), others, in the father (animalculists), and "each was content for a long time in his ideas."[54] Maupertuis replaces preformation with the experimental method (*l'expérience*). Experimentation and observation, such as that which he had performed on polydactylous members of the Ruhe family in Berlin, "has proven that one cannot accept that an infinite succession of beings has arisen from either one parent or the other."[55]

Maupertuis embraced the notion that particles residing throughout the body settle in the reproductive organs and reproduce the part of the body from which they came. The theory was not new: he merely resurrected it and gave it new life. Hippocrates, Empedocles, Democritus, Almaeon, Pythagoras, Aristotle, and Galen believed that elements circulating throughout the bloodstream of the father eventually form the offspring. Maupertuis expanded the theory to include both parents.

He believed that the reason that particles, spread throughout the body arrange themselves in an orderly fashion, some to form an eye, and others an ear, is because molecules are conscious and have the properties of desire, aversion, memory, and intelligence.[56] He defends this hypothesis by pointing out that animals exhibit intelligence: they see, understand, desire, fear, and remember. Since animals behave as if they were conscious, one must necessarily extrapolate that they are, indeed, conscious. Therefore, it would not be illogical to accord consciousness to even the smallest particles of matter. He declares that intelligence resides in the tiniest grains of sand as well as in elephants and monkeys.

Maupertuis explains the necessity of according the properties of intelligence, desire, aversion and memory to matter in order to explain reproduction. He says, "I believe I see the necessity in it. One will never explain the formation of any organized body, by the physical properties of matter alone..."[57] Thus, Maupertuis embarked upon new ground. He attempted to explain reproduction, something that he claimed that philosophers from Epicurus to Descartes had been unable to do by the motive property of atoms alone. He argued that more than motion is required: it is necessary to either allow for new properties or else acknowledge properties that do exist.[58]

As he had done in *Physical Venus*, Maupertuis reiterates that epigenesis is fully capable of explaining many phenomena that preformation cannot: resemblance to parents, teratisms lacking body parts (i.e., Cyclops), teratisms with too many body parts (i.e., polydactylous births), teratisms whose organs are reversed (i.e., the heart and stomach are on the right, the liver on the left), hybrids, and the sterility of hybrids (i.e., the mule). Parental elements that will form the offspring circulate throughout the body and finally settle in the reproductive organs of the mother and father. Each parental element is derived from the part of the body that it will form, retains a memory of its former situation, and reproduces itself as many times as it can, to form an identical part in the offspring. From this arises the conservation of the species and resemblance to parents. If some parental elements are lacking, or if they cannot unite, teratisms arise that lack some part. If some parental elements are found in too great a quantity, or if after their ordinary union, an extra particle allows another to join itself to it, there arises a teratism with extra parts.

In the case of hybrids, if the parental elements are derived from animals of different species, but there still remains a rapport among the elements, some are more attached to the father's form, other to the mother's form, there will arise hybrid animals.

If the parental elements are derived from animals that lack a sufficient analogy between them, these parental elements cannot arrive at an adequate arrangement, and generation becomes impossible. This explains sterility in the offspring of hybrids. Repeated experimentation and observation show that no animal, born from the coupling of different species, reproduces.

Conversely, if there are elements so susceptible to arrangement or in which the memory is confused, they will arrange with the greatest facility and new animals will arise as one sees in spontaneous generation with moistened flour and with other animalcules that liqueurs form. Maupertuis, like many eighteenth-century thinkers (Diderot, Buffon, Needham) believed in spontaneous generation. Since Maupertuis believed that all molecules are conscious and that all life resulted from a single prototype through the transmission of errors in the arrangement of parental elements (originally from the mineral kingdom), he did not view spontaneous generation as an impossibility.

We can also explain certain phenomena with epigenesis that cannot be explained by other theories. Sometimes resemblance skips a generation and offspring resemble grandparents more than they do parents. The elements that form certain traits may have better preserved the habit of their situation in the grandparent than in the parent.

A total forgetfulness or memory lapse of the first situation will give rise to a teratism with reversed organs.

Maupertuis asks, "Could one not explain by this how from two individuals alone, the multiplication of the most dissimilar species could have followed? They would only have owed their origin to some fortuitous productions in which elementary parts would not have retained the order they had in the father and mother animals: each degree of error would have made a new species; and by virtue of repeated digressions there would have arisen the infinite diversity of animals that we see today, which will increase still more perhaps with time, but to which perhaps the succession of centuries carries only imperceptible growths."[59] This statement constitutes a landmark in the history of biology. Maupertuis is emphasizing arrangement, pattern, and order in elements contributed by both the father and mother. When the memory of the original pattern is retained, the offspring resembles the parent who contributed the element. When the memory of the original arrangement is lost, an error occurs, and the offspring has a birth defect. These errors explain the existence of the innumerable species in the world. Further, they explain how all beings arose from a single prototype. The implications are immense: he will go on to discuss consciousness. If all beings arose from a single prototype (i.e., in the mineral kingdom), then consciousness resides, to varying degrees, in all three kingdoms.

Maupertuis hypothesizes an emergent consciousness: when particles unite, each particle loses its consciousness of self and acquires the consciousness of the larger body to which it belongs. It loses the memory and consciousness of itself and, upon uniting with other particles, acquires the consciousness of the whole. He asks, "But each element, in losing its form, & combining with the body that it is going to form, would it also lose its perception? Would it lose, would it lessen the small degree of consciousness that it had, or would it strengthen it by its union with the others, for the benefit of all?"[60] He answers the question thus: "…it seems that from all of the perceptions of the elements having been gathered together, there results a single perception that is much stronger, much more perfect than any of the constituent perceptions and which is perhaps analogous to each of these perceptions as the organized body is to the component part. Each element, in its union with others, having mingled its perception with theirs, and having lost its *consciousness of self*, we have lost the memory of the original state of the elements, and our origin must be entirely lost for us."[61] Hence, man is conscious of himself and others, but he is not aware of each molecule that constitutes his body.

Maupertuis was highly regarded by his contemporaries, Buffon and Diderot. Buffon stated that the problems inherent in the competing theories of the ovists and the animalculists "were intuited by a man of intellect, who seems to me to have reasoned better than all those who had written on this matter before him, I am talking about the author of *Physical Venus*, published in 1745; this treatise, although very short, assembles more philosophical ideas than there are in several big volumes on generation…this author is the first to have begun to approach the truth, from which we were farther than ever since we had imagined that eggs exist and animalcules were discovered."[62]

Diderot also recognized Maupertuis' genius, praised him profusely, and adopted his hypothesis that all life is derived from a single prototype. In *Thoughts on the Interpretation of Nature, Thought 12*, Diderot asks the reader whether, considering the fact that characteristics among species overlap, and also that there are beings that have the functions and body parts of two kingdoms, one would be persuaded to believe that there has only ever been a single prototype for all beings. Diderot credits Dr. Baumann (Maupertuis' pseudonym) with this hypothesis, adding that this theory should be examined more fully: "...we will not deny that it should be embraced as an essential hypothesis for the advancement of experimental physics, rationalist philosophy, and the discovery and explanation of phenomena that depend on organization."[63] In a footnote in *Thought 12*, Diderot refers the reader to Dr. Baumann's *Inaugural Dissertation on Metaphysics*.

In *Thought 50*, Diderot praises Maupertuis again and prints passages verbatim from the *Inaugural Dissertation on Metaphysics*. Diderot's laudatory words include "...the doctor of Erlangen, whose book, filled with odd and new ideas...His subject matter is the greatest to which human intelligence can apply itself...Dr. Baumann's hypothesis will explain, if you will, the most incomprehensible mystery of nature...a subject that the foremost individuals throughout the centuries have taken up...the fruit of deep meditation, the endeavour of a great philosopher."[64]

In *Thought 50*, Diderot meticulously follows Maupertuis' sequence of ideas as he had presented them in the *Inaugural Dissertation*: first, he mentions that Maupertuis had denied unverifiable hypotheses and the belief in unextended entities; he refuted the theories of plastic natures, subordinate intelligent substances, and preformation. Rather, he chose to explain life with the hypothesis that all molecules are conscious and that they have the properties of desire, aversion, memory, and intelligence. Diderot cites Maupertuis' passage on emergent consciousness in the original Latin in order to avoid the accusation that he misinterpreted him: when molecules combine to form larger entities, each molecule loses its memory of self and acquires the consciousness of the larger body that it forms. Each successive organization has its own consciousness. Diderot carries this notion to the limit and asks whether the universe, which is the composite of all nature, might be God. Maupertuis, who was not a pantheist, had no choice but to vehemently deny Diderot's conclusion.

David Beeson examines Maupertuis' reaction to Diderot's conclusion: "What most interested Diderot was the suggestion that elementary particles of matter might be assigned an elementary consciousness, and by their fusion form a whole whose consciousness would be more than the sum of its constituents and would form a mind or soul...Diderot sets out to push Maupertuis's ideas as far as he can."[65] In the *Interpretation of Nature*, Diderot pushes the notion of an emergent consciousness to the limit, positing, "I will ask him, then, if the universe, or the general collection of all conscious and thinking molecules, form a whole...the world could be infinite, this world soul, I am not saying, is, but may be, an infinite system of perceptions, and the world may be God."⁶⁶

Beeson points out that Maupertuis was to deny this charge of neo-Spinozism when he was to prepare the *Système de la nature* for inclusion in his 1756 *Complete Works*. In the *Response to the Objections of Mr. Diderot*, Maupertuis refutes Diderot by arguing against systematic reasoning, or reasoning that the universe, organisms and particles can be integrated into a single system: "Our mind, as limited as it is, will it ever find a system in which all consequences are in harmony...All our systems, even the greatest ones, embrace only a small part of the plan that the Supreme Intelligence follows; we see neither the relationship that parts have to one another, nor their relationship to the whole..."⁶⁷

Maupertuis, as Beeson points out, goes on not to "merely attack systematic reasoning in general terms; Diderot's fundamental methodology is itself, he claims, false."⁶⁸ Beeson quotes from Maupertuis, "This method of reasoning, that Mr. Diderot calls the act of generalization, and which he regards as the touchstone of systems, is merely a kind of analogy, that we can take where we want; incapable of proving either the falseness or the truth of a system."⁶⁹

Beeson advises that Maupertuis considered materialism, like Cartesianism, to be in error because "both depend on statements concerning the nature of the universe as a whole, and consequently on systematic reasoning. Such reasoning is indefensible because the human intellect is incapable of handling universal truths. Diderot's attack on Maupertuis depends on the systematic application of reasoning by analogy; Maupertuis replies by denying the possibility of applying analogy systematically, insists that we are authorized to take such arguments just as far as suits us. Diderot had explicitly declared his intention to force Maupertuis's reasoning to its logical conclusion; Maupertuis refused to be forced and denied that the conclusion was logical."⁷⁰ Hence, Maupertuis, who was an atheist, refuted Diderot's pantheist conclusion by asserting 1) the impossibility of ever understanding universal truths or the nature of the universe as a whole and 2) the impossibility of ever finding one system that fully explains the relationship that parts of the universe have to one another or to the whole.

In summation, Maupertuis' study of the recurrence of polydactly in four generations of the Ruhe family led him to arrive at some landmark conclusions in biology. First, he observed that the trait could be transmitted by either parent who had the anomaly. Secondly, he hypothesized that birth defects first come about due to an error in the arrangement or pattern of parental elements. This error is passed down from generation to generation and eventually, over time, can create new species. In this way, it is possible to explain the creation of all living things from a single prototype. Maupertuis believed that climate and nutrition can influence heredity.

Maupertuis must also be credited with devising a mathematical means (which is today called the "null hypothesis") of predicting the statistical probability that an anomaly will not recur in a family's future offspring.

Maupertuis refuted preformation and adopted Pythagoras' notion that the offspring acquires particles drawn from all over the body of its father; Maupertuis broadened the hypothesis to include particles from the bodies of both parents. The elements retain the memory of their former situation and take it up again in the offspring, thus recreating the body part from which they are derived.

In order for this to work, Maupertuis hypothesized an emergent consciousness: he posited that all molecules have the property of desire, aversion, memory, and intelligence. When molecules combine, each one loses its memory of self and acquires the consciousness of the larger entity that it forms. This explains why we are not aware of each individual molecule or organ in our body. His hypotheses were to form the basis of Diderot's transformism in 1753 and 1769.

Chapter 6

Diderot

When we see successive metamorphoses...approach one kingdom from another kingdom by gradual degrees and populate the borders of these two kingdoms...who would not be led to believe that there was not ever only one first prototype for all beings?[1]
—Denis Diderot, *Thoughts on the Interpretation of Nature, Thought 12* (1753)

This chapter will provide an overview of the salient points of *Diderot and the Metamorphosis of Species* (New York and London: Routledge, 2007). That book discusses Diderot's radical view that species metamorphose over millennia and shows how that hypothesis was influenced by contemporary sources (Buffon, Maupertuis, and La Mettrie), *Encyclopedia* articles on probability theory and fossils, Lucretius, and Needham's experiments with spontaneous generation. The material in that work is presented in six chapters that attempt to cover the expansiveness of Diderot's thought: "Chaos, Time, Flux, and Probability," "Embryology, Epigenesis, and the Metamorphosis of Species," "Spontaneous Generation," "The Chain of Beings," "The Mutability of Species," and "The Ascent of Consciousness." The most important points of each chapter will be mentioned here.

There are at least seven significant factors that contributed to Diderot's transformism:

1. the Greeks' hypothesis that atoms are in perpetual motion, continually colliding with one another, and randomly forming new combina-

tions; therefore, the universe and everything in it are the results of the motive property of atoms and random chance

2. epigenesis-William Harvey's theory that the germ is brought into existence by successive accretions and not developed from a preformed seed; this hypothesis allows for the birth of individuals that are uniquely different from either of their parents and explains the appearance of offspring that do not resemble either parent, birth anomalies, and hybrids.

3. Buffon observed that physical characteristics overlap among species: different species have similar body parts (correspondences in types of structures are called homologies). Diderot surmised that homologies provided evidence that all beings arose from a single prototype. Buffon developed the two dimensional chain of beings into a three dimensional cone in which characteristics are shared; Diderot added the fourth dimension, time, to show that physical characteristics change over millennia.

4. Buffon believed that species degenerate when they are taken out of their lands of origin and are domesticated by men. Geography, climate, food, domestication, and working an animal very hard cause minor changes in a species, but do not create new and different species. Diderot considered Buffon's idea of degeneration, and, parting company with him, concluded that these factors do, indeed, create new species.

5. Maupertuis theorized that errors occur in the arrangement of parental elements during the generative process and that these errors are transmitted from generation to generation; these errors could explain how all living things may have arisen from a single prototype. Diderot seized upon this idea and made it the foundation of his transformism for the rest of his literary career.

6. Maupertuis hypothesized that consciousness, like motion, is a property of atoms. Consciousness is inherent in all matter, at every level of organization, and this consciousness causes threads to organize into bundle of fibers, and bundles of fibers to form organs. When matter combines to form something new, the new consciousness that arises is greater than that of the sum of its parts. Hence, Maupertuis envisaged an emergent consciousness. Diderot embraced this notion and made it one of the three pivotal points on which the fulcrum of

his transformism rested (the other two being the motive property of atoms and probability theory).

As chief editor of the *Encyclopedia*, Denis Diderot was perpetually on the vanguard of science and privy to a wealth of information that, collectively, caused him to conclude that transformism is a certainty. He published articles on fossil discoveries that suggested that the earth was much older than previously thought. He was well aware that there were layers of beds of seashell fossils situated far from water that indicated that Europe was once covered with water. Moreover, they suggested that the earth and sea had changed places several times. Discoveries of intact mammoths in northern Siberia and the New World indicated that animals were once much larger than they presently were.

The *Encyclopedia* also had articles on probability theory, many of which Diderot, himself, authored, and others of which, the mathematician d'Alembert wrote. Diderot knew that studies of games of chance (ie: cards, dice) indicate that given enough time, every possible permutation will eventuate. He applied this mathematical principle to his transformist biology. Molecules have the property of motion; because molecules are in perpetual motion and are continually randomly colliding, eventually, every possible combination will come up. Given an infinite amount of time, a universe will gradually emerge, as will stars, planets, the mineral kingdom, the vegetable, the animal, and man.

Diderot's transformism was intertwined with the polemics of atheism. In 1746, while he was still a deist, he declared that the magnificence and beneficence of God could be readily discerned by observing the wonders of nature (ie: the complexity of the insect, the harmonious paths of celestial bodies, in *Philosophic Thoughts, Thought 18*). By 1749, Diderot had migrated from deism to pantheism to atheism. In his article, "Diderot and Eighteenth-Century French Transformism," Lester Crocker examines how Diderot came about to journey from deism to pantheism to atheism in three short years.[2] By 1749 Diderot had moved away from the deistic notion of a beneficent God as he was painfully aware of the human misery that is born of birth defects such as blindness. In the *Letter on the Blind* (1749), the reader experiences the pathos in Saunderson's cry, "Look at me, Mr. Holmes. I have no eyes. What have we done, you and I, to God, that one of us has this organ while the other has not?"[3] Nature, which is perpetually in flux, is continually creating monsters, as it has done since the time of Lucretius. What we perceive to be nature, is merely the outcome of a series of random events, and hence, nature, by definition, is oblivious to human suffering.

In 1749 Diderot embraced the philosophy of Epicurus, the fullest extant version of which is articulated in Lucretius' *De rerum natura*: random molecular motion, given an infinite length of time, will eventually yield life, and this includes anomalies. Because motion is a property of matter, nature is continuously creating new varieties, many of which are defective. Diderot takes Lucretius' stance that at the time of creation, many monsters were born and only those beings without serious self-contradictions survived and perpetuated their kind. Nature is still creating monsters, as there are Siamese twins, blind people, stillborn births, and false conceptions.

By 1753 Diderot had gathered more information on biology from the writings of Buffon and Maupertuis. Buffon had published the first four volumes of the *Natural History*, in which he described homologies that he had observed among the body parts of various species. Diderot understood that the chain of being is not a two dimensional line extending from God down to inanimate matter, but rather, a three dimensional cone in which physical characteristics overlap among species. Maupertuis, on the other hand, hypothesized that errors occur in the generative process and that these errors are passed on from generation to generation; this could explain how, over time, all living beings may have arisen from a single prototype. Diderot took Buffon's observations of homologies and Maupertuis' theory that inherited errors can eventually bring about new species, and extrapolated that homologies are proof that all beings, indeed, the three kingdoms, have arisen from a single prototype [*Thoughts on the Interpretation of Nature* (1753)].

By 1769 Diderot adds a fourth dimension to the three dimensional cone: time. The universe is in perpetual flux, and given an infinite time frame, species come into existence and then fall out of existence. Everything is in flux: animalcules that materialize and then die under microscopic magnification, species, planets, stars, and perhaps, even the universe itself, are all born, exist for a certain length of time, and then fall out of existence [the trilogy, *Conversation between d'Alembert and Diderot* (1769), *D'Alembert's Dream* (1769), and the *Sequel to the Conversation* (1769)].

In 1746 Diderot was still a deist, but he was just beginning to entertain atheism. In *Philosophic Thoughts, Thought 18*, Diderot affirms that it is the work of Marcello Malpighi, Isaac Newton, Pieter van Musschenbroek, Nicholas Hartsoeker and Bernard Nieuwentyt that has provided "satisfactory proofs" of "the existence of a reign of sovereign intelligence."[4] Because of their work, the world is no longer a god, in the tradition of Spinoza, but rather, a "machine with its wheels, its cords, its pulleys, its springs, and its weights" as in the convention of Newton.[5] These scientists demonstrate that there is a difference between the created and the Creator. The created universe is not a god, as in Spinoza's work, but rather, announces the existence of a Creator, as in deism.

In *Thought 19*, Diderot again proclaims deism and denies spontaneous generation. He reiterates, "...it was reserved for the knowledge of nature to make true deists."[6] He takes the classic Newtonian deistic stance that the marvels of the universe proclaim the beneficence and magnificence of God. Because only God can create life, spontaneous generation must clearly be false. He alludes to the experiments of Redi when he adds, "...all experiments agree in proving to me that putrefaction alone never produced any organism."[7] Spontaneous generation was a powerful polemical tool: atheists, relying on the experiments of John Turberville Needham, argued that living things arise from nonliving matter all the time, in the twinkling of an eye, and therefore, divine agency is not needed; the materialists cited the authority of Lucretius, who in *On the Nature of Things*, declares that random molecular motion creates everything in the universe, as is evidenced by spontaneous generation. Conversely, deists argued that God created all living things at the time of Creation and that therefore, the generative process is required from living beings to create new life. In 1746 Diderot still argued the latter: only God can create life and scientific experimentation (by Redi) shows that spontaneous generation is false.

However, although Diderot was still a deist in 1746, he was beginning to toy with atheism. He begins *Thought 21* by stating, "I open the notebooks of a celebrated professor..."[8] Paul Vernière advises that critics hold that this "celebrated professor" is Dominique François Rivard (1697–1778), who had taught philosophy and mathematics at the collège de Beauvais."[9] Vernière also notes that he introduced mathematics to the University of Paris and that Diderot was his student.[10] "Notebooks" in the phrase "I open the notebooks..." connotes authority, truth, and importance; it builds suspense by attributing these qualities to something that will follow. "Celebrated' and "professor" doubly hyperbolizes the authority connoted by "books" and doubly builds suspense by announcing that what will follow is even more important and has even more authority; it will carry the weight of the authority of the notebooks of a celebrated professor. This is the celebrated professor who introduced Diderot to the science of mathematics and hence, showed him the path to certainty and fact. Rivard also taught philosophy: therefore, he was able to make connections between philosophical speculation and mathematical certainty. He taught Diderot that philosophy and mathematics are not distinct entities, but that they overlap. The central thesis of *Thought 21* will be that probability theory sheds light on the origin of the universe. Because probability theory is mathematical in nature, it is certain and cannot be disproven.

In *Thought 21* a deist and an atheist have an argument as to whether the world is the fortuitous result of the collision of atoms. The theory was not new: it dates back to Thales (6th century BC), Leucippus (5th century BC), Democritus (460–370 BC), and Epicurus (341–270 BC). Lucretius' *On the Nature of Things* (first century BC) is the most complete statement that we have of Epicurus' work. In *Thought 21* the atheist clearly wins the argument that he has with the deist. At the beginning of the argument, the deist concedes that motion is an essential property of matter. The moment he concedes this point, he loses the argument. Motion is an inherent property of molecules that cause them to perpetually collide with one another. Probability theory shows that given an infinite time frame, every eventuality will arise (the universe, stars, planets, the three kingdoms, and man).

The language that Diderot employs is the language of games of chance and probability theory: to agree upon [*accorder*], to bring about [*amener*], analysis [*analyse*], mathematical permutations (2x) [*arrangements*], to occur [*arriver*], advantage or advantageous (3x) [*avantage* or *avantageuse*], mutual agreements [*aveux réciproques*], printer's type (2x) [*caractères*], 100,00 dice [*cent mille dés*], 100,00 sixes [*cent mille six*], chaos [*chaos*], possible combinations [*combinaisons possibles*], compensated (2x) [*compensée*], throws [*coups*], contradict oneself [*se démener*], difficulty (2x) [*difficulté*], duration [*durée*], create (2x) [*engendrer*], to follow [*ensuivre*], eternity (2x) [*éternité*], event (2x) [*événement*], all at once [*à la fois*], fortuitous (2x) [*fortuit*], fortuitously (2x) [*fortuitement*], to bet [*gager*], hypothetical [*hypothétique*], infinite (6x) [*infinie*], throw or throws (6x) [*jet* or *jets*], game [*jeu*], laws [*lois*], multitude (3x) [*multitude*], multitude of those [*multitude de ceux*], multitude of throws [*multitude de jets*], number (2x) [*nombre*], order [*ordre*], small [*petit*], no limits [*point de bornes*], possible [*possible*], possibility [*possibilité*], to propose [*proposer*], proposition [*proposition*], quantity (3x) [*quan-*

tité], quantity of throws (3x) [*quantité des jets*], real [*réelle*], result [*résultat*], definite number (2x) [*somme finie*], infinite number [*somme infinie*], chances [*sorts*], and supposition [*supposition*].[11]

The atheist in the argument relies on the certainty of the science of mathematics to understand the origin of the world. Atoms are metaphorized as dice. Just as dice are thrown and randomly yield new permutations, atoms, too, collide and form new combinations. A basic premise of probability theory is that as the number of tosses grows larger, all possible combinations eventuate. Given an eternal time frame, nothing is impossible.

Diderot employs three metaphors: atoms [*atomes*] (that perpetually collide and form new objects), printer's type [*caractères*] (given enough time, random strokes of printer's type can result in a literary masterpiece), and dice [*dés*] (the continual roll of dice will eventually yield the most unlikely patterns). The deist tells the atheist that he might as well argue that Voltaire's *La Henriade* or Homer's *Iliad* were produced by random keystrokes on printer's type; this appears, at first glance, to be a good argument that intelligence is required to produce order from chaos. However, the atheist uses the science of mathematics to prove that given an eternal time frame and random molecular motion, intelligence is not required to produce order from chaos: every possible permutation will eventuate without will or intelligence. Here, physics predominates and Divine Will is obviated.

In the conclusion to *Thought 21*, Diderot asserts, "Therefore the mind ought to be more astonished at the hypothetical duration of chaos than at the actual birth of the universe."[12] In this statement, Diderot surmises that chaos has never existed because the motive property of atoms dispels it. Hence, he takes exception to the biblical notion that God called the world into being out of chaos with His Word. The fact that the universe exists is an iconic representation of the truth of the mathematical principle that flux (of events)+time=patterns.

This is an overview of two crucial points in the fulcrum of Diderot's transformism (probability theory and the motive property of matter). A more in depth analysis, an *explication de texte* of the mathematical terminology that Diderot employs, and an enumeration of articles on probability theory that appeared in the *Encyclopedia* (1751–1765, source material for his later works), appears in my first book, *Diderot and the Metamorphosis of Species, Chapter 1*, "Chaos, Flux, Time, and Probability."[13]

Three years later, in the *Letter on the Blind* (1749), Diderot again employs probability theory to show that flux+time dispels chaos. Again the speaker is a mathematician, this time, Nicholaus Saunderson, who was in fact, Lucasian professor of mathematics at Cambridge and who lost his vision due to smallpox at the age of one. In Diderot's work, however, he is blind from birth. It is significant that the protagonists in *Thought 21* and the *Letter on the Blind* are both mathematics professors. Mathematics is the science of certainty. Theories can be proven to be true or false and there is no doubt. Hence, when mathematics is applied to philosophy (ie: speculations about the origin of the universe), it can be held up as a benchmark against which truth can be measured.

Saunderson is chosen to be the protagonist for another reason, as well: he is blind and therefore, one of the anomalies that flux has been eternally producing. He, himself, is an event, and an iconic representation of the proven mathematical fact that flux, given time, will produce monsters, birth anomalies, beings that can survive only a few hours before they die, beings that nature itself aborts because they are even more seriously defective than those that survive until birth. Saunderson, himself an anomaly, is emblematic of the fact that life is merely the result of the motive property of matter. He declares, "...if we were to return to the birth of things and times, and we perceived matter move and chaos unfold, we would encounter a multitude of formless beings for a few well organized beings."[14]

He continues, "For instance, I may ask you and Leibniz and Clarke and Newton, who told you that in the first instances of the formation of animals some were not headless and others footless? I might affirm that such an one had no stomach, another no intestines, that some which seemed to deserve a long duration from their possession of a stomach, palate, and teeth came to an end owing to some defect in the heart or lungs; that monsters mutually destroyed one another; that all the defective combinations of matter disappeared, and that those only survived whose mechanism was not defective in any important particular and who were able to support and perpetuate themselves."[15]

The terminology is the same in 1749 as it was in 1746, which indicates that Diderot is focusing on the problem of the origin of the universe from the same perspective, namely, that of probability theory. The language connotes the continual flux of events, randomness, an infinite time frame, and the eventual appearance of all permutations. He uses "birth" to connote the beginning of a string of throws: "the real birth of the universe" (1746, *Thought 21*) and "the birth of things and times" (1749, *Letter on the Blind*). He employs "admirable" to describe what we interpret to be a pattern in a string of throws: "admirable arrangements (2x, 1746) and "an admirable order" (1749). He employs "chaos" in both works: "the hypothetical duration of chaos" (1746) and "chaos unfold" (1749). Permutations from probability theory [*arrangements* and *combinaisons*] recur: "the infinite number of possible combinations" (1746) and "all the defective combinations of matter have disappeared" (1749). Flux recurs in the form of a continual, unceasing strings of events ("multitude"): "the multitude of throws" (1746) and "a multitude of formless beings" (1749). In 1746 and 1749 we have all of the components of probability theory: patterns that emerge (admirable arrangements, admirable order), the absence of patterns (chaos), permutations (combinations), the beginning of a string of throws (birth). In 1746 the events were atomic collisions; in 1749 the events are living beings, many of them deformed creatures. Hence, Diderot returned to the probability theory that he was considering in 1746, but this time, as an atheist who has concluded that the random flux of atoms (due to the motive property of matter) has dispersed chaos and eventuated living beings.

In 1746 Diderot was a deist and Newton was his authority; in 1749 Diderot is an atheist and he relies on the authority of the ancients. Lucretius, in *On the Nature of Things*, 5.418–425, hypothesizes that random molecular motion created everything in the universe. Diderot derives from Lucretius both the notion of flux (that the continual random collision of molecules has brought about the universe) and monsters (some of the events are malformed humans). Saunderson discounts Newton's "admirable order": it is merely the result of fortuitous throws. "Jusquà ce qu'ils aient obtenu quelque arrangement dans lequel ils puissent persévérer" is reminiscent of Lucretius' statement that multitudes of atoms moving through multitudes of courses through infinite time eventuate combinations that form the earth, sea, sky, and living creatures.[16] The events are random combinations of atoms that sometimes take the form of living things: many of these living things are malformed and monstrous. Saunderson concedes that Newton"s "admirable order" exists. This is ironic: Saunderson, himself, is not an example of any admirable order, he is defective (blind) and hence, he is an iconic representation of nature's flawed random events. What man interprets to be admirable, is, in fact, a fortuitous string of events.

Because Lucretius is his authority, Diderot asserts that in the beginning, nature tested every possible combination "until they obtained a permutation in which they could survive" [*jusqu'ils aient obtenu quelque arrangement dans lequel ils puissent persévérer*]. Again, we have a fortuitous permutation [*arrangement*] in a string of events. The language in the *Letter on the Blind*, like that in *Thought 21*, is the language of probability theory: permutations [*combinaisons*], number of mere possibilities [*nombre des possibles*], fortuitous string [*order*], to appear [*paraître*], occasionally [*de temps en temps*], the start of a string of throws [*commencement*], a string of repetitious events [*mes semblables étaient fort communs*], quantity [*combien*], flux [*le mouvment continue et continuera*], time [*jusqu'à*], permutation [*arrangement*], a fortuitous, random permutation [*arrangement dans lequel ils puissent persévérer*], a string of throws [*une succession rapide d'êtres*], a pattern [*une symétrie passagère, un ordre momentané*], fortuitous string [*la perfection des choses*], and time [*tout à l'heure, la durée, vos jours, éphémère(s)* (2x), *éternel* (2x), *instant, éternité, temps précis, durer*].[17]

Diderot observes that given flux (arising from the motive property of atoms), patterns randomly emerge at different levels of magnification: there is symmetry and self-similarity. The notion that symmetry and self-similarity are phenomena that arise from chaos would one day be articulated by chaos theory. While Diderot demonstrates that there must be a succession of monsters in order for a few viable creatures to arise that can survive, similarly, there are defective worlds among those that do survive. He asks how many malformed worlds in outer space ("where I do not touch and you do not see") must come into existence, last for a short period of time (because they are defective), and then disappear. Planets and stars are metaphorized to be animals (they are malformed) that can subsist only if they have no inherent self-contradictions. Diderot observes a truth that one day fractal theory would articulate: symmetry and self-similarity of parts. In the *Letter on the Blind*, he observes birth, existence and death at the magnification level of living beings [*animaux*] and planets [*mondes estropiés*]. In 1769, in *D'Alembert's Dream*, he would also observe birth, existence, and death of animalcules, stars, and the universe itself, as well. Hence, the universe has symmetry and self-similarity of its parts: the microscopic, living beings, and the macroscopic are all merely events the flux delivers and furthermore, they all have a beginning, a term of existence, and an end. Even the whole (the universe itself) is subject to birth, life, and death, as are its constituent parts.

In addition, Diderot made another astute observation of events arising from chaos: a single event may have far reaching consequences, like the butterfly effect. Saunderson asks what would have happened to the human race if the first man had had serious self-contradictions: "...what would have been the fate of the human race? It would have been still merged in the general depuration of the universe, and that proud being who calls himself man, dissolved and dispersed among the molecules of mater, would have remained perhaps forever hidden among the number of mere possibilities."[18] Diderot recognizes that one random event can spawn a series of random events: humanity may never have proceeded past the first man if he had been seriously defective.

Henri Coulet observes that during the eighteenth century, men came to terms with flux, change, impermanence, and uncertainty as to what will happen next. Coulet finds that men were ready to "tolerate contradictions" and accept the fact that "they have been immersed in the changing flux of phenomena and the inexhaustible chain of causes and effects."[19] Hence, the eighteenth century addressed mortality, unpredictability and uncertainty as it realized that man is merely an event in an eternal stream of random events.

Coulet also comments on the monsters in Diderot's writings. Coulet finds that for Diderot, the difference between monsters and normality is purely statistical-monsters appear less frequently than do normal beings. In *D'Alembert's Dream* Julie and Bordeu discuss many of the things that could go wrong in a developing embryo. Julie is surprised that the fibers that comprise the developing fetus do not get tangled more often, like the silk threads on her weaving loom, and form birth anomalies more often than they do. Coulet notes that in the section "Fœtus" in the *Elements of Physiology*, Diderot asserts that if monsters are defined as beings that do not last, then everyone is a monster because no one is immortal.[20]

Lester Crocker says that in the *Letter on the Blind* Diderot demonstrates that the "universe is not an orderly clock-mechanism, but a chaotic force in which everything is the result of blind randomness plus necessity. There is a cosmic order, but it is the transient outcome of trial and error in an endless process devoid of final causes."[21]

While in 1749 Diderot was armed with probability theory and the motive property of atoms, it would not be until 1753 that he would have the crucial third factor that would make transformism work: the conscious property of atoms. Maupertuis had devised a system that explains the transmission of inherited traits from generation to generation based on the conscious property of matter. He believed that this could explain how all living beings arose from a single source. Maupertuis first published his book in Latin in 1751 under the title *Inaugural Dissertation in Metaphysics*. Diderot cites

Maupertuis' book in *Thoughts on the Interpretation of Nature* (1753), *Thought 12, footnote 2,* and mentions that it was "brought to France in 1753." Diderot used Maupertuis' work as the foundation of his own ideas from 1753 on.

In *Thoughts on the Interpretation of Nature, Thought 12,* Diderot begins his explanation of transformism by commenting on homologies. He borrows from Buffon's observation that species share homologous parts, but he carries the notion much farther than Buffon had done: he shares Maupertuis' view that the fact that physical characteristics overlap among species is evidence that all beings arose from a single prototype. He wrote, "When we observe the successive outward metamorphoses which take place in this prototype, whatever it may be, pushing one realm of life closer to another by imperceptible stages, and populating the regions where these two realms border on each other (if they can be referred to as 'borders' in the absence of any true divisions); and, populating, as I said, the border regions of the two realms with vague, unidentifiable beings, largely devoid of the forms, qualities and functions of one region and assuming the forms, qualities and functions of the other; who, then, would not be persuaded that there had never been more than one single prototype for every being?"[22] Hence, he takes Buffon's observation of homologies that exist among species and expands it to include the shared characteristics between kingdoms. The idea that one kingdom metamorphosed into another kingdom has far reaching implications: if all of nature arose from a single entity, then consciousness, too, must have developed from a single source.

While the issue of consciousness rising up through the chain of beings from a single source is not mentioned, but implied in *Thought 12,* the mechanics for it are explained in *Thought 50.* In *Thought 50* Diderot explains Maupertuis' hypothesis that consciousness is a property of matter. All atoms have the properties of desire, aversion, memory, and intelligence. Furthermore, as atoms combine to form more highly organized entities, the new objects that are created possess desire, aversion, memory and intelligence at their level of organization, as well. In *Thought 50,* Diderot reiterates Maupertuis' assertion that when parental elements combine to form a composite, each element forgets its memory of self and acquires the consciousness and memory of the whole. Diderot quotes from Chapter 52 from Maupertuis' *Inaugural Dissertation in Metaphysics* in the original Latin: "It seems that from all of the perceptions of the elements having been gathered together, there results a single perception that is much stronger, much more perfect than any of the constituent perceptions and which is perhaps analogous to each of these perceptions as the organized body is to the component part. Each element, in its union with others, having mingled its perception with theirs, and having lost its *consciousness of self*, we have lost the memory of the original state of the elements, and our origin must be entirely lost for us."[23]

"A single perception that is much stronger and much more perfect" [*Unam fortiorem et magis perfectam perceptionem*] is Diderot's statement of emergent consciousness: as elements unite, something radically new and different emerges, different from any of its components. This made Diderot a pioneer in his time: he recognized that a random, emergent consciousness arises and that this new consciousness is not equal to the sum of its components; it is stronger and more perfect than the sum of its constituents.

Diderot and Maupertuis were pangenesists: they believed that each organ produces superfluous particles that it does not need. These particles circulate throughout the bloodstream of each parent and eventually go on to form the conceptus. When they do, they retain the memory of their former situation and go on to form the organ from which they came. When these tiny particles, which are conscious and thinking, retain the memory of their original position, the offspring resembles his parents. When these parental elements cannot unite because of a lapse of memory (*ne puisse s'unir par oubli*) of their original arrangement, birth anomalies result.

Diderot uses the term "memory" [*mémoire*] five times in *Thought 50*. Repetition of "memory" hyperbolizes the conscious property of matter. It reminds the reader that all matter is conscious, from the atom to organized matter and that this memory is responsible for heredity [*ressemblance*]. On the other hand, the absence of the memory of the arrangement of parental elements (this is an error in the generative process) is responsible for all the diversity that we see in nature. This lapse of memory creates something different from the original and could explain how all beings, in fact, even the three kingdoms, arose from a single prototype. Parental elements are like bees on a branch, each having the memory of a single position. If they retain their memory of their original position, the offspring resembles his parents. If they have a lapse of memory, a teratism results. These errors are passed on from generation to generation and sometimes skip a generation. Over time, they can create a new species or a new kingdom. Maupertuis's work had taught Diderot the terms "arrangement" [*arrangements*] and "traits" [*traits*].

Diderot recognized Maupertuis' genius and made his hypothesis his own. He was able to use it to fill a gap in the Newtonian explanation of heredity: the motive property of atoms alone does not explain resemblance to parents. The conscious property of matter does.

Furthermore, the hypothesis that each constituent element loses the memory of its former self and becomes conscious only of the new entity that is formed solves another problem: that of contiguity vs. continuity. In *Diderot and the Metamorphosis of Species* the solution to the problem is summarized thus: "Diderot is exploring contiguity vs. continuity, unity vs. heterogeneity, the particular vs. the general. By proposing a mechanical assimilation of molecular consciousness, he is resolving the conflict between contiguity and continuity. Particles may be contiguous, but they lose their sense of self and their consciousness becomes continuous. Once their consciousness becomes one, then their physical parts work as one, as well. Diderot is proposing a random, emergent ascent of conscious matter that culminates in human consciousness. He is positing that consciousness emerges in a way underivable from its constituent parts, just as the property of wetness cannot be derived from the hydrogen and oxygen alone. Diderot had a holistic view of consciousness: the conscious whole is very different from any of its constituents. This is because when conscious molecules combine, they lose memory of their former state and acquire the consciousness of the new composite that is formed. What Diderot is positing is a series of chronological events that culminate in the creation of conscious entities unlike their constituent parts. The series of events may be illustrated in the following paradigm: random motion of conscious atoms>organized matter> life>human thought."[24]

CLIMATE, GEOGRAPHY, FOOD AND WAY OF LIFE

In 1769 Diderot clearly articulates a theory of transformism (in the trilogy, the *Conversation between d'Alembert and Diderot, D'Alembert's Dream,* and the *Sequel to the Conversation*). Diderot parted company with Buffon on the question of how much influence environmental factors have on species. Buffon believed that environmental factors can cause minor changes in a species, but not create a new species because the essential character of each species is locked in by an interior molding force. Diderot, on the other hand, discarded the interior molding force. When, in *D'Alembert's Dream*, he declares that needs create organs, he is declaring adaptation to environment. He reiterates that given millions of years, climate can make a species larger, smaller, extinct, transform it into a new and different species, or even cause it to cross over into another kingdom. Furthermore, in *D'Alembert's Dream*, d'Alembert declares that if a few thousand leagues can change us into another species, then the difference of several times the earth's diameter will create even greater differences. He wonders whether beings on Saturn have more senses than we do.

D'Alembert asks why he is the way he is and he answers his own question: "I had to be like this..."[25] This immediate answer hyperbolizes the determinism of geography. There is no doubt, the answer comes right away. The quick response is an iconic representation of the certainty of science. There is a scientific basis for the physical characteristics of d'Alembert and it is indisputable. He wonders what he would have looked like if he had resided elsewhere-the North Pole, the equator, Saturn, the distance of a few thousand earth diameters. Diderot places species in the space continuum. The universe is vast and species are contingent upon location.

Then Diderot adds the fourth dimension, time: *quelques millions de siècles*. Time itself is subject to flux. Diderot compartmentalizes time into parcels (centuries) and each parcel becomes an event. We have the language of games of chance and it is applied to time: a number of throws (millions) and events (centuries). As the number of packets of time approaches infinity, every possible outcome will occur. D'Alembert asks what his species will look like in a few million centuries.[26] Nature is continually trying out new combinations-those without serious self-contradictions will survive. We do not know what the wild polar man will look like with time. Buffon thought that the wild polar man is an example of degeneration. For Buffon, God created species perfectly at the time of Creation, and therefore, they can only degenerate, not ameliorate. However, for Diderot, nature randomly and blindly creates variations. If anything is created perfectly, it is by chance. Therefore, it may be that the wild polar man is on his way to perfection, rather than to extinction. One should not make the mistake of Fontenelle's roses and draw erroneous conclusions about what happens over an eternal time frame. In his sleep d'Alembert warns against the fallacy of the ephemeral, alluding to the error that Fontenelle's roses made in thinking that their gardener is immortal.[27] The character Bordeu explains that the fallacy of the ephemeral is made by an ephemeral being who believes in the immortality of things.[28] Diderot is instructing the reader that we, who, like Fontenelle's roses, have a limited lifespan, cannot tell what happens over an infinite length of time.

DEGENERATION AND PERFECTION

Buffon thought that animals degenerate when taken out of their lands of origin, are poorly fed and overworked by men. He believed that the deleterious effects that a harsh climate and domestication have on a species could be reversed by crossbreeding varieties of the same species from different locations in order to introduce healthy new characteristics. Diderot carried the idea of degeneration and perfection much farther, to its utter extreme. In *D'Alembert's Dream*, Bordeu states that the tiny worm that grovels in the dirt may be on its way to becoming a large elephant and that the large elephant, that frightens us by its enormous size, may be on its way to becoming a tiny worm. Animals were not in the past what they are at present. We have no idea what they will become with time. Degeneration and perfection have no meaning for Diderot-they are useless, judgmental terms. Nature is in continual flux and we should not conclude that nature is taking a direction, either towards perfection or degeneration. There is no design in nature. Nature improves itself haphazardly, only to undo the improvements.

The fact that living beings randomly come into and out of existence is evidenced by the spontaneous generation of animalcules under microscopic magnification. Diderot engaged the theory of spontaneous generation to prove the veracity of transformism. Buffon and Needham had conducted experiments (albeit flawed) that showed spontaneous generation to be a fact, and he used them as his authority. Hence, he reversed his opinion of 1746 when he argued that experimentation has shown that putrefaction alone produces nothing (*Philosophic Thoughts, Thought 18*), relying upon the work of Redi. During the eighteenth century the issue was not clear cut, as there were experiments that proved both sides of the controversy. Diderot added spontaneous generation to his arsenal and argued that since it proves that life arises from inanimate matter all the time, divine agency is unnecessary. Furthermore, animalcules form from Buffon and Needham's infusions of vegetable broth, showing that the animal kingdom arises from the vegetable kingdom.

Animalcules come into existence in the twinkling of an eye, endure a short time, and fall out of existence. They are an iconic representation of the flux that nature delivers. Species are like animalcules: species, too, emerge, last for a duration, and fall out of existence. Entire worlds, planets, stars, the universe itself, are like animalcules-they, too, are born, last for a length of time, and disappear. Since the universe is comprised of molecules in motion, it, too, may have a finite duration and then disappear. Diderot returns to symmetry and self-similarity, phenomena arising from chaos, that would one day be articulated by chaos theory. Animalcules, animals, species, planets, stars, the universe, are events in an endless continuum of flux. Each is hatched (*éclore*), lives, and dies. Even planets and stars are hatched, as are animals. The universe is hatched [*éclore*]: "I conjecture, then, that in the beginning, when matter in a state of ferment brought this world into being, creatures like myself were of very common occurrence."[29] Each succeeding level of magnification reveals that objects at that level metamorphose into other objects over time; many of them are defective, at every level of organization. There are defective worlds as there are defective animals. Those without serious self contradictions survive, and those are the ones that we presently observe. The flux of events is observed at the microscopic, visible and macroscopic levels. Animalcules pass in and out of existence in the twinkling of an eye. Species take centuries or millennia. At the macroscopic level, entire star systems and planets pass in and out of existence. Our sun would not be the first star to lose its light.

Symmetry and self-similarity (the whole resembles its parts) is implied in the metaphorization of earth as an atom: "Endless succession of animalcules in the fermenting atom, the same endless succession of animalcules in the other atom that we call Earth."[30] During the eighteenth century, the primary definition of *atome* was a "Body that is considered to be indivisible because of its smallness. *Democritus and Epicurus claimed that the world was comprised of atoms, that the body was formed by the fortuitous convergence of atoms.*"[31] The secondary definition was "that tiny dust that is seen flying in the air in the rays of the sun."[32] Hence, *atome* is a clever play on words: the earth is metaphorized as a speck of dust that flies through space in the rays of the sun. The sentence, "Endless succession of animalcules...," shows that flux exists at the level of the microscopic (animalcules), the visible (species are events), and the macroscopic (the earth is an event). The parcels of time here are the twinkling of an eye and centuries. Elsewhere Diderot shows that flux exists at the level of the macroscopic, as well. In the *Conversation between d'Alembert and Diderot*, the character Diderot asks, "Do you consent to my extinguishing our sun?"[33] The character d'Alembert replies, "The more readily, since it will not be the first to have gone out."[34] The sun undergoes the processes of birth, life, and death, and thus, it is established that the formation of stars is an event in a continual stream, just as are molecules and living beings. The extinction and reillumination of the sun is representative of flux at the level of stars. This concurs with fractal theory: the larger body resembles its components. For Diderot, the larger and smaller undergo the same processes and are comprised of ever smaller, similar, constituent parts. This is a reiteration of Saunderson's statement in *Letter on the Blind*: "How many faulty and incomplete worlds have been dispersed and perhaps form again, and are dispersed at every instant in remote regions of space which I cannot touch..."[35] Since the random collision of molecules can bring about life on this planet, it probably can on others, as well. Diderot considers the metamorphosis, over millennia, of star systems, planets, species, and animalcules.

HEREDITY [RESSEMBLANCE]

In *D'Alembert's Dream* (1769) Diderot reiterates Maupertuis' theory of heredity that he had discussed in *Thoughts on the Interpretation of Nature* (1753). He holds epigenesis, not preformation, to be the correct explanation of the generative process. Both the father and mother contribute particles that circulate in the bloodstream of each parent. The offspring is formed by epigenesis, a process that William Harvey defined as "partium superexorientium additamentum," holding that the germ is brought into existence by the addition of parts that bud out of one another or by successive accretions.[36] This allows for the formation of a unique new entity that is different from either parent. In the *Conversation between d'Alembert and Diderot*, Diderot explains d'Alembert's origins this way: "...the molecules that were necessary to form the first rudiments of my geometer were scattered throughout the young and frail machines of one and the other."[37] Epigenesis can explain self-replication: because it shows how a unique being is formed, different from either parent, it can explain teratisms and hybrids.

Heredity is transmitted because each parental element retains its memory of its former situation that it had in the parent and takes it up again to form the same part from whence it came. When it retains its memory, the offspring resembles the parent. When it does not, a birth defect occurs. Because matter is conscious, particles organize into threads, threads into bundles of fibers, and bundles of fibers into organs. Each successive entity has a consciousness of its own and functions as a whole. When matter combines to form a new entity, it loses its memory of its former self and assumes the memory of the new entity it has formed. Each of the fibers in the bundle is transformed solely by nutrition and according to its conformation into a particular organ; an exception is made for those organs in which the germs themselves are reproduced.

Fibers [*brins*] are the key to transformism: man is a complicated machine that proceeds towards perfection through countless, successive stages, whose formation depends on delicate fibers that can be mutilated: "where the least important fiber cannot be broken, ruptured, moved from its original position, without a problematic consequence for the whole, is bound to get twisted, entangled even more often in the place of its formation than my silks on my skein-holder."[38] It was Maupertuis who conceptualized that errors that occur in the generative process cause birth anomalies. It was Diderot's attendance at autopsies and dissections that exposed him to the myriad anomalies in miscarried fetuses. When the character Julie says that she is amazed that errors do not occur more often, it is really Diderot, who has witnessed many dissections of miscarried fetuses, who is speaking. Diderot cites examples of people whose fibers were ruptured, moved or missing during gestation: there are hunchbacks, cripples, Siamese twins, and Jean-Baptiste Macé, the man who was born with his heart and stomach on the right side and his liver on the left.

Diderot mentions that if Jean-Baptiste Macé had lived, some descendant in a hundred years would have his deformities because such irregularities make jumps in generations. We see the influence of Maupertuis here, who, in the *Inaugural Dissertation on Metaphysics* discusses the fact that certain traits skip generations.[39] Diderot surmises that perhaps one of the parents fixes the defect that the other has and that the defective network does not recur until the next generation, when the descendant in the family with the monstrosities predominates and determines the formation of the network.[40] Diderot derives the terminology "predominates" [*prédomine*] from Maupertuis. It is a testimony to the genius of Diderot and Maupertuis that they were able to relate errors that occur in the arrangement of parental elements to the idea that over time, these errors could create new species and even cross over into other kingdoms.

Diderot envisaged that one day science would be able to control the process of generation and give mankind the offspring of his choice. He believed that one day man could breed improved offspring that excel at certain occupations or have specific talents: "A hot room, covered with small flasks, and on each of these flasks, an identification tag: warriors, magistrates, philosophers, poets, flask of courtiers, flask of harlots, flask of kings."[41] He also advocated investigating whether crossbreeding could result in improved species that would be useful, perhaps as servants to man. While experiments had shown that crossbreeding results in the production of infertile offspring, he advised that more experimentation was needed. The purpose of crossbreeding would be to create more perfect beings, to improve what nature cannot improve by itself. In the *Sequel to the Conversation* (1769), the character Bordeu suggests that crossbreeding men and goats might yield an intelligent, fleet-footed race of beings that could be servants to man. However, Julie wisely articulates the possibility that tampering with nature may be dangerous and that it may produce unexpected, horrific results: the goat-men might turn out to be an immoral race of terrible lechers and society may degenerate to the point that there might be no safety for women.

Furthermore, tampering with science raises religious issues. Bordeu and Julie joke about hybrid men: would they have a soul that requires redemption from original sin? The reader is led to wonder how highly organized a being must be before it is endowed with an immortal soul that requires salvation. It also implies that animals are conscious; Diderot argues that animals are indeed conscious in the article "Animal." Since the chain of beings is a continuum of infinite gradations, experiments may show that it is possible to crossbreed man with lower species.

WHAT CRITICS HAVE WRITTEN ABOUT DIDEROT'S TRANSFORMISM

Mary Efrosini Gregory's book, *Diderot and the Metamorphosis of Species*, fills a gap that exists in literary criticism. Much has been written about individual elements of Diderot's biology, but there is an absence of material discussing how he synthesized a variety of elements from different sources to create a composite transformist biology. *Diderot and the Metamorphosis of Species* examines how Diderot derives his ideas from various contemporary sources (Buffon, Maupertuis, La Mettrie, *Encyclopedia* articles, attendance at dissections) and combines them to arrive at something new. It is also the only book that focuses specifically on the influence that probability theory and games of chance had on Diderot's notion that species metamorphose over millions of years.

Lester Crocker's article, "Diderot and Eighteenth-Century French Transformism" is the only article that provides an overview of Diderot's thought from 1746–1769. In his article Crocker mentions the most important contributions that Leibniz, La Mettrie, Maillet, Buffon, Maupertuis, Robinet, and Bordeu, made to Diderot's transformism. He also discusses the factors that caused Diderot to migrate from deist to pantheist to atheist during the three year period 1746–1749. Crocker concludes that it was La Mettrie who had convinced Diderot that "Nature had to be conceived of as a self-creating, self-patterning force, as an experimenting-and a blindly experimenting-force."[42]

Several scholars have written articles on Diderot's monsters: Emita Hill's two articles, "Materialism and Monsters in Diderot's *Le Rêve de d'Alembert*"[43] and "The Role of 'Le Monstre' in Diderot Thought,"[44] Gerhardt Stenger's, "L'ordre et les monstres dans la pensée philosophique, politique et morale de Diderot,"[45] Aurélie Suratteau's, "Les hermaphrodites de Diderot,"[46] Johan Werner Schmidt's, "Diderot and Lucretius: The *De Rerum Natura* and Lucretius' Legacy in Diderot's Scientific, Aesthetic and Ethical Thought,"[47] and Christine M. Singh's, "The *Lettre sur les aveugles*: Its Debt to Lucretius."[48] The critics agree that monsters are iconic representations of the random events that flux provides and evidence of the metamorphosis of living things that occur all the time. Nature continually produces monsters; monsters are not something that has ceased at the beginning of time. Henri Coulet, in his article, "Diderot et le problème du changement," discusses the impact that the notion of flux had on contemporary thought. He observes that during the eighteenth century, men were ready to "tolerate contradictions" and accept the fact that "they have been immersed in the changing flux of phenomena and the inexhaustible chain of causes and effects."[49] Because nature is random, the terms "normal" and "anomaly" have no meaning. Nature is continually producing errors that result in variations without regard to their outcome. Coulet observes that for Diderot the difference between monsters and normalcy is purely statistical-in the *Dream*, Julie is surprised that the formative fibers of the embryo do not get mixed up or mutilated more often, like the silk threads on her spindle. Coulet adds that in the article "Fœtus" in the *Elements of Physiology*, Diderot shows that if monsters were those that do not last, then everyone is a monster because no one is immortal.[50]

There are also a few articles on the influence that Maupertuis had on Diderot. Maupertuis had studied polydactyly in a Berlinese family and noted its recurrence over several generations. He extrapolated that parental elements contributed by both parents, statistical probability of recurrence, and birth anomalies were all interdependent. He was even able to calculate, with exactitude, the statistical probability that polydactyly would appear in a population of 100,000 (he calculated it to be 8,000,000,000,000:1)[51] Bentley Glass offers an outstanding criticism of Maupertuis and Diderot in his article, "Maupertuis, Pioneer of Genetics and Evolution,"[52] as does Aram Vartanian, in his article, "Diderot et Maupertuis."[53]

There is also ample criticism of the influence of that Buffon's *Natural History* had on Diderot: there exists articles by Michèle Duchet, "L'anthropologie de Diderot,"[54] Jean Ehrard, "Diderot, l'*Encyclopédie*, et l'*Histoire et théorie de la Terre*,"[55] Arthur O. Lovejoy, "Buffon and the Problem of Species,"[56] Jacques Roger's book, *Buffon: A Life in Natural History*,[57] Roger's article, "Diderot et Buffon en 1749,"[58] Roger's book, *The Life Sciences in Eighteenth-Century French Thought*,[59] and Aram Vartanian's article, "Buffon et Diderot,"[60]

Several articles have also been written on La Mettrie's influence on Diderot: Jean E. Perkins, "Diderot and La Mettrie,"[61] Ann Thomson, "La Mettrie et Diderot"[62] and "L'unité matérielle de l'homme chez La Mettrie et Diderot,"[63] Aram Vartanian, "La Mettrie and Diderot Revisited: An Intertextual Encounter"[64] and "Trembley's Polyp, La Mettrie, and Eighteenth-Century French Materialism,"[65] and Marx W. Wartofsky, "Diderot and the

Development of Materialist Monism."[66] In this last work by Wartofsky, the critic addresses the Diderot's monism and the causality between the motive property of atoms and the metamorphosis of inanimate matter to animate matter. Wartofsky says that La Mettrie taught Diderot that consciousness is the product of motion. Diderot concluded that consciousness arises in the egg after the motion of atoms has successively caused various levels of organization to occur. Because the motion in atoms eventually leads to consciousness, all life could have resulted from the mineral kingdom.

In summation, Diderot's originality is seen in the fact that he viewed species as mutable, rather than fixed. His transformism rests on a fulcrum with three pivotal points: probability theory, the motive property of matter, and the conscious property of matter. Flux+time dispels chaos. Atoms are continually in motion and randomly colliding, testing every possible combination (Lucretius). They are also conscious: they have desire, aversion, memory, and intelligence. When they combine, they lose the memory of their former situation and acquire the consciousness of the new entity that they form. Hence, all matter is conscious at every level of organization. Carrying the notion to the limit, perhaps the universe is conscious, and is therefore God. In the developing embryo, conscious fibers form bundles of threads, which in turn, form organs. Errors occur in the generative process when particles suffer from a lapse of memory and do not go on to take up the exact arrangement that they had before. Parental elements are like bees sitting on the branch of a tree: they have a specific arrangement and memory of their original position in that arrangement. When they fail to remember their original situation, there arises an anomaly. Over time, these anomalies can create new species and cross over into other kingdoms. There errors can explain how all living things arose from a single prototype, and how the three kingdoms arose from one.

Chapter 7

Rousseau

I shall suppose his conformation to have been at all times what it appears to us at this day; that he always walked on two legs, made use of his hands as we do, directed his looks over all nature, and measured with his eyes the vast expanse of Heaven.
—Jean-Jacques Rousseau, *Discourse on the Origin of Inequality* (1755)[1]

Jean-Jacques Rousseau embraced anthropological (intraspecies) change and sociological progress, but he rejected biological transformism (the notion that man metamorphosed from species of a lower organization). As an observer of nature, he held that there was no evidence that man had ever been a quadruped. As a deist, he agreed with Buffon that man left the hands of God as a biped, not a quadruped, in the same anthropomorphic form as we see him today. God differentiated man from the animals by giving him intellectual potential or intellectual perfectibility (that is, the ability to learn and improve himself as he grows older) and free will. There were several factors that influenced Rousseau's thought:

1. He had enormous respect for Buffon and Buffon rejected transformism. Buffon hypothesized that an interior molding force shapes the essential physical characteristics of each species and that therefore, species do not undergo significant changes over time. Rousseau accepted the permanence of man's anthropomorphic characteristics.
2. Rousseau was a deist and he believed that God created all species perfectly at the time of Creation. In the first sentence of *Emile, Part*

1, he says, as Buffon had done before him, "Everything is good as it leaves the hands of the Author of things, everything degenerates in the hands of man."
3. There was an absence of scientific proof that man had metamorphosed from quadrupeds and this lack of evidence was significant.[2] Rousseau demanded evidence.
4. It was highly improbable that man had metamorphosed from quadrupeds and he explains why: the placement of eyes in front of the head, placement of his feet, and high hind quarters would make him vulnerable to prey if he had been a quadruped.[3]

While Buffon argues that the physical constitution of species degenerate when they are taken out of their natural environment and domesticated, Rousseau attempts to show that the morals of man have degenerated when man left his original environment, the woods, to form civilized society. Rousseau's originality lies in the fact that he applied Buffon's notion of degeneration, environment, food, soil, time and space, to the deterioration of man's spiritual self.

In the *Discourse on the Origin of Inequality* (1755), Preface, Paragraph *1*, Rousseau asks, "And how shall man hope to see himself as nature made him, across all the changes which the succession of place and time must have produced in his original constitution? And how can he distinguish what is fundamental in his nature from the changes and additions which his circumstances and the advances he has made have introduced to modify his primitive condition?"[4] Here Rousseau is not referring to changes in physical characteristics, but to changes in man's psyche. While Buffon examines the details of physical changes that species undergo with the passage of time and a change in environment, Rousseau's goal will be parallel, but will examine instead the changes in man's soul. Hence, he does recognize the impermanence of nature and that it continually produces a flux of events. He uses the terms "changes," "succession," "times," "original," and "progress."

Rousseau uses the terms "changes" [*changements*], "succession" [*succession*], "times" [*des temps*], "things" [*des choses*], "original" [*originelle*], "circumstances" [*circonstances*], and "progress" [*progrès*]. We have the language of probability theory here: events (changes, times, things, circumstances), parcels of time (time itself if divided into periods, which themselves, are events), the beginning of a string (original), and flux (succession).

When he says "original constitution," he is not talking about the human body. He does not believe that the human body has changed significantly since God created it. He is referring to the spiritual constitution-morality, values, passions, likes and dislikes. The attempt to "distinguish what is fundamental in his own nature" will have to do with the psychic part of man, rather than his physical self.

Rousseau compares the spiritual degeneration that man has suffered when he left the woods to enter civilization to the erosion weathered by the statue of Glaucus: "Like the statue of Glaucus, which was so disfigured by time, seas and tempests, that it looked more like a wild beast than a god, the human soul, altered in society by a thousand causes perpetually recurring, by the acquisition of a multitude of truths and errors, by the changes happening to the constitution of the body, and by the continlual jarring of the passions, has, so to speak, changed in appearance, so as to be hardly recognizable. Instead of a being, acting constantly from fixed and invariable principles, instead of that celestial and majestic simplicity, impressed on it by its divine Author, we find in it only the frightful contrast of passion mistaking itself for reason, and of understanding grown delirious."[5]

Rousseau metaphorizes the spiritual deterioration of man as the physical decay of the statue of Glaucus. He hypothesizes that changes in place and time must have caused man's psyche to degenerate just as the statue was so disfigured by time, seas and tempests, that it looked more like a wild beast than a god. *Tous les changements que la succession des temps et des choses a dû produire dans sa constitution originelle* indicates the importance that Rousseau gives to time and the events that flux delivers to man's decline into immorality and vice. It appears that Rousseau was influenced by Buffon, who declared that everything left the Creator's hands perfectly at the time of Creation and that afterward it degenerated. Rousseau, a deist, believed the same thing. In fact, the first sentence of *Emile, Book 1*, testifies to that notion: "Everything is good as it leaves the hands of the Author of things, everything degenerates in the hands of man." [*Tout est bien sortant des mains de l'Auteur des choses, tout dégénère entre les mains de l'homme.*].

The decay of the statue of Glaucus is a striking reversal of the journey up the chain of beings and demonstrates contempt for civilization: instead of a beast gradually transforming into a god over time, the image of a god degenerates, with time, into the figure of a wild beast. This is the opposite of Montesquieu's early Troglodytes, who resembling animals, eventually assumed a human appearance, as well as a reversal of man in the writings of Aristotle, Herodotus, and Pliny, who have portrayed the effect of time on man as an ameliorating one. In the works of the ancients and in Montesquieu's Troglodytes, there was a time when man was indistinguishable from any other animal; he was disfigured and monstrous, half human, half animal; it took time for him to acquire his present human form. However, in Rousseau, we have the opposite: he compares man to a likeness of a god that slowly metamorphosed into that of a wild beast. Rousseau tells us that man started out always acting from fixed and invariable principles, having a celestial and majestic simplicity that God had impressed upon him, and having understanding, but with time, he developed a frightful passion mistaking itself for reason, and understanding grown delirious. These are marred personality traits and Rousseau is talking about the fact that civilization and time have warped and twisted man's soul. He metaphorizes the warping and twisting of the spiritual as the marred physical appearance of Glaucus' statue.

Again we have the language of probability theory: time (time, forever) [*temps, sans cesse*], events (sea, storm, causes, knowledge, errors, changes, passions) [*mer, orages, causes, connaissances, erreurs, changements, passions*], number of throws (thousand) [*mille*], flux (sea, storm, lap of society, recurring, continual) [*mer, orages, sein de la société, sans cesse, continuel*]. What Rousseau shows is that flux+time=degeneration of man's psyche. He compares man's mind to the statue of Glaucus; he likens ownership, the arts and sciences, and society to the sea and storms that mutilated the statue. The beginning of the string is "as nature made him." Time passes. There is a flux of events (discoveries, landmarks in human progress, the arts and sciences). The long term effect is the degeneration of men's souls. While Buffon described the degeneration of the physical characteristics of animals, Rousseau also uses the terms "disfigured" [*défigurée*], "ferocious Beast" [*bête feroce*], "altered" [*altérée*], "changes" [*changements*], "changed in appearance" [*changé d'apparence*], "unrecognizable" [*méconnaissable*], and "deformed" [*difforme*]. The difference between Buffon and Rousseau is this: Rousseau applies these terms to man's psyche, not his body. It is men's passions and understanding that are both deformed: "deformed contrast of passion" [*difforme contraste de la* passion] and "understanding grown delirious" [*l'entendement en délire*].

In the *Preface, Paragraph 2*, we have yet another reversal: "It is still more cruel that, as every advance made by the human species removes it still farther from its primitive state..."⁶ "Cruel" [*cruel*] connotes animality, ferocity, carnivorousness, all of the brutal qualities of animals in the jungle. However, Rousseau associates cruelty with "every advance" [*tous les progrès*] and "the more new knowledge" [*plus...nouvelles connaissances*]. Progress and civilization take man farther from his original state, cause him to forget what he once was, and make it more difficult to retrace his steps back to his true self, that of natural man. Hence, there is a reverse transformism, a degeneration in the Buffonian sense, but not of physical characteristics, but of morality.

In *Paragraph 3* there is another Buffonian analogy: as species differentiate into varieties, man differentiated into various social classes and inequality began. There is an analogy between "physical causes had introduced those varieties which are now observable among some of them" and "the origin of those differences which now distinguish men." Rousseau mentions that when man's character began to ameliorate or degenerate, men "were acquiring various good or bad qualities not inherent in their nature." Here we see that it is Rousseau's thesis that man is born neither good nor evil-he is a *tabula rasa* upon which society imprints. The fact that men "were acquiring various good or bad qualities not inherent in their nature" indicates that good and evil are learned and acquired.

In the *Exordium* (the prefatory material immediately preceding *Part 1*), *Paragraph 2*, Rousseau conceives of two kinds of inequality: the physical (comprised of differences in age, health, body strength, and quality of the mind or soul) and the moral or political inequality. The latter is contingent upon convention and man's consent and consists of various privileges that men have, some more than others: wealth, honor, and power. Rousseau states his goal: it is to examine how inequality arose among men, when originally, when man lived in the woods and before he joined civilization, all men were equal, did not own property, but the earth belonged to everyone. In order to pinpoint the origin of inequality among men, it is necessary to go back to the state of nature, before it arose. When man existed in a state of nature, he did not understand the concept of "belong to" and "ownership." Need, greed, oppression, desires and pride were vices that came with society. The authority of the strong over the weak is another convention that arose by mutual agreement.

In *Part 1, Paragraph 1*, Rousseau shows contempt for Aristotle's inclusion of hairy monsters in the great chain of being. Rousseau summarizes the

Aristotelian viewpoint thus: "I shall not ask whether his long nails were at first, as Aristotle supposes, only crooked talons; whether his whole body, like that of a bear, was not covered with hair; or whether the fact that he walked upon all fours, with his looks directed toward the earth, confined to a horizon of a few paces, did not at once point out the nature and limits of his ideas. On this subject I could form none but vague and almost imaginary conjectures. Comparative anatomy has as yet made too little progress, and the observations of naturalists are too uncertain to afford an adequate basis for any solid reasoning."[7] Here Rousseau clearly articulates the materialists' transformist hypothesis that he would go on to refute: that man walked on all fours, had claws, was hairy like a bear, gazed down at the ground, and because he looked down at the ground, his perspective, and hence his intellect, were necessarily limited. This statement indicates that Rousseau was aware of the tranformist theory and that he had carefully read Buffon, who had previously articulated it and then, too, went on to refute it.

Rousseau rejects the transformist view of man's origins and he articulates several reasons why. First, he points out that no one has any idea of man's origins and that therefore, the transformist hypothesis is purely speculative, not factual. His statement, "On this subject I could form none but vague and almost imaginary conjectures" undercuts Aristotle, Herodotus, and Pliny, as well as contemporary materialists. The terms "vague" [*vagues*], "imaginary" [*imaginaires*], and "conjectures" [*conjectures*] are tautological and hyperbolize the unknown or speculative nature of the subject. He reiterates this further: comparative anatomy has made "too little progress" [*trop peu de progrès*] and the observations of naturalists are "too uncertain" [*trop incertaines*]. "Too little progress" and "too uncertain" are also tautological and further undercut contemporary scientists and the transformist point of view. He demonstrates contempt for science by hyperbolizing its limitations.

In his endnotes, Rousseau explains in detail why he rejects interspecies transformism. In *Note 3, Part 2*, he refutes the notion that originally man walked on all fours; he asserts that man must always have been a biped for at least six reasons.[8] First, even if one could show that he could originally have been structured differently than he presently is, hypotheses based on possibilities are not enough-one would have to prove not that it is possible, but that it is probable. Possibility [*la possibilité*] is not enough, one must show probability, likeliness, likelihood [*la vraisemblance*]. Here we see the influence of probability theory. Rousseau relies on probability to show whether man was originally structured differently than he presently is. Hence, he cleverly uses mathematics and probability theory, a tool of the materialist transformists, against them. Secondly, if man had originally walked on all fours, then the way his head is attached to his body would cause his gaze to be directed downwards, rather than straight ahead, and this is antithetical to survival. Rousseau stresses the all importance of survival and self-preservation and points out that a downward gaze is a position scarcely favorable to the preservation of the individual. Thirdly, man has no tail because bipeds do not need a tail. Quadrupeds find a tail useful and none of them is without it. Fourthly, a woman's breast is well placed for a biped holding a child in her arms and is so poorly placed for a quadruped, that none has it so placed. Fifthly, the hindquarters is inordinately high in relation to the forelegs, which is why we drag ourselves around on our knees when we walk on all fours. This, too, would be antithetical to survival in a quadruped. Sixthly, if man were originally a quadruped, he would not be able to lay his foot flat on the ground the way animals do. Lastly, Rousseau dismisses the argument that originally man walked on all fours because children crawl on the ground. He points out that puppies also crawl on the ground for several weeks after birth and then eventually walk on all fours. He surmises that the only reason that babies crawl is because of the weakness of their limbs. What infants do does not prove the metamorphosis of species.

Rousseau accepted Buffon's stance against transformism and offered his own ideas to refute it. It appears that Rousseau adopted Buffon's opinion about an original molding force that make species what they are and because of it, they do not change radically. Furthermore, Rousseau's deism also caused him to adopt Buffon's view that God made everything perfectly and that it did not need to be improved in physical form.

Rousseau was contemptuous of the materialists who proposed transformism: regarding the question of whether man originally walked on all fours, Rousseau says, "On this subject I could form none but vague and almost imaginary conjectures. Comparative anatomy has as yet made too little progress, and the observations of naturalists are too uncertain to afford an adequate basis for any solid reasoning." He dismisses conjectures, then, as futile and speculative and asserts, "…I shall suppose his conformation to have been at all times what it appears to us at this day; that he always walked on two legs, made use of his hands as we do, directed his looks over all nature, and measured with his eyes the vast expanse of Heaven."

Hence, Rousseau was not a transformist. He did not believe man ever walked on all fours and enumerates a list of reasons why he believes that God made man a biped, not a quadruped that, with time, became a biped. He takes his cue from Buffon and, in fact, cites him in the first sentence of *Emile, Part 1*, "Everything is good as it leaves the hands of the Author of things, everything degenerates in the hands of man."

In *Part 1, Paragraph 2*, Rousseau says that man in a state of nature is an animal: "…we behold in him an animal weaker than some, and less agile than others; but, taking him all round, the most advantageously organized of any."[9] Here we see the influence of Buffon: "the most advantageously organized of all" [*organisé le plus avantageusement de tous*] reiterates the importance that Buffon gave to man's organization. Buffon said that man stands at the head of all created beings and that the chain of beings extends from the most organized animal to the crudest matter…"[10] From Buffon Rousseau had gleaned that man is also an animal, but he is one of an organization of a higher order. In *Paragraph 2* Rousseau demonstrates that because man is an animal, nature filled all of savage man's needs and he lacked nothing. In a state of nature savage man easily acquired everything his heart desired: he was able to satisfy his hunger beneath an oak tree, drink from a stream, find his bed beneath the same tree that supplied his meal, and thus, all of his needs were satisfied.

In *Paragraph 3* Rousseau declares that men learned to use nature's provisions by observing and imitating the behavior of animals. The earth and forests offer storage and shelter to all animals: by copying the behavior of lower species, men, who had no instinct of their own, raised themselves up to the level of the animals' instincts. Here man originally had no instincts, but intellect instead. Having no instincts of his own, he learned by observing the behavior of animals. Rousseau repeats that fact that originally man was devoid of instinct: "attain even to the instinct of the beasts" and "man, who perhaps has not any one peculiar to himself." This is more evidence that man rejected transformism: God created man as a distinct species at the time of Creation and the fact that originally he had no instinct is evidence of this. By observing and imitation various animals, man learned to eat a variety of different foods. Each species of animal instinctively searches for and consumes a particular food. Man observed them all and learned to eat a variety of foods. Thus, by observing, imitating, and having new experiences, he developed his intellectual potential or his intellectual perfectibility and this intellect was much more useful than animal instinct and made man superior to animals. It is man's intellect that puts him at the head of the chain of beings.

In *Paragraph 4* Rousseau mentions the rugged constitution of men living in the state of nature: "Accustomed from their infancy to the inclemencies of the weather and the rigour of the seasons, inured to fatigue...," natural man had a robust and almost unalterable constitution.[11] We see the influence of Lucretius here: all those that have weak constitutions perished and only the strong survived: "Nature in this case treats them exactly as Sparta treated the children of her citizens: those who come well formed into the world she renders strong and robust, and all the rest she destroys..."[12] Thus nature dealt with men as with all other animals, according to Lucretian law. He makes an analogy between the physical and the moral realms: the converse is true of moral man. The law of nature in which only the robust survive in the opposite of social law: "...differing in this respect from our modern communities, in which the State, by making children a burden to their parents, kills them indiscriminately before they are born."[13] Civilized society kills children morally by instituting inequality, slavery, vice, and crime. Men become a burden to others because man must bear the burden of criminals (it must have a legal system to mete out justice) and must take care of the poor (there is no poverty in the natural state).

In *Paragraph 11* Rousseau returns to Buffon's opinion that domestication brings about degeneration and he again applies it to the deterioration of man's morals: "The horse, the cat, the bull, and even the ass are generally of greater stature, and always more robust, and have more vigour, strength and courage, when they run wild in the forests than when bred in the stall. By becoming domesticated, they lose half these advantages; and it seems as if all our care to feed and treat them well serves only to deprave them."[14] When they are domesticated, "they lose half these advantages." Similarly it is with the character of social man: "It is thus with man also: as he becomes sociable and a slave, he grows weak, timid and servile..."[15] When man enters society, his character changes along with his physical constitution: he learns to be afraid of others, to value others more than he does himself, and to consent to being subservient to others. There are no servants in nature: slavery is a convention instituted by the mutual agreement of men. Thus Rousseau takes Buffon's notion that the domestication of animals cause physical degeneration and applies it to the destruction of man's morals, the introduction of inequality, vice, greed, murder, slavery, honor, and war, into the psyche of civilized man.

In *Paragraph 15* Rousseau says that man is a machine with free will: "I see nothing in any animal but an ingenious machine, to which nature hath given senses to wind itself up, and to guard itself, to a certain degree, against anything that might tend to disorder or destroy it. I perceive exactly the same things in the human machine, with this difference, that in the operations of the brute, nature is the sole agent, whereas man has some share in his own operations, in his character as a free agent. The one chooses and refuses by instinct, the other from an act of free-will."[16] Unlike animals, that have only instinct, man has perfectibility and free will. While animals always behave in the same way by instinct, man has the ability to choose his course of action. As an example, Rousseau says that man can select the food he eats from what is available; on the other hand, certain species eat only certain foods and do not diversify their meals. Hence, a pigeon would starve to death next to a bowl of meat and a cat, by a dish of fruit or grain; by observing animals, man has learned that meat, fruit, and grain can all be consumed. Man has perfected his intellect by observing animals behaving from instinct.

In *Paragraph 17* Rousseau points out another difference between animals and man: the faculty of self-improvement or perfectibility. An animal is, at the end of a few months, all that he will ever be during the course of his lifetime; at the end of a thousand years, the species will not change. Man, however, has the capacity to learn and improve his condition. He examines why men grow senile and animals do not: animals acquire no intellect through the course of their lifetime and therefore, have nothing to lose. Man perfects himself and then loses the intellectual faculties that he has acquired.

Natural man lived from moment to moment. He was destitute of intelligence and desired only what he needed at the moment. The only things that he feared were pain and hunger. He feared pain and not death because he was an animal. Animals live from moment to moment and do not contemplate death. Natural man did not contemplate death either until he left the natural state and entered society.

The notion that man is an animal is derived from Buffon's *First Discourse*: "he must put himself in the class of animals whom he resembles by everything material he is."[17] Placing man in the category of animals dates back to Aristotle, who had classified man among social animals who engage in common activities, along with bees, wasps, ants, and cranes.[18] In 1735 Linnaeus, in the *Systema naturæ*, had placed man in the category of anthropomorphic quadrupeds with monkeys and sloths.

Rousseau was a disciple of Buffon, who based his views on Cartesian dualism. Buffon articulated the importance of "recognizing the nature of the two substances of which we are made."[19] The two natures are the unextended, the immaterial, the immortal, and the other is the extended, the material, and the mortal. Buffon went the way of Descartes, who posited, "I think, therefore I am" [*Cogito, ergo sum* or *Je pense, donc je suis.*]. Buffon agreed with Descartes that we have a soul and that our existence is proven by our ability to think: "The existence of our soul is proved to us, or rather we only form a unity, this existence and us: to be and to think are the same thing for us; this truth is intimate and more than intuitive, it is independent of our senses, our imagination, our memory, and of all our other related faculties."[20] Rousseau also declares that we have a soul. In the *Discourse on the Origin of Inequality, Preface*, he compares man's soul to the statue of Glaucus: as the statue is disfigured by time, sea, and storm, so is "the human soul, altered in society." It was "contemplating the first and most simple operations of the human soul" that caused Rousseau to see that 1) well-being and self-preservation and 2) a natural repugnance to seeing others suffer, are natural characteristics of the soul that antecede the acquisition of reason. Before he learned how to reason, man's soul, in the state of nature, sought survival and did not want to see other men or animals suffer.

In the *Exordium, Paragraph 2*, Rousseau mentions natural inequalities of man: differences in age, health, body strength, and qualities of the mind or soul. When he says that man's morality was poisoned when he left the state of nature to enter society, he is saying that man's mind was corrupted. It was man's mind that learned inequality, slavery, vice, greed, crime, murder.

Roger summarizes Rousseau's description of natural man and then explains why Buffon had no choice but to challenge him: Rousseau's man of nature roamed through the woods alone and did not need other human beings; it was natural disasters and catastrophes such as droughts and floods that forced man to leave his state of solitude and enter society.[21] Even Rousseau, himself, admits that his account is a moral allegory and that it cannot be proven. In the *Preface* he acknowledges, "I have here entered upon certain arguments...For it is by no means a light undertaking...to form a true idea of a state which no longer exists, perhaps never did exist, and probably never will exist..."[22]

Roger points out that it was these suppositions and absence of evidence that caused Buffon to reject Rousseau's natural man. Buffon wanted to explain the present with known facts of the past, not allegory. Roger explains that Buffon considered Rousseau's natural man to be allegorical and fabulous, not scientifically proven fact. Naturalists who have traveled around the world have never seen any natural men such as those described by Rousseau: "We do not find, 'in traveling over all the isolated places of the globe, human animals lacking words, deaf to voices as well as signs, dispersed males and females, abandoned children...' 'Children would die if they were not helped and cared for over several years...it is not possible to maintain that man has ever existed without forming families.' And Rousseau's natural man was no more than a myth."[23] Thus Buffon rejected Rousseau's natural man. Roger demonstrates that Buffon concluded that the distance between natural man and the modern primitive "such as Rousseau had established it, did not exist."[24] Buffon rejected the notion that natural man had ever existed. Buffon had traveled all over the world, he had seen the Hottentots and other people living in extremely primitive conditions, and none of them fit the picture of Rousseau's natural man: the Hottentots had a language, families and society. There is no evidence anywhere on the earth that Rousseau's natural man, without language, wandering solitary in the woods, ever existed. Buffon shows that man cannot exist apart from a family because children must be protected by their parents for many years before they can survive on their own. Therefore, the most primitive and elementary form of a society, the family, must have existed as long as man has. Language must have always existed as long as families have, because family members must communicate among themselves for the purpose of survival.

Roger summarizes the reasons for Buffon's rejection of Rousseau's natural man: human children take many years to mature and they must be protected by their parents longer than do animal offspring; families remain intact for a longer time than do animal families in order to raise the children; language and intellect are functions of the soul and only men have souls, imparted by the divine breath of God.[25]

Furthermore, there is an insurmountable distance between the most primitive man and the ape. Humans living in the most primitive conditions have thoughts and words; apes cannot speak and there is no evidence that

Rousseau 155

they can think, either. If apes had thoughts, they would have spoken by now and become man's rival. This has not happened and therefore, man's superiority over all the animal kingdom is permanent.

Hence Buffon rejected Rousseau's natural man as fictional because:

- he traveled all over the world and never found a solitary man wandering through the woods
- families must have always existed in order to protect the children
- language must have always existed for the purpose of survival
- God exists and He endowed man with a soul
- language is evidence of the insurmountable chasm between man and other animals

Otis Fellows also compares the works of Buffon and Rousseau. He finds it significant that Rousseau greatly admired Buffon and held him in high esteem: "Marguerite Richebourg, in her *Essai sur les lectures de Rousseau*, has found in Jean-Jacques's voluminous correspondence, his *Confessions*, and other writings, some fifty references to the author of the *Histoire naturelle*."[26] Fellows goes on to examine how Rousseau derived his ideas from Buffon's work. First, there are similarities between Buffon's account of the origin of the universe and that of Rousseau in his *First Discourse, Part 1*: neither writer mentions Genesis or God, but rather, events unfold and things change independently of God's will. Fellows observes, "…Rousseau opened the "Première Partie" of his essay with a daring, majestic sweep which ignored the account of man's origin in *Genesis* and stressed his rise from primeval nature. In this he was following, wittingly or not, in the footsteps of Buffon…Jansenists of the *Nouvelles ecclésiastiques* and the Faculté de Théologie of the Sorbonne were scandalized to find no evidence of belief in any particular act of creation. The thought immediately arises that here Rousseau found more than fleeting inspiration for the development of his own early speculations on evolving man in a changing universe."[27] Fellows goes on to show how Rousseau also derived his notion of natural man's virtue and innocence from Buffon. In 1749 Buffon had written, "virtue belongs to the savage man more to civilized man, and…vice originated in society."[28] Fellows amusingly points out that while Rousseau embraced the notion and made it the foundation of his second discourse, Buffon "was to take Jean-Jacques to task for adopting precisely the same supposition."[29]

Fellows advises that there are numerous "instances when Rousseau openly calls on Buffon's authority in support of his own arguments and, by so doing, furnishes irrefutable proof of influence."[30] For example, there is the lengthy quotation of Buffon in the *Discourse on the Origin of Inequality, Note 2*.[31] Fellows refers the reader to Jean Morel's "Recherches sur les sources du *Discours de l'inégalité*" for an examination of Rousseau's use of Buffon in his work. Fellows advises that "Jean Morel has proved the early volumes of the *Histoire naturelle* to have been a primary source of the 'information scientifique' scattered throughout the Second Discourse."[32]

In the *Discourse on the Origin of Inequality, Note 10*, Rousseau discusses species of apes that eighteenth-century naturalists were confronted with classifying: orangutans, pongos, enjokos, beggos, and mandrills. The question arose whether these anthropomorphic species were men or animals. Rousseau finds that orangutans "occupy something like a middle position between the human species and the Baboons."[33]

In *Note 10* Rousseau mentions Dapper's statement that the orangutan is so similar to a man, it is thought that it might have been the offspring of a woman and a monkey. Rousseau goes on to discuss the striking similarities between apes and men and he even goes so far as to say that there are fewer differences between apes and men than there are between one human being and another. Rousseau notes that while naturalists hypothesize that these beasts cannot possibly be men because they are stupid and have no language, one must take note of the fact that speech is not natural to man and that it is something that he acquired when he first entered society. Hence, he is reasoning from his own premise that man never spoke before he entered society and is thus open to classifying orangutans and pongos as men. Rousseau maintains that if scientific experiments were to show that these apes have the faculty of perfectibility, then it would be proven that there are multiple species of man. Naturalists had performed experiments with monkeys and they knew that monkeys are not men because they lack perfectibility. However, experiments had not been done with orangutans and pongos to ascertain whether they can be taught language and whether they have the faculty of perfecting themselves, a characteristic which is the hallmark of the human species. Hence, pending scientific investigation, Rousseau would not judge whether or not orangutans and pongos are a variety of the human species. This distinguishes him from Buffon, who held that the great apes are not men.

The question arises as to whether Rousseau thought that man metamorphosed from the apes. It should be noted that even though Rousseau observes all of the similarities between apes and men, and even though he is open to the possibility that experimentation may prove that orangutans and pongos can learn language and have the faculty of perfectibility, and that they may, indeed, be men, nowhere does he indicate a transformist biology, as say, Diderot does in *D'Alembert's Dream*. Rousseau maintains faithful to the fixity of species, as Buffon had done. What Rousseau is proposing is anthropological and sociological transformism, rather than biological. The metamorphosis that he sets forth is intraspecies (anthropological), and not interspecies (biological).

Buffon, Rousseau's role model, clearly denies the possibility of biological transformism. Buffon articulates the materialist hypothesis in order to go on to refute it: "…if one had to judge only by its form, the monkey species could be considered to be a variety of the human species" is immediately followed by "whatever resemblance there may be between the Hottentot and the monkey, the interval that separates them is immense because inwardly he is filled with thought and outwardly, with the spoken word."[34] Hence, Buffon maintained the insurmountable chasm between man and animal. Rousseau, at the end of *Note 10*, says that what is required is for men of the stature of Montesquieu, Buffon, Diderot, Duclos, d'Alembert, and Condillac to travel across the world to every country, and ascertain whether these creatures are men or animals. When such great men arrive at a conclusion about this matter, we must believe them: "I say that when such Observers assert about a given Animal that it is a man and about another that it is a beast, they will have to be believed…"[35]

Nineteen years after Rousseau wrote his second discourse, Lord Monboddo examined Buffon's influence on Rousseau and tried to ascertain whether either of these men thought that orangutans were men. Monboddo considered Buffon's statement, "…if one had to judge only by its form, the monkey species could be considered to be a variety of the human species" to which he adds, "whatever resemblance there may be between the Hottentot and the monkey, the interval that separates them is immense because inwardly he is filled with thought and outwardly, with the spoken word."[36] Monboddo then carefully examined Rousseau's *Note 10* and saw that Rousseau allowed for the possibility that orangutans might be of the same species as man. Because Rousseau was open to the idea, Monboddo surmised that there is a difference of opinion between Buffon and Rousseau as to whether orangutans are men. Monboddo concluded, "It is from these facts that we are to judge, whether or not the Orang Outang belongs to our species. Mr. Buffon has decided that he does not, Mr. Rousseau inclines to a different opinion."[37]

Otis Fellows examines the implications of this. He observes that in the *History of the Ass* (1753), Buffon summarizes the materialist hypothesis and then denies it: "...if the ass were to be considered of the same line as the horse, then one could also add that the ape was of the same family as man, and furthermore that both man and ape had the same origin. In fact, all animals may have descended through eons of time from a single source. This was indeed bold speculation for the day, and two paragraphs further on Buffon hastens to add: 'But no: it is certain, by revelation, that all animals have equally participated in the grace of Creation.'"[38] Rousseau, on the other hand, was open to the possibility that orangutans and man were of the same species. In *Note 10* he says, "Dapper confirms that…This Beast…is so similar to man that it has entered into the mind of some travelers that it might have been the offspring of a woman and a monkey…one finds in the description of these supposed monsters striking conformities with the human species, and smaller difference than might be pointed to between one human being and another."[39]

The point of contention between Rousseau and Buffon is whether or not apes are men, not whether man or apes metamorphosed from animals of a lower organization. Fellows reiterates that for Rousseau, the chasm between man and ape was not insurmountable as it was for Buffon.

In *Note 10* Rousseau argues that there is a strong possibility that orangutans and men belong to the same species: pongos gather around campfires abandoned by men, sit by them until the embers die out, and then leave. He criticizes Andrew Battel and Samuel Purchas for concluding that these animals, *as dexterous as they are, do not have the sense to keep the fire going by adding wood to it*. Rousseau italicizes Battel and Purchas' conclusion in order to ridicule it: *Car avec beaucoup d'adresse, ils n'ont pas assez de sens pour l'entretenir en y apportant du bois*. Rousseau takes them to task for extrapolating that apes are stupid. He demands to know how they can tell that the apes' departure from the campfire is an indication of their stupidity or their will. Here we see that Rousseau attributes free will to apes and in so doing, erases the chasm between men and apes. For Rousseau, free will is a hallmark of natural man, as is his perfectibility: these two characteristics distinguish natural man from the animals. Here, Rousseau is clearly ascribing free will to apes. He argues that apes do not need the warmth of a campfire in a hot climate and that therefore, the flames must entertain them. When they get bored from sitting in one location, they move on and forage, which requires more time to do than if they ate flesh. Rousseau observes that animals are amused and cheered, they get bored, they need new stimulation, and they are aware that foraging takes time, and they choose not to waste time by sitting in one place. Furthermore, the great apes, like natural man, are not carnivorous (Rousseau specifies that foraging takes more time than if they ate meat), but rather, they, too, are vegetarians.

Rousseau goes on to say that pongos bury their dead and make roofs out of branches. This clearly indicates volition and intelligence. Therefore, he doubts that they would not know how to push embers into fires. Rousseau declares that he has seen a monkey push embers into a fire, an operation that Battel and Purchas claim that pongos cannot do. Rousseau admits that while it is certain that a monkey is not a man because he lacks the faculties of speech and perfectibility, experiments have not been done with pongos or orangutans to ascertain whether or not they are men.

Fellows points out that it is obvious that Rousseau parts company with Buffon to allow for the possibility that apes are men. Does this make him a transformist? Nowhere does Rousseau say that man or apes metamorphosed from animals of a lower organization. All he is articulating is the possibility of an anthropological (intraspecies), not biological (interspecies) relationship. Because Rousseau believed that natural man had once existed, perhaps he thought that he had found evidence of this. He is open to the possibility that men and the great apes are of the same species and this would explain the will and intelligence of apes. Fellows notes that Arthur O. Lovejoy, who is "more prudent," points out that Rousseau is a transformist "in the anthropological and sociological sense rather than in the biological."[40]

Francis Moran also argues that Rousseau was not a transformist and contends that the orangutans are used to make a political statement, not a biological one. First, Moran points out that Rousseau takes advantage of the eighteenth-century debate on the possibility of multiple human species as an opportunity to criticize contemporary civilized Europe: "The significance of his speculation has less to do with his special insights in human descent than with the political point to be made were it true. Rousseau uses *orangs-*

outang like the pongo to construct a viable model for criticizing his contemporary Europe and to defend his claim that the kind of political inequalities associated with late European society do not issue from God or nature but are accidental events in the life of the species."[41] Moran reminds the reader that the title of Rousseau's work is *Discourse on the Origin and Foundations of Inequality Among Men* and that it is essentially a political statement. Moran sees *Note 10* as a tool to bolster his suppositions about natural man, which, in turn, are merely an entrée to the real dish, a criticism of inequality in European civilization.

Moran agrees with Gourevitch that the notion of natural man is conjectural and not factual.[42] Rousseau, himself, admits, in the *Preface*, "I have hazarded some conjectures…For it is no light undertaking…to know accurately a state which no longer exists, which perhaps never did exist, which probably never will exist…" The purpose of his conjecture is to criticize contemporary civilization, as the title of the work indicates.

Secondly, Moran points out that the relationship (either proximate or distant) between humans and apes was an important issue among eighteenth-century naturalists because they were trying to gain a better understanding of the chain of being. The human-ape relationship was not a concern about descent, but rather, about the chain of being, its corollaries, plenitude and continuity, and most of all, man's place on the great chain. Moran explains, "Throughout the Enlightenment we find naturalists preoccupied with arranging species along a continuum descending from God and linking each part of His creation. The meaning of the claim that *orangs-outang* might be some sort of mid point is lost unless we place it in the context of this tradition. The discovery of the *orangs-outang* was important for eighteenth-century naturalists not for what these animals could tell us about human descent, but for what they revealed about the viability of the chain of being and the position of human beings within that chain…the apparent uniqueness of human consciousness and spirituality seemed to raise special difficulties for a theory which presented nature as a continuum. For instance, Rousseau, Voltaire, and Buffon all refer to this gap in rejecting the idea of a natural continuum."[43]

Moran finds that Rousseau "is extending the range of the human species as it currently exists," something which was not unusual to do in the eighteenth century. He points out that Maillet, in *Telliamed*, also ends his discussion on orangutans by saying "if we could not say that these living creatures were men, at least they resembled them so much that it would have been unfair to consider them only as animals."[44]

Rousseau concedes that more accurate accounts of the apes are needed, as well as experimentation to see if they are of the same species as man. He acknowledges that men of the caliber of Montesquieu, Buffon, Diderot, Duclos, d'Alembert, and Condillac, are needed to travel abroad, observe the apes, and describe them for others.

Moran concludes that Rousseau raises the question as to whether apes can be perfected (since they seem to have volition and intelligence) in order to make a political point. What Rousseau is declaring is the unity of the human race and that racial and social equality is a basic tenet of natural law. He was condemning the strict divisions among social classes in Europe. Furthermore, Europeans had discovered the existence of other races living under very primitive conditions. Rousseau, in widening the umbrella to allow for the possibility of multiple human species, was declaring that all humans, regardless of their social status, culture and living conditions, belong to one species, namely, the human race; again, the title of the work indicates that his treatise is a condemnation of the evils of inequality and prejudice.

Moran summarizes the point that Rousseau was making with his orangutans thus: "If his discussion of natural man is to become a foil for criticizing contemporary Europe, and if that description is culled from European reports of the world's primitive populations, then it is essential that these people be included in the same species as Europeans. If this were not the case, he could have been vulnerable to questions about their relevance to the kind of European political problems that he was addressing. He therefore needed a definition of the human species which not only recognized a single species but also accommodated the diversity within the species. In other words, he needed an account of human natural history which could explain how *orangs-outang* might become Europeans (or vice-versa).[45]

Francis Moran, in another article, "Between Primates and Primitives: Natural Man as the Missing Link in Rousseau's *Second Discourse*," examines Rousseau's objective to understand man's placement in the chain of beings. This job involved assessing the gap between man and the next higher being on the chain of beings. In the *Discourse on the Origin of Inequality, Note 10*, Rousseau cites the abbé Prévost's description of orangutans as a "sort of middle point between the human species and the baboon."[46] In *Part 2* he puts natural man "at equal distances from the stupidity of brutes and the fatal enlightenment of civil man."⁴⁷ Moran advises that Rousseau had devised natural man to fill a gap on the chain of beings between man and orangutans. Moran interprets Rousseau's analogy between natural man and the orangutan according to the eighteenth century:

> By referring to natural man as a mid-point between animals and human beings, Rousseau is providing his audience with a recognizable framework for understanding the kind of creature he will be describing. In the context of mainstream eighteenth-century thought, this reference would probably have been read as an allusion to the chain of being rather than as an indication of human descent, for unlike the later evolutionists, eighteenth-century naturalists who suggested a possible relationship between primates and human beings were generally uninterested in tracing the genealogy of these populations. Instead, their claims were meant to establish the relative position of each in the chain of being.
>
> Because there was some concern that the human species represented a possible break in the natural hierarchy of the chain of being, those naturalists interested in preserving the chain began to search for possible "missing links" which would reunite human beings with other animals. This search focused primarily on the (alleged) anatomical, morphological, and behavioral similarities of the populations presumed to be closest to the break...i.e., primates (as the highest animal) and the native populations of Africa, the South Pacific and the Americas (as the lowest human beings).⁴⁸

Jean Starobinski also discusses Rousseau's anthropologic transformism and Buffon's influence on his thought in his article "Rousseau and Buffon."⁴⁹ Starobinski begins by observing that Rousseau adopts Buffon's method of explaining man by beginning from the most elementary form of existence and proceeding to the most complex. They both embrace Cartesian dualism (body and soul) modified by Locke (sensations). The body works according to the "physical properties of matter" and man's spirituality lies in "the activity of the 'reasoning soul.'"⁵⁰

Starobinski finds that Rousseau adopted Buffon's view that man alone is capable of comparison and also, "that man alone is capable of anticipating the future and remembering the past. Further, animals, even at their most ingenious, infallibly obey their instincts, repeating the same actions without modification; only man has the power to perfect himself, to progress..."⁵¹ Starobinski mentions that this is found in Buffon's *Natural History* and in Rousseau *Discourse on the Origin of Inequality* and the *Profession of Faith*.

He point out that there is a difference between them: Buffon believes that man's spirituality lies in his understanding and Rousseau says that it is in his freedom.[52] Starobinski notes that for Rousseau, the animal "is nothing more than 'an ingenious machine,' whereas man is endowed with freedom."[53] Starobinski observes, like Moran, that Rousseau's inclusion of the great apes in the human species is a historical commentary on how European civilization has caused man to degenerate morally: "By including creatures so different from civilized man in the human race, Rousseau pointed to the existence of a huge gap between primitive man and the disciplined European. This gap could only be explained by history, which altered and transformed if not man's nature then at least his 'Constitution.' This made man a particularly eloquent example of the restricted transformism whose stages Buffon had so ably described for those species modified by human husbandry. The opening sentences of *Emile* make clear that Rousseau saw no essential difference between man's transformation of himself and his transformation of such natural species as dogs and horses."[54] Here Starobinski recognizes the parallel between the moral degeneration of natural man when he entered society and the physical degeneration of Buffon's domesticated animals.

Starobinski also examines natural man's consciousness of self. For Rousseau, "natural man can be aware of his existence without forming any ideas."[55] Furthermore, natural man is more aware of his existence than is civilized man because he is not distracted by what others think of him, he is not preoccupied with pleasing others, he is not living outside of himself. Starobinski explains: "In fact, the less man reflects, the more aware he is of his existence: 'His soul, which nothing disturbs, dwells only in the sensation of its present existence, without any idea of the future, however close that might be.'"[56] and "Rousseau holds that we are turned away from self-knowledge by active contemplation of the past and, even more, by concern for the future. When we reflect, we compare objects and moments of our experience and as a result distinguish between ourselves and others and look to others to confirm our sense of ourselves. In other words, reflection is alienating. Rousseau therefore maintains that 'the savage lives within himself; social man lives always outside himself; he knows how to live only in the opinion of others. It is, so to speak, from their judgment alone that he derives the sense of his own existence.' If living within the sense of present existence is also living within oneself, then natural man spontaneously attains an ideal of independence that civilized man can attain only after lengthy philosophical exertion..."[57]

Leonard Sorenson agrees with Jean Starobinski that the purpose of the *Discourse on the Origin of Inequality* is political: Sorenson says, "Rousseau apparently addressed the theme of inequality in his *Discourse on the Origin of Inequality* in order to defend the fundamental principle of human equality."[58] Sorenson observes that while Rousseau's natural man is equal, inequality is born of "accidental, historical conditions": "Inequality is caused by human relations-social, economic, and political-but all relations and even most if not all human capacities are artificial products of accidental historical circumstances. The root of all evil, inequality, has its source in an accidental history that disjoined man from nature."[59]

However, Sorenson points out that inequality is not just a result of history and social convention. Rousseau concedes on the first page of his *Discourse on Inequality* that there is also natural inequality that is established by nature: there are differences in age, health, strength, and mind. Rousseau recognize the natural inequality in intelligence and faculty of the mind (*esprit* or *âme*). How does Rousseau propose to reconcile natural equality with natural inequality for the good of all? After examining the question in 21 pages, Sorenson capsulizes the answer thus: "One of the tasks of political philosophy, intrinsic to its own success, is to discover or invent ways to benevolently reconcile in so far as possible, 'natural inequality' with 'natural equality,' the requirement of rule by the naturally superior for the good of the naturally equal with the requirement of self-rule by the naturally equal: to reconcile wisdom and consent. As has been recognized by many, Rousseau himself recognized and addressed this dilemma."[60] The naturally superior, those with superior minds, can cultivate philosophy for the good of all. The goal of philosophy is to restore self-rule to the naturally equal.

David Gauthier, in *Rousseau: The Sentiment of Existence*, identifies the transformation of natural man to civilized man not as biological, but rather, as psychological.[61] He describes the transition as a passage from an inward focus to an outward focus, from that of the inward man, out to the world outside of himself. Gauthier explains the passage as the conversion of *amour de soi* (self-preservation) into *amour propre* (vanity): "Rousseau represents this as the conversion of *amour de soi* into *amour propre*. *Amour de soi (-même)* is no more than the case each person-indeed, each animal-has for its own preservation. It is a love centered on the self and addressed to its natural needs; it involves no awareness of others, much less comparison between self and others. But as awareness of others develops, this self-love is transformed into *amour propre*, a love centered on the relation between the self and others and addressed to comparative advantage."[62]

Gauthier points out that self-preservation (*amour de soi*) is related to the sentiment of our existence, while vanity (*amour propre*) is linked to the concern of what others think of us-whether they think we are powerful or weak. We want others to think that we are powerful, and there is an enslavement here-a concern for the opinion that others have of us, that takes over all our thoughts, and so we forget our sentiment of existence. Gauthier puts it this way: "*Amour de soi* is linked to our sentient of existence. As long as it alone holds sway, each person unreflectively senses his existence in himself. But as it comes to be transformed into *amour propre*, each senses his existence not in himself, but in his relation to those whom he perceives as other. It is the regard that others have for me, their concern with my power, or their contempt for my lack of power, their valuing or disdaining my assistance, their fearing or ignoring my opposition, that form the basis of my own self-conception. I am no longer psychologically self-sufficient, and so no longer free; I seek the recognition of the others that confers prestige. But this loss of freedom depends on distinguishing self and other..."[63]

Gauthier states that Rousseau contrasted liberty with its opposites: slavery, illusion and prestige: "The slave fails to gain the recognition of his or her fellows and can survive only by literal abasement, by being a person entirely for another and not for himself."[64] Similarly, the person who is regarded as powerful, as having gained the esteem and recognition of others is also enslaved: he is concerned with what others think of him. Gauthier says, "He cannot be free, since the powers requisite to meet his needs and desires are not his own; rather, they are the powers he is believed to possess, for it is these that affect the responses of others, and so determine whether he will gain satisfaction."[65]

Gauthier cites Rousseau, who in the *Discourse on the Origin of Inequality*, says, "having been formerly free and independent, behold man, due to a multitude of new needs, subjected so to speak...especially to his fellows, whose slave he becomes in a sense even in becoming their master; rich, he needs their services; poor, he needs their help."[66] Gauthier concludes, "The master's condition is one of psychological slavery."[67]

In summation, Jean-Jacques Rousseau proposed anthropological (intraspecies) and sociological change, not biological (interspecies), as did the materialists Diderot and Maupertuis. He recommended that more experimentation be conducted, in the way of hybridization, to ascertain whether orangutans belong to the human species. He also recommended that minds of the caliber of Montesquieu, Buffon, and Diderot travel around the world and relate their observations of how the great apes behave in their natural environment. Rather than proposing transformism, he was declaring a political statement as to the unity of the human race. The eighteenth century was confronted with the discovery of humans living under the most primitive conditions, as well as the discovery of the great apes, which appeared to have free will and intelligence (i.e., orangutans bury their dead, put roofs on their houses, and exhibit boredom and restlessness). Because he was a deist and a follower of Buffon, he articulated numerous reasons why he did not believe that man had ever been a quadruped. His purpose then, was to make a scathing social commentary on civilized Europe that had divided human beings into different social and economic classes and demonstrate that inequality is not natural, but an anthropological and historical artifice based on consent.

Chapter 8

Voltaire

People may tell me that porphyry is formed of bears' bristles; I will believe them when I find that white marble is made of ostrich feathers.[1]
—François-Marie Arouet de Voltaire, *A Defence of My Uncle* (1767)

Voltaire did not believe that the earth is more than 6,000 years old; nor did he believe that one species can metamorphose into another. His satirical jabs at contemporary naturalists are iconic representations of the enormous extent to which investigation into man's beginnings was polemicized in the eighteenth century. Voltaire was a fervent deist, and so, he vigorously opposed transformism because it threatened to bring deistic cosmogony to the ground.

Déisme is derived from the Latin *deus*, god. During the eighteenth century "deism" was defined as "System of those, who, not having any particular cult, and rejecting every kind of revelation, believe only in a sovereign Being. *To be suspected of deism*."[2] "Deist" was defined as "He or she who recognizes a God, but who does not recognize any revealed Religion. *He is a deist.*"[3] The *Oxford English Dictionary* defines "deism" as "usually, belief in the existence of a Supreme Being as the source of finite existence, with rejection of revelation and the supernatural doctrines of Christianity; 'natural religion.'"[4]

Because Voltaire believed in God the Creator, he recognized the danger inherent in random creationism: it obviated the need for a Prime Mover. Furthermore, if everything in the universe is the result of the motive property of matter, then random molecular motion is determinative and not God's will; God's will is rendered subordinate to Newtonian physics.

Since Voltaire was an impassioned deist, he defended only those hypotheses that he could use as polemic tools and spent the last thirteen years of his life attacking science: he zealously promulgated preformation theory and the notion that the earth is only 6,000 years old and vociferously attacked the metamorphosis of species, epigenesis, spontaneous generation, the notion that random molecular motion brought about the universe and everything in it, the chain of being, along with its corollaries, plenitude and continuity, fossils, and the hypothesis that the earth is much older than 6,000 years old. Because he did not believe in transformism, he had to necessarily deny the unity of the human race: all the different colors of the various races could not have proceeded from one man because physical characteristics had not changed since Creation. Therefore, he defended the notion that at the time of Creation, God created men of different colors as separate acts of Creation.

OPPOSITION TO RANDOM CREATION

In the article "Atheism," *Section 2*, in the *Philosophical Dictionary* (1764), Voltaire declares that conscious, intelligent beings cannot have arisen either by spontaneous generation or by random chance: "We are intelligent beings; well, intelligent beings cannot have been formed by a crude, blind, insensate being: there is certainly some difference between the ideas of Newton and the dung of a mule. Newton's intelligence comes, therefore, from another intelligence."[5] Here Voltaire focuses on the inability of the materialists, who embraced creation by random chance, to prove that consciousness arises from random molecular motion. Maupertuis recognized the insufficiency of the hypothesis, and so, he had to necessarily attribute aversion, desire, memory, and intelligence to molecules. However, Maupertuis' hypothesis was merely a hypothesis, and not a proven fact. Therefore, Voltaire jumps on this lack of evidence, and points out the difference between consciousness (the ideas of Newton) and insensate matter (mule dung). Another contemporary, Diderot, had also pondered the difference between living matter and dead matter in 1753 and wondered whether it is possible that the two states oscillate back and forth *ad infinitum*. The answer to this question was crucial because it meant that either God exists or He does not.

This sentence is also an attack on spontaneous generation (the notion that living things are born of non-living material). He attacks spontaneous generation because if animalcules are continually born of inorganic mater, then there is no need for God. This sentence is an attack on Buffon and Needham, whose flawed experiments indicated that spontaneous generation was a fact, and the materialists (Maupertuis, Diderot, and La Mettrie) who found spontaneous generation useful in proving that life is the result of random molecular action.

A few paragraphs later, under the subtitle, "Reasons of the Atheists," Voltaire goes on to say, "Nevertheless I have known roguish individuals who say that there is no creative intelligence at all, and that motion alone has by itself formed all that we see and all that we are. They impudently tell you: The combination of this universe was possible, hence the combination exists: therefore it was possible that motion alone arranged it."[6] Here Voltaire clearly states his opponents' argument before refuting it: random molecular motion, given an eternal time frame, will eventually bring about the universe. He proffers his response beneath the subtitle, "Answer," and it is a repetition of what he has previously said: intelligence and consciousness exist and random molecular motion is insufficient to produce either. As is his style, Voltaire focuses on a point, magnifies it, and repeats its, all the while excluding everything else his opponents offer. He cites Spinoza, who proclaimed universal consciousness, but visibly ignores Maupertuis and Diderot, who both hypothesize that molecules are conscious and that consciousness exists at every level of organization.

Having reiterated the insufficiency of motion to bring about consciousness, he goes on to pounce on the argument of Diderot, La Mettrie, and Maupertuis that parts of animals conform to their needs not because of Divine planning or final causes, but because all animals that were born with self-contradictions have perished; only the ones that random chance brought about without self-contradictions have survived. Voltaire's response to this argument is the same reply that he has given to antecedent arguments, namely, that the combinations that random chance produces cannot create consciousness or intelligence.

At this point, Voltaire goes further and identifies by name the materialist *philosophe* that he is attacking: the next subtitle reads, "Maupertuis' Objection." In this discourse, the atheist interlocutor, who has finally been identified, argues that deists see God's beneficence in the folds of a rhinoceros (which permit the thick-skinned animal to move), but one could equally deny

His existence because of the shell of the tortoise. In his response to this argument, Voltaire sounds like a scientist: the physical structures of both the rhinoceros and the tortoise demonstrate the infinite variety of different species and the goal of nature, which is preservation, generation and death. As is his style, Voltaire searches deeply and calls up the facts: there is infinite variety in nature and the legitimate question does arise, "Is it due to the fact that nature is in perpetual flux and is nature continually creating new variations?" Voltaire's position is the deistic one: the magnificent complexity of a fly's wing and a snail's organs suffice to bring the atheist to the ground.[7] The infinite variety in nature announces God's beneficence and genius. The beauty of God's infinitely varied creatures, their proportion, their uses, all exist, even if we do not fully comprehend them.

In the dialogue Maupertuis continues to argue that the snake appears to have no purpose in nature and does only harm-this is an attack on final causes-what purpose could God have had in creating an animal that has no use and does only harm? Voltaire's response to this is comical and he scolds Maupertuis: snakes are venomous and so is he; snakes do harm and so has he; men are worse than snakes.

REFUTATION OF THE CHAIN OF BEINGS

We have seen that Voltaire attacks random creation via molecular motion by arguing that there is no evidence that it leads to consciousness. What remains is to show that nature is not in perpetual flux, that the time of Creation has long passed, and that no new species have been created since Genesis. In the article, "Chain of Beings," in the *Philosophical Dictionary*, Voltaire launches an attack to demonstrate that the chain of beings is pure fiction: when one examines the great chain carefully, it vanishes;[8] he points out that there is no chain because the supposed gradation no longer exists in vegetables and animals;[9] there are species of plants and animals that have become extinct;[10] we no longer have murex;[11] Jews were forbidden from eating griffin and ixion: these two species have probably disappeared from this world;[12] he asks, "Where then is the chain?";[13] it apparent that species can be destroyed: the lion and the rhinoceros are beginning to be exceedingly rare;[14] it is likely that there have been races of men that are no longer found;[15] he asks, "Is there not a visible gap between the monkey and man?"[16] In Voltaire's cosmogony, God created a fixed number of species at the time of Creation and there have been no new species since then. Nature is not continually producing new creations. Some may be extinct because men have hunted and killed them. For Voltaire, only God can create species and the time of Creation has passed.

Arthur O. Lovejoy examines the article, "Chain of Beings," and advises that Voltaire makes three cogent arguments that the chain of beings is fictional. Lovejoy says, "First, some species which once existed have disappeared; others are in the process of extinction; and yet others might be or may yet be destroyed by man, if he should so desire…It is probable also that there have been races of men which have vanished."[17] Lovejoy mentions that Voltaire had observed that there is a visible gap between the monkey and man. Voltaire recognized that man can destroy species and render them extinct; he did not believe that new species can metamorphose into existence from existing species. Hence, the number of species is always getting smaller and there is no chain of beings that exists today. Therefore, we cannot hypothesize that there ever was a chain of being.

Secondly, Lovejoy points out that Voltaire argues that because we can imagine intermediary species between existing ones shows that there are gaps between species. Voltaire asks, "Is there not visibly a gap between the ape and man? Is it not easy to imagine a featherless biped possessing intelligence but having neither speech nor the human shape…"[18] Voltaire posits that the fact that man can imagine an intermediary creature between himself and the monkey shows that it is visibly absent from nature.

Thirdly, Lovejoy maintains that because Voltaire believed in God, "…the supposition of the completeness of the chain of beings requires the existence of a vast hierarchy of immaterial beings above man."[19] In an apostrophe Voltaire cries out to Plato, "But you, what reason do you have for believing in it?"[20] The inferred answer is simply because the chain of beings, by definition, requires a complete ladder from the atom all the way up to God Himself. Hence, imaginary celestial beings are intrinsic to the chain by its definition.

Voltaire declares that there is no chain of beings in the inanimate world either: he points out that there is no gradation among the planets or moons, either in their magnitude or in their orbits.[21] Voltaire also mentions the vacuum that exists in outer space and points out that not everything in nature is connected: "And then how can you expect there to be a chain that links everything in the great empty spaces?"[22]

Lovejoy researches and cites several works in which Voltaire uses Newton's vacuum in space to refute the notion that everything in nature must be connected. In *We Must Take Sides, or the Principle of Action* (1772), Voltaire reiterates the fact that the vacuum that exists in outer space negates the notion that continuity must exist in all of nature. Newton had proven the existence of a vacuum in space: since this vacuum exists, it can exist anywhere, even in the hierarchy of species: "Why should, and how can, existence be infinite? Newton demonstrated the reality of a vacuum. If in nature there can be a void beyond nature, wherein lies the necessity that entities should extend to infinity? What would an infinite extension be? It could no more exist than an infinite number."[23]

Lovejoy also cites a footnote that Voltaire appended to the *Poem on the Lisbon Disaster* (1756), in which he says, "It is demonstrated that the heavenly bodies perform their revolutions in a non-resistant space. Not all space is filled. There is not, therefore, a series (*suite*) of bodies from an atom to the most remote of the stars; there can therefore be immense interims between sensible beings, as well as between insensible ones. We cannot, then, be sure that man is necessarily placed in one of the links which are attached to one another in an unbroken sequence."[24] In that same footnote Voltaire reiterates his denial that a continuous gradation exists that links all beings. He declares that there is probably an immense distance between man and animals, between man and superior (celestial) beings, and there must be infinity between God and all created beings. He repeats his statement of 1765 that the planets that revolve around the sun do not exhibit any gradual gradations in their sizes, distances, or their moons.[25]

REFUTATION OF SPONTANEOUS GENERATION

In the *Questions on Miracles* (1766), the 4th *Letter, Foreword to the 5th Letter, 5th Letter, 6th Letter,* and 7th *Letter,* Voltaire ridicules John Turberville Needham and warns worshippers of God to steer clear of spontaneous generation because it is a tool of deception that materialist atheists will use to shake their faith. Voltaire's association of spontaneous generation to atheism is a personal jab at Needham, who was a Catholic priest. In the *Foreword to the 5th Letter,* Voltaire ridicules Needham for thinking that he has discovered that spurred wheat flour steeped in water is transformed into small animals that look like eels. He declares that the claim is false because the Italian scientist Lazzaro Spallanzani has proven it to be false and furthermore, it is false because it is impossible to do. Spallanzani was also a Catholic priest, but Voltaire significantly omits this fact because of his opposition to Catholicism; however, Christians benefited from debunking the materialists' tool just as much as the deists did, and in this rare moment, Voltaire is in the camp of a priest. Voltaire argues that if animals could be born without seed, generation would be unnecessary and a man could be born from a clump of earth as easily as an eel could be born of a piece of dough. This dangerous belief system leads to atheism and the materialist *philosophes* use it to show that God is not necessary for the greatest miracle of all, the generation of living beings.

In the *5th Letter* Voltaire challenges the veracity of Needham's experiments: Needham has made quite a little reputation for himself among atheists by claiming that flour produces eels and that thus, all men, starting from the first one, may have been born the same way. Voltaire makes fun by asking a comical question: "The only difficulty that remains is to know how there could have been flour before there were men."[26]

Voltaire repeatedly calls Needham an atheist. In the *Foreword to the 5th Letter* he says, "This system…would lead…to atheism" and "Needham's microscope…was considered to be the atheists' laboratory." In the *5th Letter* he says, "You got yourself a little reputation among atheists," "Atheist that you are," and "…you have shaken their faith." Aside from the fact that he was insulting a Catholic priest, Voltaire clearly understood that if life springs forth from dirt all the time, than there is no need for Divine agency.

In the *Defence of My Uncle* (1767), *Chapter 19*, Voltaire again refutes spontaneous generation by associating it with another hypothesis that could not be proven: Descartes' globular, penetrating, channeled, striated matter. Voltaire makes the analogy between the two hypotheses, one biological and the other pertaining to physics, one purported to explain the origin of man, and the other, the origin of the universe: "The seed is useless; everything will grow spontaneously. Upon this supposed experiment a new universe is constructed, in the same manner as a new world was formed a hundred years ago, with a penetrating, globulous, and channeled matter."[27] This is a dig at Descartes, who in the *Principles of Philosophy, Part 3, Sections 46–48 and 86–87*, tries to explain the genesis of the stars in the universe by molecular motion. Descartes hypothesizes that tiny globules of matter seek to penetrate the vortex of a whirlwind of matter, and in doing so, become channeled or striated in shape. These fluted globules are visible as sunspots. For Voltaire, the attempt to explain either the origin of man or the origin of the stars in the universe through Newtonian mechanics is dangerous to deism because it threatens to replace God as a causative agent. Newtonian physics is a two-edged sword: it both proves the brilliant engineering of a beneficent Creator, but also threatens to eliminate the need for Him.

Two years later, Diderot would respond to Voltaire in *D'Alembert's Dream*:

> Mademoiselle de l'Espinasse: "Voltaire can joke all he wants, but the eel-monger is right; I believe my eyes; I see them: how many there are! How they go! How they come! How they jump about!"[28]

As far as Diderot was concerned, scientists did see animalcules under microscopic magnification; they had no reason to doubt spontaneous generation as it was a belief system that dated back to antiquity. If Francisco Redi and Lazzaro Spallanzani were able to perform experiments that disproved the hypothesis, Jan Baptista van Helmont, John Needham and Georges-Louis Leclerc de Buffon were able to uphold it.

Voltaire's assault on spontaneous generation and its chief proponent, Needham, is endless. In *The Life Sciences in Eighteenth-Century French Thought*, Jacques Roger advises that during the thirteen year period 1765–1778, Voltaire dedicated twenty-five of his works to denouncing atheism and ten works to ridiculing Needham and his eels.[29] Roger copiously researches and enumerates the sources for readers to explore further: Voltaire unrelentlessly assails Needham in the *Questions sur les miracles* (1766), 4th letter, "Avertissement," 5th letter, 6th letter, and 7th letter; *La Défense de mon oncle* (1767), ch. 19 "Des montagnes et des coquilles"; *Singularités de la nature* (1768), ch. 20 "De la prétendue race d'anguilles formés de la farine et de jus de mouton"; *Homme aux quarante écus* (1768), ch. 6 "Nouvelles douleurs occasionnées par de nouveaux systèmes"; *Les Deux Siècles* [the *Siècle de Louis XIV* (1751) and *Supplément du siècle de Louis XIV* (1753)]; *L'ABC* (1768), 17th discourse; *Les Colimaçons du révérend père L'Escarbotier* (1768), 3rd letter "Dissertation du physicien de Saint-Flour"; *Précis du siècle de Louis XV* (1768), ch. 43; *Questions sur l'Encyclopédie*

Voltaire

(1770–1772) "Dieu" §4; and *Dialogues d'Evhémère* (1777), 9th dialogue ("Sur la génération").[30]

OPPOSITION TO FOSSILS

During the seventeenth and eighteenth centuries there were many fossil discoveries: mammoth bones and teeth in Siberia and the New World, stones containing fossilized figures of small animals and plants (hence, the term "figured stones" were given to fossils), and seashells on the peaks of very high mountains. While pondering what these strange artifacts might be, scientists began to move away from the Aristotelian notion that fossils were of inorganic origin and began to accept the view that they were the remains of living things. However, there were conflicting schools of thought as to how old the fossils might be. At first, scientists, accepting the Genesis chronology of Creation, believed that these articles, of organic origin, had been deposited by the Noachian flood; they believed, therefore, that the fossils could not be more than 6,000 years old. Soon, however, forward thinkers like Buffon and the encyclopedists extrapolated that they could not possibly be 6,000 years old, and that therefore, they were proof that the earth was much older than that.

In addition, there were discoveries of heaps of seashells in beds in the vicinity of Paris, atop very high mountains, and all of Europe, in areas far from the sea. The question arose as to why these shells were found in very large quantities in areas far away from oceans and seas. Buffon and Diderot vehemently denied that deposits of fossils and seashells had anything to do with Noah's flood, and they were not the first to do so. Leonardo da Vinci, c. 1500, intuited that the seashell beds were unrelated to the Flood: "And if you wish to say that it was the Deluge which carried these shells hundreds of miles from the sea, that cannot have happened, since the Deluge was caused by rain, and rain naturally urges rivers on towards the sea, together with everything carried by them, and does not bear dead objects from sea shores toward the mountains. And if you would say that the waters of the Deluge afterwards rose above the mountains, the movement of the sea against the course of the rivers must have been so slow that it could not have floated up anything heavier than itself."[31]

During the seventeenth century, Steno (Niels Stenson) and Robert Hooke corroborated da Vinci's thesis with scientific investigation, as did Antonio Vallisneri during the first half of the eighteenth century. In 1666 Steno inspected the mouth of a shark and discovered that its teeth were similar to fossils called *glossopetræ* (tongue stones), which had previously been thought to be snake tongues or dragon tongues. Comparing the two, Steno surmised that the *glossopetræ* fossils were, in fact, shark teeth, and therefore, they were the remains of organic organisms. What remained was to ascertain how the figures got encased in the stone. Steno posited that the sandstone must have once been loose sand that enveloped the fossils and then became petrified. Hence, he extrapolated that the fossils inside must be older than the stone encasing it. In 1669 Steno articulated his hypotheses in *Prodromus a Dissertation Concerning Solids Naturally Contained with Solids*.[32]

Robert Hooke arrived at the same conclusions as Steno and went a step further: because many fossils have no living counterparts, species must pass in and out of existence. Hence, species must be mutable, not fixed. In 1668 Robert Hooke read his "Discourse of Earthquakes" to the Royal Society in which he noted that some fossils seem to have no living equivalents and that "there may have been diverse species of things wholly destroyed and annihilated, and diverse others changed and varied..."[33] This raised eyebrows as it was still commonly held that God created all species during Creation and that they were still in existence, unchanged.

Antonio Vallisneri also broke away from the Noachian flood theory. In 1721 Vallisneri hypothesized that fossils are of organic origin and are the remains of animals who lived in previous ages; their location on mountains is unrelated to the Noachian flood.[34]

Voltaire, because he was an impassioned deist, could not help but become embroiled in the controversy surrounding the origin of fossils and the location of seashells far from water. He denied that they had anything to do with Noah's flood because he was fighting a war simultaneously on two fronts. First, he was afraid that the fossils would be used as proof to substantiate that there really was a flood and that therefore, everything in the *Old Testament* was true (as Johann Scheuchzer had done in 1726), and secondly, the atheist materialists could use them to prove random creation, that species come into and pass out of existence all the time, without any particular purpose.

In the *Dissertation Sent by the Author in Italian to the Academy of Bologna and Translated by Himself into French on the Changes that Have Taken Place on Earth and on Petrifications that are Pretended to Still Bear Witness* (1746), Voltaire maintains that fossils found on mountaintops were carried there by men: a stone that appears to bear the impression of a turbot (species of flatfish) has been found in the mountains of Hesse (in Germany), and a petrified pike was found in the Alps; he mocks scientists who have concluded that the sea and rivers must have flowed around and around over mountains. He declares that it would have been more natural to suppose that "these fish, carried by travelers, having spoiled, were thrown away, and petrified over time; but this idea was too simple and too systematic." A boat anchor was discovered atop a Swiss mountain: men probably carried it with them to serve as an anchor for the loads they carried in case of a rockslide. *Glossopetræ* are merely the shells of a genus of contemporary mollusks called "Venus" and they were not left by Noah's flood.

The foreword to the first complete edition of Voltaire's works, the Kehl edition (1784–1789), advises that the *Dissertation* had appeared anonymously and that for a long time no one knew that Voltaire was the author. Kehl states that Buffon was unaware that Voltaire had written the piece when he severely criticized the work and scolded its author in the *Natural History, General and Particular, Volume 1, The Theory of the Earth* (1749).

In the 1829 edition of Voltaire's *Works*, the editor, Beuchot, advises that Voltaire had written the piece in three languages and circulated it through diverse venues: Voltaire sent a letter to G.-Fr. Muller on June 28, 1746 indicating that he had sent his piece in English to the Royal Society of London and that he proposed to translate it into Latin to send it to the Academy of Saint-Petersburg. A French translation of the Italian was published in the *Mercure* in July 1746. It was in the 1748 edition of his works (Dresden), that Voltaire added his own translation into French, which, for the majority of readers, was preferable to the original in Italian. Moreover, Voltaire made several additions and corrections in various editions. The *Digression*, which follows the *Dissertation*, was published in 1751.

Buffon was in the midst of correcting the proofs to the *Natural History* when the *Dissertation* crossed his desk. The naturalist could not help but comment on the piece and he felt it necessary to admonish its anonymous author: "While reading an Italian letter on the changes that have taken place on earth, published in Paris this year (1746)...petrified fish are only, in his opinion, rare fish, thrown off the table of Romans, because they were not fresh; and regarding shells, he says that it was Syrian pilgrims who carried them back from eastern seas during the times of the Crusades, that are now found petrified in France, Italy and in other Christian countries; why didn't he add that monkeys carried the shells to the tops of high mountains and all the places where men cannot inhabit, that would have spoiled nothing and would have made his explanation even more probable. How is it possible that enlightened men, who even take pride in philosophy, still have such erroneous ideas on this subject? Therefore, we will not be satisfied to have said that petrified shells are formed in nearly every place on earth that has been excavated, and to have reported the testimony of authors of the *Natural History*; as one could suspect them of perceiving, in view of some systems, shells where there are none, we believe it necessary to cite travelers besides, who have noticed them by chance and whose eyes, being less trained, could recognize only whole and well preserved shells; their testimony will be perhaps of a greater authority next to people who do not have thee ability to ascertain the truth of facts by themselves, and those who do not know either shells or petrifications, and who not being in a position to make a comparison, could doubt that the petrifications were in fact true shells, and that these shells are found heaped together by the millions in all climates on earth. Everyone can see with their eyes the banks of shells that are in the hills in the vicinity of Paris..."[35]

Impervious and unyielding to the opinions of naturalists, Voltaire, in *A Defence of My Uncle* (1767), *Chapter 19* ("Of Mountains and Shells"), devotes himself to proving that fossils have nothing to do with any flood and to deny that there ever was a Noachian flood. He enumerates nine reasons why he disagrees with "a great naturalist" (Buffon) "who has imagined that mountains were formed by the sea" and that the sea deposited petrified fish on mountain peaks through "flux and reflux."[36] "Flux and reflux" is a scatological pun and he uses it several times to make fun of the hypothesis. During the eighteenth century the primary definition of "flux" was diarrhea: "FLUX also means the flow of excrement that has become too fluid, and signifies loose bowels."[37] "Flux" and "reflux" strung together, to form the phrase "flux and reflux," connote the ebb and flow of the ocean's tide. However, Voltaire takes care to use the terms "flux" and "reflux" separately,

Voltaire

and as often as possible, to ridicule the image of the ocean depositing fossilized objects all over the place, and especially, in very high spots.

Voltaire proffers nine reasons why seashells, discovered on mountain peaks, could not have possibly been deposited there by Noah's flood:

1. If the sea had formed a small mountain by its flux, it would destroy what it built by its reflux.[38]
2. The flux of the ocean can create heaps of sand, but it cannot create rocks.[39]
3. If it takes 6,000 years for the ocean to raise hills of sand 40 feet perpendicular, it would require 30 million years to form the highest mountain of the Alps, which is 20,000 feet high.[40]
4. How could the flux of the sea, which never rises more than eight feet high at the coasts, have formed mountains 20,000 feet high? How could there be enough water to cover them to leave fish at the summits?[41]
5. How could the tide and currents form chains of mountains, almost circular, such as those that circumscribe Kashmir, Tuscany, Savoy, and Vaud?[42]
6. If the sea were above the mountains, all the rest of the globe must also have been covered with water equal in height, otherwise the waters would have fallen again by their own weight. An ocean that high would contain the water of forty of our present oceans; therefore, thirty-nine oceans must have necessarily vanished.[43]
7. If the earth were covered with water, it would have been inhabited only by fish. It is difficult to comprehend how porpoises could produce men (This is a little dig at Maillet's *Telliamed*, in which the author posits that all life originated in the sea).[44]
8. If the oceans covered the mountains for a long time, there would have been no fresh water for two-footed animals and quadrupeds to drink. The Rhine, the Rhone, the Soane, the Danube, the Po, the Euphrates, and the Tiber, derive their water from snow and rain.[45]
9. Nature never contradicts itself. All species remain the same forever. Animals, vegetables, minerals, metals, everything is invariable. Everything preserves its essence. The essence of the earth is to have mountains, without which there would be no rivers. The steep banks of some rivers and lakes are embroidered with seashells. I have never seen that they were the remains of sea-monsters. They look

like the torn coats of muscles and other small shellfish that inhabit rivers and lakes. There are some which are apparently nothing but talc, which have taken different forms. There are a thousand earthly productions that are mistaken for marine productions.[46]

Voltaire also offers an explanation as to why oyster shells can be found a hundred leagues from the sea: because he has seen Roman money buried twenty feet deep and rings of knights more than 900 miles from Rome, he must necessarily extrapolate that these shells, coins and rings were not made there, but rather, that travelers carried them to those locations.[47]

His response to the news that Syrian shells have been discovered in the Alps is that "these shells were probably brought there by Pilgrims, upon their return from Jerusalem."[48] He adamantly declares, "People may tell me that porphyry is formed of bears' bristles; I will believe them when I find that white marble is made of ostrich feathers."[49] It is thus that Voltaire acquired his reputation for believing that fossils were carried to their present locations by the Crusaders returning from Jerusalem, picnickers who threw away their trash, and conchologists who misplaced their samples.

In *On the Singularities of Nature* (1768), Voltaire devotes 102 pages to refuting the findings of contemporary scientists and the impact that their hypotheses had on deism. In *Chapter 12* he denies that the discovery of seashell beds in Touraine, the suburbs of Paris, and all over Europe, prove that the earth is more than 6,000 years old: "We find in some places in this globe heaps of shells; we see in some others petrified oysters: from this it has been concluded that, despite the laws of gravity and those of fluids, and despite the depth of the ocean bed, the sea has covered all the earth millions of years ago."[50] He reiterates his argument that the sea, whose tide reaches a maximum of fifteen feet, could not have possibly, through the action of its flux, formed tall rocks 18,000 feet high.

What Voltaire feared the most about this hypothesis was that if the earth was covered with water, life must have originated in the sea, and perhaps Maillet was right. This was insupportable. The hypothesis had to be refuted, but since it could not be, with eighteenth-century science, then it would suffice to ridicule it to the fullest extent possible.

In *Chapter 13* Voltaire makes note of the fact that although "a thousand places are filled with the remains of testaceans, crustaceans, and petrifications...they are almost never found either on the tops or the sides of mountain chains."[51] Rather, they are found "several leagues from mountains, in the middle of the ground, in caves, in places where it was very likely that there were small lakes that have disappeared, small rivers whose courses changed...but true marine bodies, those you never see. If there were any, why have we never seen the bones of dog-fish, sharks or whales?"[52] The answer that comes immediately to the reader's mind is that dog-fish, sharks and whales can swim out to sea in the event of a cataclysmic event, but turtles, muscles, snails, and small river crustaceans cannot. It is obvious that Voltaire is grasping at straws here. It gets worse in other passages, where he appears to be reverting to the Aristotelian notion that fossils are of inorganic origin. He concludes *Chapter 13* by asking, "Again, I am not denying that we find, a hundred miles from the sea, some petrified systems, sea-shells, univalves, products of nature that perfectly resemble marine productions; but are we assure that the soil of the ground cannot produce these fossils? Mustn't the formation of arborized or herborized agates make us postpone our judgment? A tree has not produced the agate that perfectly represents a tree; the sea may also not have produced shells that resemble the dwelling places of small marine animals."[53] Voltaire attempts to substantiate his thesis that marine fossils have originated in the soil with an eyewitness account in *Chapter 14*.

In *Chapter 14*, entitled, "Important Observations on the Formation of Rocks and Shells," Voltaire proves that seashells do not necessarily come from the sea: a certain Mr. Le Royer de la Sauvagère claims that near his castle, a part of the earth metamorphosed into a bed of soft stone twice in eighty years. He built with this stone, which became very hard when used. The small hole in the ground that he created when he removed the stone began to take form again. Shells reappeared there that at first, could be distinguished only with a microscope, and which grew with the stone. These shells were of different species: there were ostracites, gryphites, which are not found in any sea; cams, telines, hearts, whose seeds developed imperceptibly until they were a half inch thick. Voltaire asks, "Isn't that enough to at least surprise those who maintain that all shells that are found in some places one earth were deposited there by the sea?"[54]

In 1778 Buffon reveals that he had been horrified when he discovered that the author of the *Italian Letter* was Voltaire. Being the courteous gentlemen that he is, he tries to make amends and soften the harsh criticism that he had given twenty-nine years earlier. However, he remained adamant that the heaps of seashells discovered buried in the earth in diverse places in

France and other parts of Europe prove that these places had once been covered by water. In his *Supplement* to the *Natural History*, he admitted, "Concerning what I have written, *page 281*, on the subject of the Italian Letter, where the author states *that it is pilgrims and others, who, in the times of the Crusades, have brought back shells from Syria that we find in the bowels of the earth in France*, etc. one could have found, as I find myself, that I did not treat Mr. Voltaire seriously enough; I admit that I would have done better to drop the matter, rather than to jokingly take it up again, insofar as it is not my style, and that it is perhaps the only instance of it in my writings. Mr. Voltaire is a man, who by the superiority of his talents, merits the greatest esteem. This Italian Letter was brought to me at the same time I was correcting the proofs to my book that is at issue; I read this letter only in part, believing that it was the work of some scholar in Italy, who, according to his historical knowledge, only followed his prejudices, without consulting Nature; and it was only after the publication of my volume on the Theory of the Earth, that I was apprised that the Letter was by Mr. Voltaire; therefore, I regret my language. This is the truth, I declare it as much for Mr. Voltaire, as for myself and for posterity to which I would not want to leave any doubt as to the high esteem that I have always had for such an extraordinary man who has done such credit to his century."[55]

Having shown Voltaire the respect that he was due, Buffon goes on to reiterate that seashells are found everywhere, and in such great quantity in certain places, and they are arranged in such a way, that one must necessarily extrapolate that spots on the earth that are presently land must have formerly been covered by water. The seashells suggest that where there is now ground, there must have once been a seabed, and that by some revolution of the earth, the sea pulled back and left its productions. Buffon politely introduces the work of a contemporary naturalist, P. Chabenat, who differs with Voltaire. Buffon diplomatically says, "Because Mr. Voltaire's authority had made an impression on some people, there were a few who wanted to verify for themselves if the objections against the shells had some basis, and I believe that it is necessary to provide here the extract of a Memoir that was sent to me..."[56]

Buffon cites the authority of naturalist Chabenat, who had traveled extensively throughout France and Italy, and who concluded, "I saw figured stones everywhere, and in some places in such great quantity, and arranged in a way, that one could not help but believe that these parts of the Earth were a seabed in former times. I have seen shells of every species and which perfectly resemble their living counterparts. I have seen them in the same shape and the same size: this observation seemed sufficient to persuade me that all these individuals were of different epochs, but they were of the same species. I have seen ammonite horns from a half inch up to almost three feet in diameter. I have seen scallops of all sizes, other bivalves and univalves as well. Other than that I have seen belemnites, sea-mushrooms, etc. The form and quantity of all these figured stones prove to us almost invincibly that they were once animals that lived in the sea. Above all the shell that covered them seems to leave no doubt because in certain ones, it is as shiny, as fresh and as natural as in the living ones; if it were separated from the stone, one would not believe that it was petrified...All this seems to tell me very intelligibly that this country was formerly a seabed, which by some sudden revolution, receded and left its productions there as in many other places. However, I suspend my judgment because of Mr. Voltaire's objections. To answer him, I wanted to combine experimentation with observation."[57]

Buffon concludes, "Mr. P. Chabenat reports afterwards many experiments to prove that shells which are found in the bowels of the earth are of the same nature as those of the sea; I do not report them here, because they teach nothing new, and no one doubts the identity between fossils shells and seashells. Finally, Mr. P. Chabenat concludes and ends his Memoir by saying: 'Therefore, one cannot doubt that all these shells that are found in the bowels of the earth are true shells and the remains of marine animals that formerly covered all these regions, and that consequently, Mr. Voltaire's objections are ill-founded.'"[58]

Martin J.S. Rudwick demonstrates how Voltaire, who insisted that there was no flood and was compelled to deny the mutability of species, had painted himself into a corner and was forced to argue that Crusaders brought fossils back after they had ransacked Jerusalem: "Yet if the evidence was sometimes strained to provide support for the reality of the Deluge, it could be equally strained in the opposite direction by those who, in the name of Enlightenment, wished to deny that any such inexplicable event had ever occurred. For example, Voltaire, whose first-hand knowledge of fossils was probably minimal, nevertheless felt himself qualified to assert that they gave no evidence of any interruption of the Newtonian regularity of the universe. To reach this conclusion he was obliged to dismiss fossils variously as inorganic productions, as the relics of freshwater lakes, and as shells dropped on land by pilgrims: but these were arguments hardly calculated to persuade naturalists who knew that many fossils closely resembled marine organisms and yet were embedded within strata. On the whole, therefore, it was the diluvialists whose work most encouraged the acceptance of an organic interpretation of fossils."[59]

OPPOSITION TO EPIGENESIS

Voltaire embraced preformation and he refuted epigenesis, which he believed was dangerous and would lead to atheism: epigenesis could be used as stepping stone to spontaneous generation and if inanimate matter brings forth living things, then Divine agency is unnecessary. Shirley A. Roe explains why deists regarded preformation critical in their polemics: "The belief that the clockwork universe had been created by an intelligent and all-powerful God demanded that God's involvement in the creation of living creatures play a fundamental role in explanations of their generation. For there was always a danger in the mechanist world view that God as Creator might be a superfluous entity and that matter and motion might themselves be responsible for all of the phenomena of the universe, including the creation of life and the existence of the human soul. Preformation undercut these dangers by making both God and mechanical laws essential to the explanation of generation."[60]

Therefore, Voltaire denounced epigenesists (ie: Maupertuis, Buffon, and Needham) going all the way back to William Harvey. In his review of Charles Bonnet's *Considerations on Organized Bodies* (1762) that appeared in the *Literary Gazette of Europe* on April 14, 1764, he takes the opportunity to refute the work of Harvey and Maupertuis. According to Voltaire, even the legendary William Harvey, who dissected dogs taken from the royal parks of England, did not succeed in disproving preformation.

Voltaire's contemporary, Charles Bonnet, believed in preexistent germs and therefore, he devotes eight pages to defending the hypothesis in his review of Bonnet's *Considerations*. Bonnet, on the first page of his book, declares his purpose, which is to prove that preformation is a fact: "Philosophy, having understood the impossibility of explaining the formation of organized beings through mechanical means, has fortunately imagined that they have already existed in miniature, in the form of germs or organic corpuscles."[61]

In *Chapter 8* Bonnet summarizes his defense of encasement thus: "I will content myself in reminding my readers of the astonishing apparatus of *fibers, membranes, vessels, ligaments, tendons, muscles, nerves, veins, arteries*, etc. that enter into the composition of an animal's body. I will ask them to consider with great care the structure, the relationships and the functioning of all these parts. Then I will ask them if they imagine that a whole as assembled, as interrelated, and as harmonious, could have been formed simply by the collision of molecules in motion, or directed, following certain laws unknown to us. I will ask them to tell me if they do not feel the necessity we are in of admitting that this admirable machine was first designed in miniature by the same hand that traced the design of the universe."[62]

In his review of Bonnet's work, Voltaire begins by admitting that he does not believe that *Considerations* "can shed much light on the great and mysterious question of generation, which is the despair of both ancient and contemporary philosophers, but at least his book reveals a very wise and very enlightened mind."[63] He remarks that "the ancients had to guess as we do the secrets of nature, but they had no thread to guide them in the roundabout ways of this immense labyrinth."[64]

The assistance of microscopes, comparative anatomy, and two centuries of observation have offered a little enlightenment, but not much. For Voltaire there is no point in trying to unravel the mystery of generation with a microscope because it cannot be done. Harvey, Malpighi, Leeuwenhoek, Vallisneri, they all solved nothing, not even after copious investigation: "They concluded by remaining in doubt; which is what always happens when you want to go back to first causes."[65]

Voltaire takes Maupertuis to task for defending epigenesis in *Physical Venus*: Maupertuis pretended that "the left leg of the fetus attracts the right leg without being mistaken; one eye attracts an eye while leaving the nose between them, one lobe of lung is attracted by the other lobe, etc."[66]

Having ridiculed Maupertuis' theory that embryonic particles attract, Voltaire supplies the answer: it is necessary to return to preformation: "It seems necessary to return to the ancient opinion that all germs were formed at the same time by the hand that arranged the universe; that each germ contains within itself all those that must be born of it; and, whether the germs of animals are contained in the males, or in the females, it is likely that they have existed since the beginning of things, just like the earth, the seas, the elements, the celestial bodies."[67]

In the article "Atheist, Atheism" in the *Philosophical Dictionary* (1764), Voltaire relates preformation to eradicating atheism from the planet: "Above all, let me add that there are fewer atheists today than there have ever been, since philosophers have perceived that there is no vegetative being without germs, no germ without design, etc., and that grain is not produced by putrefaction. Unphilosophical mathematicians have rejected final causes, but true philosophers accept them; and as a well-known author has said, a catechist announces God to children, and Newton demonstrates him to wise men."[68] The well-known author to whom Voltaire refers is himself (see his article, "Theism," published in 1742).

In *On the Singularities of Nature* (1768), Voltaire reiterated his defense of preformation: "And today it has been shown to the eyes and to reason that there is neither vegetable, nor animal that does not have its germ. We find it in the egg of a chicken as in the acorn of an oak. A formative power presides over all these developments, from one end of the universe to the other."[69]

REFUTATION OF TRANSFORMISM

Voltaire thought that it was preposterous that one species could metamorphose into another over millennia. He believed that God had created all species in perfect condition at the time of Creation and that He had not created any new species since then. Moreover, he considered transformism to be a dangerous tool used by atheists to promote their philosophy. It was the fulcrum of the random creation of the materialists and if it could be successfully refuted, the mechanics of their argument would disappear. It was also dangerous to deism because it negated final causes: if transformism is true, then God contradicts Himself all the time and eradicates something that originally must have had a purpose.

In the *Treatise on Metaphysics* (1734) Voltaire calls upon believers in God to reject transformism because it negates final causes: each species is unique and distinct and God made each one with a special purpose in mind. Voltaire begins his argument by declaring that there are two ways to arrive at the notion of a Being that presides over the world. God is demonstrated through the order of the universe and the purpose that each being seems to have.[70] The world's order implies a creative intelligence and final causes. Species do not vary or metamorphose. At the end of *Chapter 1* he declares,

"…a moderately instructed man has never advanced the notion that unmixed species degenerate"[71] and "Therefore, it seems to me that I am quite justified in believing that it is with men as with trees; that pear trees, pines, oaks, and apricot trees do not come from the same tree, and that bearded white men, woolly haired black men, yellow men with manes, and beardless men do not come from the same man."[72] Here he is denying the unity of the human race: this is because the unity of the human race presupposes transformism and a common origin. Voltaire denied that there have been any physical changes since Creation, and so he was forced to deny the common origin of all men. He thought that the white, black, and yellow races are different species of men and that they have been created separately by God. In the 1784–1789 edition of Voltaire's *Works*, Kehl was compelled to add a footnote to Voltaire's statement to remind the reader of Linnaeus and Buffon's definition of species: "All these different races of men produce together individuals capable of reproduction, which we cannot say about trees of different species; but has there been a time when there existed one or two individuals of each race? That is what is completely unknown to us."[73]

In *Chapter 2*, Voltaire reiterates that each species has a purpose or final cause: "We must, above all, reason in good faith, and not seek to fool ourselves; when we see a thing that always has the same effect, that uniquely has only this effect, that is comprised of an infinite number of organs, in which there are an infinite number of movements, all of which cooperate to meet the same end, it seems to me that we cannot, without some hidden reluctance, deny a final cause. The seed of all vegetables, of all animals, is in this category: wouldn't we be a bit rash to say that all this is unrelated to any end?"[74]

Voltaire restates his belief that the different races of men are different species in *Relation touchant un Maure blanc amené d'Afrique à Paris en 1744* (1744). This piece was written after an albino Negro had been exhibited in Paris in 1743 and as a response to Maupertuis' *Dissertation physique à l'occasion du nègre blanc* (1744). Voltaire's comment on the phenomenon of the white Negro was, "Here at last is a new wealth of nature, a species that no more resembles ours than spaniels do greyhounds."[75] Voltaire thought that the albino Negro was a species of man different from that of the other blacks in Africa and from that of white Europeans. He uses the term "species" [*espèce(s)*] eleven times in the piece and each time he employs it, he indicates that he believes that there are multiple species of men. Some of the phrases are "This species is scorned by blacks" [*Cette espèce est méprisée des nègres*], "To blacks they seem to be an inferior species made to serve them" [*Ils paraissent aux nègres une espèce inférieure faite pour les servir*], "We knew about the existence of this species a few years ago" [*Il y a quelques années que nous avons connu l'existence de cette espèce*], "there was a species of men as black as moles" [*il y avait une espèce d'hommes noirs comme des taupes*], "a species that no more resembles ours than" [*une espèce qui ne ressemble pas tant à la nôtre que*], and "There is probably yet still some other species towards Australian lands" [*Il y a encore probablement quelque autre espèce vers les terres australes*].[76]

It is significant that Voltaire finishes the piece by declaring the equal worth of all men: "…if we think that we are worth much more than they, we are very mistaken.[77] This climax to the four page essay reaffirms the deistic ideal, which is to recognize the equal worth of all men and to treat them with dignity and respect, regardless of how they originated. Voltaire devoted his life to fighting intolerance of all kinds, including racial. In the article entitled, "Theism," in the *Philosophical Dictionary*, Voltaire defines the theist as one who says, "I love you," to all of humanity's races: "What is a true theist? It is someone who says to God: *I adore you, and I serve* you; it is someone who says to the Turk, to the Chinese, to the Indian, and to the Russian: *I love you.*"[78] In the article entitled, "Theist," he reiterates that the theist "has brothers from Peking to Cayenne, counting all wise men as his brothers."[79]

In the *Defence of My Uncle* (1767), Chapters 18 and 19, Voltaire again denies the common origin of the various races. In *Chapter 18*, entitled, "Of Men of Different Colours," he defends the Abbé Bazin, whose views questioning the unity of man had been attacked in a journal entitled the *Œconomique*. Voltaire says, "He never thought that English oysters were engendered by the crocodiles of the Nile; or that gilliflowers of the Molucca islands derived their origin from the firs of the Pyrénées."[80] Voltaire repeats what the *Œconomique* journalists had to say about the Abbé: his thesis that there are multiple species of men is as absurd as the supposition of ancient philosophers who thought that there are black atoms and white atoms that have produced black men and white men. Voltaire supports the multiple species theory by observing that autopsies show that the *reticulum mucosum* of the African is black, from head to toe, while that of the European is white; he also cites the example of four red men who traveled to France in 1725; the natives of the New World have skin the color of red copper. This leads one to concur with Bazin that there are varieties of men and that one does not proceed from another.

In *Chapter 19* Voltaire makes another statement that implicitly denies the unity of the human race: "Never lose sight of that great truth that nature never belies itself. All species ever remain in the same situation. Animals, vegetables, minerals, metals, everything is invariable in this great variety. Everything preserves its essence."[81] Again, the implication is that each of the races was created separately.

Jacques Roger advises, "What Voltaire defended was a stable creationism, which could admit only that nature had changed only minimally since its creation. Buffon could not escape his criticism here. In 1767 Voltaire published the *Défense de mon oncle* [Defense of My Uncle], which attached, first of all, the unity of the human species. Such a unity was impossible by definition, since it presupposed that the current races had appeared progressively and therefore that nature had changed. For the same reasons, he then attacked the 'Theory of the Earth' and, in passing, organic molecules."[82]

In *The Man of Forty Ecus* (1768), Voltaire takes the opportunity to ridicule Benoît de Maillet's *Telliamed* (published posthumously in 1748) for its transformist views. Maillet, in the chapter entitled, "Sixth Day. Of the Origin of Man and Animals, and of the Propagation of the Species by Seeds," proffers much transformist material that Voltaire was able to use as propaganda against science. Maillet expounds on how life began in the sea, the interrelatedness of all species, and how flying fish left the water and over time, were transformed into land creatures. Fish developed wings that enabled them to fly and then "the little Wings they had under their Belly, and which like their Fins helped them to walk in the Sea, became Feet, and served them to walk on Land."[83] Land creatures come from their counterparts in the sea: men and women come from mermen and mermaids, apes come from sea monkeys. Intermediary species between men and fish (mermen and mermaids) have been sighted by seafarers and the *Telliamed* recounts many such events. There are different species of men, some with tails and some without. Maillet asks the ultimate question as to whether men could have metamorphosed from apes with tails: "To return to the different Species of Men. Can those who have Tails, be the Sons of them who have none? As Apes with Tails do not certainly descend from those which have none, is it not also natural to think, that Men born with Tails are of a different Species from those who have never had any?"[84]

Voltaire went wild and pounced on Maillet with delight: he would use *The Man of Forty Ecus* as propaganda to ridicule the transformist hypothesis. Maillet's book provided an excellent opportunity to ridicule the wild imagination of contemporary transformists, which he heaped all together. The fact that Maillet was not a transformist, but believed that the ocean carried human eggs to land where they hatched (this made him a panspermist), was irrelevant: transformism was dangerous and now was the time to expose it for the lie that it is. Voltaire mocks Telliamed, "who taught me that mountains and men are made by sea waters. First, there were handsome mermen that later became amphibians. Their beautiful forked tail transformed itself into buttocks and legs. I was full of Ovid's Metamorphoses, and a book where it was shown that the human race was the bastard of a race of baboons. I liked descending from a fish about as much as from a monkey. With time I had a few doubts about this genealogy, and even about the formation of mountains."[85]

In *On the Singularities of Nature* (1768), *Chapter 10*, Voltaire makes it clear that everything has a purpose. An iconic representation of final causes is a great mountain range because it "seems to be an essential part of the world machine, just like bones are to quadrupeds and bipeds. It is around their summits that clouds and snow congregate, which, emanating from there unceasingly, form all the rivers and all the fountains, whose source we have for such a long time and so erroneously attributed to the sea...These mountain ranges that cover both hemispheres have a more sensible use. They strengthen the ground; they serve to irrigate it; they enclose at their bases all the metals, all the minerals. We should notice here that all the pieces of this world machine appear to be made for one another."[86]

In *Chapter 11* he goes on to reiterate his refutation of transformism and takes a jab at Maillet: "The great Being who made gold and iron, trees, grass, man and the ant, has made the ocean and mountains. Men were not fish, as Maillet says; everything has been probably what it is by immutable laws."[87]

In *Chapter 12* he discusses what he believes to be erroneous theories of the metamorphosis of man from marine creatures that have found new life thanks to the discoveries of seashell beds. He compares the unearthing of seashells to sightings of Needham's eels under microscopic magnification. He ridicules Maillet thus: "The same thing has happened with shells as has happened with eels: they have hatched new systems...If the sea were everywhere, there was a time when the world was populated only by fish. Little by little the fins became arms; the forked tail, becoming longer, formed buttocks and legs; finally fish became men, and all that has happened according to shells that have been unearthed. This theory is about as good as the horror of the void, substantial forms, globular, penetrating, channeled, striated matter, the denial of the body's existence, Jacques Aimard's divining rod, pre-established harmony, and perpetual motion."[88] In this Proustian stream of consciousness, Voltaire unleashes every imaginable insult that he can muster at the moment.

"The horror of the void" [*l'horreur du vide*] is a little jab at Pascal, who, in his *Thoughts*, addresses the universal issue of man's terror when confronted by the inevitability of death. Pascal declares, "When I see the blindness and the wretchedness of man, when I regard the whole silent universe and man without light, left to himself and, as it were, lost in this corner of the universe, without knowing who has put him there, what he has come to do, what will become of him at death, and incapable of all knowledge, I become terrified, like a man who should be carried in his sleep to a dreadful desert island and should awake without knowing where he is and without means of escape."[89] Pascal employs the phrases "without knowing...what he will become upon dying" [*sans savoir...ce qu'il deviendra en mourant*] and "I enter into horror" [*j'entre en effroi*]-this is Voltaire's "horror of the void." The tautology of "blindness," "wretchedness," "silent," "without light," "left to himself," "lost," "without knowing," "death," "incapable," "terrified," "sleep," "dreadful," "without knowing," and "without means" hyperbolizes the horror and the void. Pascal forces the reader to confront the issue of horror and dread when faced with the corruption of the human body, as well as the great unknown regarding a spiritual end, if any. Pascal observes that as a result, men, who are wretched and lost beings, seek escape from their terror with diversions. Pascal invites the reader to wager that God exists to extricate himself from this abyss of terror. When the reader believes in Christ, he will undergo a transformation of the spirit: he will find peace in the midst of despair, just as Saint Paul had discovered the presence of God in the midst of turmoil. Voltaire ridiculed Pascal not only because Pascal was a follower of Christ, but also because he invited the unbeliever to wager that God exists. As far as Voltaire was concerned, a wager is unnecessary because the complexity and harmony of the Newtonian universe suffices to prove that God exists.

For Voltaire, the horror of the void is fictional, as is everything else in his enumeration, "horror of the void, substantial forms..." However, Voltaire places Pascal's horror of the void at the beginning of his list of notions that he wants to ridicule because of the association between Pascal and the Catholic Church. For Voltaire, Pascal represented not only a religion that he did not embrace, but a church that tortured its critics on the rack and burned them at the stake. There is no doubt that the execution of the young Chevalier de La Barre on July 1, 1766 was still troubling Voltaire: when the teenager did not raise his hat at a passing procession of Capuchins, he was tried, convicted of blasphemy, his hands were cut off, he was beheaded, and then his body was burned along with Voltaire's *Philosophical Dictionary*. Voltaire tried to get La Barre reinstated, but he was unable to do so. La Barre was finally reinstated by the National Convention of the 25th of Brumaire, Year II (November 15, 1794). Paul Johnson explains, "What made Voltaire hate Pascal was not the latter's awareness of the limitation of reason, for he shared it, but the way in which Pascal was used to defend a Christianity still capable of monstrous cruelty. In 1766 there was a further outrage, when the young Chevalier de la Barre failed to doff his hat in respect while a Capuchin religious procession passed through the streets of Abbeville. (It was raining.) He was charged and convicted of blasphemy, and sentenced to 'the torture ordinary and extraordinary,' his hands to be cut off, his tongue torn out with pincers, and to be burned alive. This atrocious case haunted Voltaire for the rest of his life, and indeed it was a reminder to the European intelligentsia that Catholic Europe, despite the apparent triumph of reason, was still basically unreformed."[90] Hence, Voltaire's hatred of Pascal, the Jansenist sect to which Pascal had belonged, and everything Catholic, far outweighed the horror of the void that Pascal had used as an entrée to the wager that God exists.

The analogy that Voltaire makes between Maillet's transformism and substantial forms is also intended to deride Maillet's metamorphosis of species. The *Oxford English Dictionary* defines "substantial forms" as "the nature or distinctive character in virtue of possessing which a thing is what it (specifically or individually) is.[91] The *OED* also defines the term as "In the Scholastic philosophy: The essential determinant principle of a thing; that which makes anything (*matter*) a determinate species or kind of being; the essential creative quality."[92] *Chambers Cyclopædia* (1728) defines "substantial forms" as "Forms independent of all Matter; or Forms that are Substances themselves."[93] La Mettrie devotes *Treatise on the Soul, Chapter 7*, to substantial forms and clarifies the definition greatly. La Mettrie explains that ancient philosophers distinguished between two kinds of substantial forms in living bodies-those that comprise the organic or physical parts, and those that are considered to be their principle of life, or the soul; the term "soul" was assigned to the latter, and Aristotle distinguished among the vegetative soul, the animal soul, and the rational soul. Voltaire had contempt for metaphysical discussions on the nature of the soul and for scholasticism, which reveled in it. In the article "Soul" [*Ame*] in the *Philosophical Dictionary*, he ridicules Aristotle's tripartite soul and declares the vegetative, the animal, and the rational soul to be false. It is evident that he regarded the subject of substantial forms, and their divisions, the tripartite soul, so preposterous, that he considered the analogy to transformism a great insult to the latter.

The analogy between transformism and "globular, penetrating, channeled, striated matter" [*matière globuleuse, subtile, cannelée, striée*] is an assault upon Descartes, Newton, and science. The *Encyclopédie* defines globulous matter [*la matière globuleuse*] as "matter comprised of detached parts that have the form of small globes."[94] It defines "penetrating matter" [*la matière subtile*] as "the name that Cartesians give to a matter that they imagine crosses and freely penetrates the pores of all bodies, and fills these pores so as not to leave any void or interstice among them."[95] The *Encyclopédie* points out that if such matter existed (perfectly solid and penetrating all things), it would necessarily be heavier than gold and it would contradict the laws of physics.[96]

The *Encyclopédie* mentions that Newton also incorporated penetrating matter into his physics: "Nevertheless, Mr. Newton accepts the existence of a penetrating matter, or of an environment much finer than air, that penetrates the densest bodies & which thus contributes to the production of several of nature's phenomena."[97]

In the *Principles of Philosophy, Part 3*, entitled, "Of the Visible Universe," Descartes attempts to explain the genesis of the universe and the formation of stars with globulous matter. He posits that the universe is full of matter that is perpetually in motion and colliding with other matter. This creates many circular movements and vortices or whirlwinds of tiny particles. In *Sections 46–48 and 86–87*, Descartes posits that small globules of matter circle around and because of centrifugal force, fly away from the center of the vortex towards its circumference. These small globules of matter give rise to the phenomenon of light emanating from a star. Other matter, seeking to penetrate the vortex, becomes channeled or striated in shape and goes on to form spots in the sun.[98] Voltaire thought it utterly preposterous that anyone would try to explain the formation of stars and again, because the physics is purely hypothetical and cannot be proven, he felt that the comparison to Maillet's transformism is à propos and demeaning. Thus, the enumeration "globular, penetrating, channeled, striated matter" is intended as an affront to Descartes, Newton, and all of speculative science.

Voltaire had used the same enumeration, "globular, penetrating, channeled, striated matter," earlier in his literary career, in the *Defence of My Uncle* (1767), *Chapter 19*. Here, he uses analogy to ridicule Needham's animalcules, which are as fantastic and as farfetched, as Descartes' globules: "The seed is useless; everything will grow spontaneously. Upon this supposed experiment a new universe is constructed, in the same manner as a new world was formed a hundred years ago, with a penetrating, globulous, and channeled matter."[99]

In summation, Voltaire viewed science as a death threat to deism. The science of the eighteenth century revolved around random creation, spontaneous generation, transformism, and the unity of all living things which have proceeded from a single prototype; this was insupportable to Voltaire and he devoted the last thirteen years of his life, right up until his death, to denouncing atheism and its partner in crime, science. Jacques Roger sums up, "Voltaire would write more than twenty-five works, treatises, dialogues, putative letters, fictive questions, verse and prose satires, meditations, or tales, all of them directed in whole or in part against atheism and its scientific underpinning…Voltaire was to avenge the honor of God on the backs of eels, mountains, and shells. He would even perform some hasty 'experiments' in order to persuade himself more firmly that the lime pits of Touraine were not the result of fossilized shells and that the polyp was not an animal. For he had understood that these wretched little details called into question the existence of God."[100]

Chapter 9

The Controversy over whether Apes Can Be Taught to Speak

In a word, would it be absolutely impossible to teach the ape a language? I do not think so.[1]

—Julien Offray de La Mettrie, *Man a Machine* (1747)

As travelers voyaged around the world and returned to Europe with stories about strange beasts that resembled men (ie: men with tails), interest in the plenitude and continuity of the great chain of being increased. As naturalists began to take notice of the physical similarities between man and apes, and to note that apes, too, have a larynx and pharynx, the question arose whether they can be taught language. Hence, the debate over whether apes can be taught language arose from the effort to determine which animals were right beneath man on the chain of beings (*scala naturæ*). Materialists such as La Mettrie and Diderot, who held that only one substance exists (matter), and that the universe is the result of random molecular organization, did not see any reason why apes could not be taught to speak: since apes possess the physical apparatus required for speech, all they require is instruction. Rousseau found it significant that the great apes exhibit intelligent behavior and he did not rule out the possibility that perhaps they are a variety of the human species. He recommended that scientific experiments be performed with hybridization to ascertain whether orangutans belong to the same species as man and to find out exactly how intelligent they are. He allowed for the possibility that perhaps apes are like natural man, who did not speak simply because he had no reason to do so.

The question also involved theological polemics and sides were drawn between the monist materialists and the Cartesian dualists. Cartesians argued the notion of *homo duplex*, that is, that man is comprised of two substances, body and soul, matter and mind; intellect arises from the activity of the soul and only man has a soul. Therefore, apes will never be taught to speak. On the other hand, the materialists, who rejected the notion of *homo duplex*, argued that consciousness is the product of the activity of the brain and organs, education, environment, climate, and food. Rejecting the notion of the immortal soul as fictional, they argued that apes, who possess a larynx, pharynx, and trachea, could probably be taught to speak. The answer to this question would determine just how great the gap actually was between man and the other animals, and if it were to be shown that apes could be taught to speak, then it would show that man does not have a special place on the great chain of beings as the head of all animals and that the distance between men and other animals is not as great as had been previously thought. It would indicate that the difference between men and other animals is caused by environmental factors that can sometimes be remedied with instruction. It would raise the question of whether apes, too, have an immortal soul that requires forgiveness and redemption. It would also pose the more basic question of exactly how highly organized a being must be before it could be considered to have a soul. This theological question is cleverly noted by Diderot, who, in the *Sequel to the Conversation* (1769), humorously asks, "Have you seen in the King's Garden, in a glass cage, an orangutan that looks like Saint John preaching in the wilderness?...Cardinal Polignac said to him one day, 'Speak, and I will baptize you.'"[2]

THE PHYSICAL SIMILARITIES BETWEEN MAN AND THE APES

In 1597 an English translation of Philippo Pigafetta's account of apes seen by the Portuguese sailor Odoardo Lopez was published under the title, *A Report of the Kingdome of Congo, a Region of Africa*. Pigafetta's work briefly alludes to the similarities between human and apes that Lopez had noted. In *Chapter 9*, entitled, *The Six Provinces of the Kingdome of Congo, and First of the Province of Bamba*, Pigafetta writes, "*Apes, Monkeys*, and such other kinde of beasties, small and great of all sortes there are many in the Region of *Sogno*, that lyeth upon the River *Zaire*. Some of them are very pleasant and gamesome, and make good pastime, and are vied by the Lordes there for their recreation and to shew them sport. For although they are venerable Creatures, yet will they notably counterfait the countenances, the fashions, & the actions of men. In every one of these Regions abovenamed, there are some of the aforesaid Creatures, in some places mo, and in some places fewer."[3] The salient point of this quote is that multitudes of apes were discovered that delighted the nobility by imitating human gestures. Pigafetta's book also contains drawings of the De Bry brothers in which apes are depicted as resembling chimpanzees-the apes in the drawings have no tail, long arms, large ears, and look anthropomorphic.

In 1613 Samuel Purchas published *Purchas his Pilgrimage, or Relatons of the World and the Religions Observed in all Ages and Places Discovered, from the Creation unto this Present*. In *Book 6, Chapter 1*, entitled, *Of Africa, and the Creatures Therein*, he relates the descriptions of apes that Andrew Battell had seen during his voyages to the Congo. Purchas writes, "But more strange it seemed which hee told mee of a kinde of Great Apes, if they might so be termed, of the height of a man, but twice as bigge in feature of their limmes, with strength proportionable, hairie all over, otherwise altogether like men and women in their whole bodily shape. They lived on such wilde fruites as the Trees and woods yielded, and in the night time lodged on the Trees."[4]

Purchas discusses the resemblance between pongos and men in much greater detail in his subsequent work, *Purchas his Pilgrimes* (1625). In *Book 7, Chapter 3, Section 6*, entitled, *Of the Provinces of Bongo, Calongo, Mayombe, Manikesocke, Motimbas: Of the Ape-monster Pongo, their Hunting, Idolatries; and divers other Observations*," he writes, "The greatest of these two Monsters is called, *Pongo*, in their Language: and the lesser is called, *Engeco*. This *Pongo* is in all proportion like a man, but that he is more like a Giant in stature, than a man: for he is very tall, and hath a man's face, hollow-eyed, with long haire upon his browes. His face and eares are without haire, and his hands also. His bodie is full of haire, but not very thicke, and it is of a dunnish colour. He differeth not from a man, but in His legs, for they have no calf...They sleepe in the trees, and build shelters for the raine. They feed upon Fruit that they find in the Woods, and upon Nuts, for they eate no kind of flesh. They cannot speake, and have no understanding more than a beast. The people of the Countrie, when they travaile in the Woods, make fires where they sleepe in the night; and in the morning, when they are gone, the *Pongoes* will come and sit about the fire, till it goeth out: for they have no understanding to lay the wood together...When they die among themselves, they cover the dead with great heaps of boughs and wood, which is commonly found in the Forrest."[5]

The salient features of Purchas' account deserve summary, as they will be discussed by the abbé Prévost in *Histoire générale des voyages* (Paris, 1746–1789), by Buffon, Rousseau, and in numerous articles in the *Encyclopedia*. Purchas' main points are:

- Pongos look like men, except they have no calves: they are tall, have a man's face, they are hollow-eyed, have eyebrows, the face and ears are hairless, the hands are hairless, and they walk upon their legs.
- They build shelters from the rain. This implies roofs over their shelters and is therefore significant.
- They are vegetarian.
- They cannot speak because they do not have more intelligence than other animals.
- They are attracted to campfires left by humans and they remain by the campfires until the embers die out.
- They do not have the intelligence to lay wood together to keep a campfire going.
- They cover their dead with great heaps of boughs and wood.

After Purchas' account, seventeenth and eighteenth-century naturalists began to pay attention to the physical similarities between man and the apes. Francis Moran, in his article, "Between Primates and Primitives: Natural Man as the Missing Link in Rousseau's Second Discourse," has researched and cited many primary sources.[6] Moran mentions that François Leguat (1708) and Daniel Beeckman (1718) compared apes to human populations. Moran cites François Leguat, who found that the ape resembles the Hottentot: "Its Face had no other Hair upon it than the Eyebrows, and in general it much resembled one of those *Grotesque* Faces which the Female *Hottentots* have at the Cape."[7] Similarly, Daniel Beeckman agreed that the orangutan was "handsomer I am sure than some Hottentots that I have seen."[8] Other writers observed that apes walk like human beings: besides Samuel Purchas (1625), there were Edward Tyson (1699), François Leguat (1708), Daniel

Beeckman (1718), William Smith (1744), Benoît de Maillet (1748), and Buffon (1766).[9] Edward Tyson also gave attention to the "human face" of an ape that he called a "pygmie."[10] Likewise, William Smith discusses the "boggoe" or "mandrill" that bore the "near resemblance of a human creature, though nothing at all like an Ape."[11] Similarly, Rousseau, in the *Discourse on the Origin of Inequality, Note 10*, examines Purchas' account in great detail and reiterates that the pongo "is said to have a human face" and to "resemble man exactly."

Moreover, Moran points out that when artists drew pictures of monkeys and apes for illustrations to be incorporated into naturalists' books, they often embellished upon the human qualities of simians by depicting them standing with a walk stick or cane. Because most Europeans had never come into contact with these primates, they believed, from the drawings in books, that apes had the ability to use human tools and that they were very close to humans on the chain of beings. Moran surmises that it was drawings of apes with walking sticks that suggested to Enlightenment audiences that perhaps the gap between man and other animals was not as great as previously thought: "These drawings help us to appreciate Rousseau's uncertainty over whether such creatures were animals or primitive human beings who had been misidentified by careless observers. Recall that in Note X Rousseau wonders whether "various animals similar to men (i.e., *orang outangs*), which travelers have without much observation taken for Beasts…might not indeed be genuine Savage men whose race…had not acquired any degree of perfection, and was still in the primitive sate of nature." Indeed, this speculation becomes all the more plausible when we take into account European description of primate ethology."[12] Moran also mentions Tyson's observation that when orangutans have the choice of associating with men or monkeys, they prefer the company of men.[13] Tyson concluded that "the *orang outangs* themselves recognized their proximity to human beings."[14]

There were also many articles on apes in the *Encylopedia*. The encyclopedists used the articles on the apes as propaganda to further their materialist agenda. They emphasized the great physical similarities between apes and man in order to narrow the distance between man and the other animals on the great chain of beings and to suggest that the chasm was not insurmountable. In *Man a Machine* (1747), La Mettrie would take the next logical step, which was to declare that it is possible to teach apes to speak.

The unsigned article, "Ape" [*Singe*] (1765), describes and differentiates among 5 species of monkeys, incorporating 38 varieties, paying attention to distinguishing physical characteristics.[15] The author emphasizes and continually reiterates the great similarity between the apes and man: "Most of these animals have more in common with man than do other quadrupeds-above all in their teeth, ears, nostrils, etc. They have eyelashes on both eyelids and two breasts at the chest. Females for the most part have menses similar to women. The front feet are very similar to man's hands; the hind feet also have the form of hands, since the four fingers are larger than those on the front feet and the thumb is large, fat and far removed from the first finger; they also use their hind feet as they do their front to grab and point."[16] A few lines below, describing the monkey, the author reiterates, "they have a great resemblance to man in their face, ears and nails."[17] Next he identifies the "man of the woods" orangutan thus: "It resembles man more than does any other species of ape; its hair is short and very soft."[18]

In the article "Pongo" (1765), Jaucourt describes the pongo according to Purchas and Battell's account: The pongo is "more than five feet tall; it has the height of an ordinary man, but is twice as heavy. Its face is hairless and resembles a man's, it has rather large, sunken eyes, and its hair covers the head and shoulders...they build types of shelters in the trees against the rains with which this country is inundated during the summertime. They live only on fruit and plants; they cover their dead with leaves and branches: the blacks consider this as a kind of sepulchre. In the morning, when pongos find campfires that Africans lit at night, as they travel across the forests, they are seen to approach them with pleasure. Nevertheless, they have never imagined kindling them by throwing wood into them. The Africans assure us that pongos have no language, and that they are not considered to have any degree of intelligence that would make them superior to animals."[19] The important points that Jaucourt captures are that pongos resemble men, they built roofs on their shelters (the build types of shelters against the rain), they cover their dead, and they exhibit pleasure (*on le voit s'en approcher avec une apparence de plaisir*).

THE DEBATE OVER WHETHER APES CAN BE TAUGHT TO SPEAK

Robert Wokler, in his article, "The Ape Debates in Enlightenment Anthropology," traces the history of the controversy over whether apes can learn language back to the seventeenth century.[20] Wokler says, "To be sure, the Enlightenment controversy about apes, men and language stemmed in large measure from an ancient tradition of speculation about the chain of being and about the distinction between the natural and cultural determinants of human behavior which had attracted the interest of both Plato and Aristotle among a host of other thinkers."[21] Wokler points out that during the latter half of the seventeenth century, scholars were trying to determine which animals were right beneath man on the chain of beings. William Petty held that elephants, rather than monkeys, were right beneath man because they were more intelligent. In 1676 Claude Perrault said that although monkeys have a larynx, pharynx, and other speech apparatus, they did not have the intellect that God intended for man alone.[22] In 1699 Edward Tyson, an anatomist who was interested in the physical similarities between chimpanzees and humans, agreed that chimpanzees have all of the essential body parts of man, but that they do not have man's intellect.[23]

The eighteenth century saw more diverse opinions on the subject. The doctor and materialist *philosophe*, La Mettrie, author of *The Natural History of the Soul* (1745) and *Man a Machine* (1747), argued that consciousness is the result of the organization of matter alone: levels of awareness, intelligence, behavior, and a penchant for criminal behavior, are determined by the brain, nervous system, heredity, education, climate, and food, in short, biological and environmental factors. Regarding language, he astutely observed that language does not provide evidence of the existence of the immortal soul because the deaf-mute cannot speak and they are still men. Since La Mettrie did not believe in an immortal soul, and because he noted that man and apes have similar physical apparatus, he contended that apes can be taught to speak.

In *Man a Machine* (1747), La Mettrie compares the size of man's brain to those of other animals. Because he is a doctor and a materialist, he attributes behavior and intelligence to how highly organized the brain is. He observes that in general, the form and structure of the brain of quadrupeds are almost always the same as those of the brain of man. He finds that of all animals, man has the largest brain, followed by that of the monkey, the beaver, the elephant, the dog, the fox, and the cat. Next, after the quadrupeds, birds have the largest brains; fish have very little brain, and insects have none.

La Mettrie establishes a causality between the size of the brain and intelligence and he also observes an inverse relationship between the size of the brain and ferocity and the size of the brain and instinct. He draws the following conclusions: 1) the less brain that animals have, the fiercer they are; conversely, the larger the brain, the gentler the animal, and 2) the more intelligence that an animal has, the less instinct. La Mettrie also establishes a causality between brain size and the ability to learn. He observes that man is not the only animal that can learn things. He points out that the brain and vocal apparatus of birds permit them to vocalize: "Among animals, some learn to speak and sing; they remember tunes, and strike the notes as exactly as a musician."[24] He concludes that since apes have been shown to be unable to learn music, it must be due to some defect in their organs.[25] He is optimistic that one day apes will be taught to speak: "But is this defect so essential to the structure that it could never be remedied? In a word, would it be absolutely impossible to teach the ape a language? I do not think so."[26] Hence, La Mettrie takes the position that apes can be taught language.

He bolsters his thesis by noting the great physical similarity between man and ape: "The ape resembles us so strongly that naturalists have called it 'wild man' or 'man of the woods.'"[27] He is confident that the task could be accomplished if an excellent teacher, someone of the caliber of Amman, took charge of the responsibility, if the subject selected were a young ape (one not too young or too old), and if one selected the ape with the most intelligent face, one that appeared to live up to its look of intelligence by performing a thousand small operations. La Mettrie considers Johann Conrad Amman, a Swiss/Dutch physician, to be the best teacher for the job. Amman was celebrated in his time for teaching the deaf-mute to speak. His method of instruction was to have his pupils observe the motions of his lips, feel the vibrations in his throat with their fingertips, and then to imitate his movements until they repeated distinct letters, syllables, and words.

La Mettrie places the burden of teaching apes to speak squarely on the shoulders of the teacher. La Mettrie admits, with reference to the ape, "Finally not considering myself worthy to be his master, I should put him in the school of that excellent teacher whom I have just named, or with another teacher equally skillful, if there is one."[28]

La Mettrie finds that it is significant that although Amman's pupils could not hear, their teacher was able to instruct them via sight and touch. From this La Mettrie extrapolates that perhaps it might be easier to teach apes to speak than deaf-mutes because they do hear: "…but apes see and hear, they understand what they hear and see, and grasp so perfectly the signs that are made to them, that I doubt not that they would surpass the pupils of Amman in any other game or exercise. Why then should the education of apes be impossible? Why might not the ape, by dint of great pains, at last imitate after the manner of deaf mutes, the motions necessary for pronunciation. I do not dare decide whether the ape's organs of speech, however trained, would be incapable of articulation. But, because of the great analogy between ape and man and because there is no known animal whose external and internal organs so strikingly resemble man's, it would surprise me if speech were absolutely impossible to the ape."[29]

La Mettrie reiterates his confidence that the ape can learn language a few paragraphs later: "Not only do I defy anyone to name any really conclusive experiment which proves my view impossible and absurd; but such is the likeness of the structure and functions of the ape to ours that I have very little doubt that if this animal were properly trained he might at last be taught to pronounce, and consequently to know, a language. Then he would no longer be a wild man, nor a defective man, but he would be a perfect man, a little gentleman, with as much matter or muscle as we have, for thinking and profiting by his education."[30]

For La Mettrie, it is language that differentiates man from the apes, and this difference can be overcome through education. He points out that before man learned the artifice of language, he was indistinguishable from the ape. As a materialist, La Mettrie was a transformist: he did believe that species metamorphose into other species; he believed that with education, apes could become men. The continuum between man and ape is evident: "The transition from animals to man is not violent, as true philosophers will admit. What was man before the invention of words and the knowledge of language? An animal of his own species with much less instinct than the others. In those days, he did not consider himself king over the other animals, nor was he distinguished from the ape, and from the rest, except as the ape itself differs from the other animals, i.e., by a more intelligent face...Words, languages, laws, sciences, and the fine arts have come...Man has been trained in the same way as animals."[31]

Conversely, there were eighteenth-century thinkers, such as Buffon, who were Cartesians, and who held that consciousness is a uniquely human phenomenon and that animals are merely insensate automata. Leonora Cohen Rosenfield explains Cartesian dualism in an excellent study, *From Beast-Machine to Man-Machine; Animal Soul in French Letters from Descartes to La Mettrie*. In this work, Rosenfield examines the genealogy of the Cartesian view that animals have no soul, no intelligence, no free will, and that they are merely machines. She identifies an early statement that Descartes made regarding the nature of animals, one that far antecedes his well known exposition on animals in the *Discourse on Method* (1637): "From the very perfection of animal actions we suspect that they do not have free will."[32] Rosenfield advises that this passage, from *Private Thoughts* [*Cogitationes privatæ*], is Descartes' earliest reference to animals. He jotted it down in his private notebooks some time between 1619 to 1621. He addresses the subject more fully in 1637 in *Chapter 5* of the *Discourse on Method*. Rosenfield cites this quote: "...though there are many animals which manifest more industry than we in certain of their actions, the same animals are yet observed to show none at all in many others: so that the circumstance that they do better than we does not prove that they are endowed with mind, for it would thence follow that they possessed greater Reason than any of us, and could surpass us in all things; on the contrary, it rather proves that they are destitute of Reason, and that it is nature which acts in them according to the disposition of their organs: thus it is seen, that a clock composed only of wheels and weights, can number the hours and measure time more exactly than we with all our skill."[33] This often quoted statement was the foundation of the seventeenth-century view that animal are insensate machines devoid of consciousness, intelligence, and feeling. It would also influence Buffon, who metaphorized animals as machines and clockwork.

Descartes goes on to declare that only man has a soul, which is real and distinguishable from the body: "...we much better apprehend the reasons which establish that the soul is of a nature wholly independent of the body...and finally, because no other causes are observed capable of destroying it, we are naturally led thence to judge that it is immortal."[34] Hence, Descartes recognized the existence of two substances, the extended (that which occupies space) and the unextended (that which does not occupy measurable space, namely, the soul).

Buffon, who embraced Cartesian dualism, held that animals will never be taught to speak and that there is an insurmountable chasm between man and ape. He maintained that animals are mindless, soulless machines that are unteachable because they lack the intelligence that only a soul can bring. In the *Discourse on the Nature of Animals* (1753), he continually employs the metaphor "machine": *la méchanique vivante* (the living machine), *mis en mouvement* (set in motion), *la machine animale* (the animal machine), *cette méchanique* (this machine), *tous les ressorts de la machine animale* (all of the springs in the animal machine), *une machine moins compliquée* (a less complex machine), *en mouvement* (in motion), *le mouvement du corps et des membres* (motion of the body and members), and *le mouvement continuel* (continual movement), are a few examples. Buffon is a Cartesian and believes that animals are purely machines without a soul. He further objectifies animals by reducing them to machines whose springs exhibit either kinetic energy or potential energy: "An animal is distinguished by two modes of existence, that of motion, and that of rest...In the former, all the springs of the machine are in action; in the latter, all is at rest, excepting one part..."[35]

Besides the machine metaphor, Buffon frequently employs the phrase "animal economy" [*économie animale*]. He states that he will examine the nature of the animal world and "study the animal economy in general," and also, "Sleep...serves as a basis to the animal oeconomy," "we must not understand the animal economy, per se," "the animal economy of the oyster," "one is the fundamental part of the animal economy," "the action of the heart and lungs...is the first part of the animal economy," and "if we imagined beings to whom nature accorded only this first part of the animal economy." These are a few examples taken from the first few pages of the *Discourse on the Nature of Animals*. The phrase "animal economy" hyperbolizes the mechanical nature of animals and reduces them to objects governed by Newtonian physics: the universe operates according to laws of economy and the least action required to accomplish a motion. We should recall that Maupertuis was embattled in a controversy over who wrote *The Law of Least Action*. The "principle of least action" states, "When some change occurs in nature, the quantity of action used for this change is always the smallest possible."[36] The "principle of least action" also posits that "in all the changes that take place in the universe, the sum of the products of each body multiplied by the distance it moves, and by the speed with which it moves, is the least that is possible." Hence, Buffon metaphorizes animals as mechanical devices whose cogs, wheels, springs, and pulleys operate according to the principle of least action.

Having established that animals are soulless, mindless machines that operate according to Newtonian physics, Buffon goes on to distinguish man from animals: only man has a soul. He says, "With regard to man, whose nature is so different from that of other animals...the soul participates in all our movements; and it is not easy to distinguish the effects of his spiritual substance from those produced solely by the material part of our frame...But, as this spiritual substance has been conferred on man alone, by which he is enabled to think and reflect, and, as the brutes are purely material, and neither think nor reflect, and yet act, and seem to be determined by motives, we cannot hesitate in pronouncing the principle of motion in them to be perfectly mechanical, and to depend absolutely on their organization."[37] The salient points of this citation are that God has conferred the soul only upon man, that the soul participates in our movements, that it permits man to think and reflect; and that conversely, animals do not have a soul and that that is why they can neither think, nor reflect.

Buffon goes on to point out that "animals never invent, nor bring anything to perfection; of course, they have no reflection; they uniformly do the same things in the same manner"[38] and "may all the actions of animals, however, complicated they appear, be explained, without the necessity of attributing to them either thought or reflection. Their internal sense is sufficient to produce every motion they perform."[39] On what is this internal sense based? Buffon explains, "The internal sense of the brute, as well as its external senses, are pure results of matter and mechanical organization."[40] Man also has the internal and external senses of brutes, but in addition, he has something superior, the soul: "Like the animal, man possesses this internal material sense; but he is likewise endowed with a sense of a very different and superior nature, residing in that spiritual substance which animates us, and superintends our determinations."[41]

Buffon surmises that animals do have a consciousness of their existence and that that awareness is only of the present moment: "Now the power of reflection being denied to brutes, it is obvious, that they cannot form ideas, and consequently, that their consciousness of their existence must be less certain and less extensive than ours; for they have no idea of time, no knowledge of the past, or of the future. Their consciousness of existence is simple; it depends solely on the sensations which actually affect them, and consists of the internal feelings produced by these sensations."[42] Later on he reiterates this point: "...brutes have no knowledge of past events, no idea of time, and, of course, no memory."[43] He goes on to declare that brutes are incapable of comparing sensations and forming ideas from them; animals cannot compare ideas, they cannot form a chain of reasoning, and hence, they cannot deduce abstract truths. Buffon concludes that animals lack these faculties because they have no understanding.

Conversely, Buffon declares that man is *homo duplex*: he is comprised of two natures that are at war with one another because they are opposite in their actions: the mind, or principle of all knowledge, is opposed to the body; man has two natures, body and soul, matter and mind. His intelligence arises from his soul. Jacques Roger explains Buffon's view thus: "The soul was 'thought, reflection,' 'understanding,' 'the mind, memory,' 'reason.' It 'was given to us to know,' not 'to feel.' It constituted the 'principle of knowledge,' what we would call today the whole of our 'cognitive faculties.' Essentially it was 'the power that produces ideas,' that is capable of 'comparing' sensations and therefore of 'judging.'"[44] In 1755 Buffon declared the unbridgeable chasm between man and other animals. God had given intellect to man alone.[45] Hence, because he believed that God gave a soul to man alone, Buffon denied that animals could ever be taught to speak.

Buffon continually reiterated the insurmountable chasm between man and ape: in 1753 in the *Discourse on the Nature of Animals*, in 1758 in *Carnivorous Animals*, and in 1766 in the *Nomenclature of Apes*. Jacques Roger explains, "In all these texts, he wanted to show that man is not an animal, even when the physical resemblance is striking, as with the orangutan. And he always repeated the same necessary sequence: slow growth, therefore long education; durable families, therefore language and thought. In addition, he always maintained his two major assertions: the soul is a 'living immortal force,' a 'divine light,' and although the Creator chose to give man an 'animal form similar to that of the ape,' He suffused this animal body with His divine breath.'"[46]

In 1766 Buffon flatly denied that apes can ever be taught to speak. In the *Nomenclature of Apes*, he declares that the soul, thought, and speech do not depend on the organization of the physical body; they are gifts given to man alone and this explains why orangutans do not speak. When one considers the physical appearance of the orangutan, one might supposes that this animal is the first among apes or the last among men. However, it lacks a soul, thought, intellect, and language, and therefore it cannot be considered to be the last among men: "In the history of the orang-outang, we shall find, that, if figure alone be regarded, we might consider this animal as the first of apes, or the most imperfect of men; because, except the intellect, the orang-outang wants nothing that we possess, and, in his body, differs less from man than from the other animals which receive the denomination of *apes*. Hence mind, reflection, and language depend not on figure, or on the organization of the body. There are endowments peculiar to man. The orang-outang, though he neither thinks nor speaks, has a body, members, senses, a brain, and a tongue perfectly similar to those of man: He counterfeits every human movement; but he performs no action characteristics of man...the Creator has not formed a man's body on a model absolutely different from that of the mere animal. He has comprehended the figure of man, as well as that of all other animals, under one general plan. But, at the same time that he has given him a material form similar to that of the ape, he has penetrated this animal body with a divine spirit. If he had conferred the same privilege, not on the ape, but on the meanest, and what appears to us the worst constructed animal, this species would soon have become the rival of man; it would have excelled all the other animals by thinking and speaking. Whatever resemblance, therefore, takes place between the Hottentot and the ape, the interval which separates them is immense; because the former is endowed with the faculties of thought and of speech."[47]

Jacques Roger finds that it is significant that Buffon distinguishes between people living under the most primitive conditions, the Hottentots, and apes. Having examined the physical characteristics and living standards of the Hottentots in detail, Buffon concludes that there is an insurmountable chasm between the most primitive man and the smartest ape. Jacques Roger summarizes, "...despite everything, 'the distance separating them is immense, since in the inside the savage is filled with thoughts and on the outside with words.' If an ape or any other animal species had possessed these powers belonging only to man, 'this species would have soon become man's rival; enlivened by the mind, it would have prevailed over the others, it would have thought, it would have spoken.' Nothing like this has ever occurred. Even though his origin was purely natural, man's superiority is absolute."[48]

Buffon goes on to solidify his position by addressing the argument that materialists raise, namely, that apes do not speak because they are like imbeciles who are defective in the organization and functioning of the brain. Buffon, a dualist, maintains that it is more than the physical organization of the brain that distinguishes imbeciles from apes. He asks, "Who will ever be able to ascertain how the organization of an idiot differs from that of another man?"[49] He concedes that it is physical organization that defines the imbecile, the delirious man, the healthy man who sleeps, the newborn who does not think yet, and the senile old person who no longer thinks. However, animal offspring learn in a few weeks everything that their parents know, while human offspring require years of education and protection from their families to acquire their knowledge. Hence, there is a qualitative difference between the nature of animals and that of man, that difference is perfectibility, and it is a function of the soul.

Rousseau agreed with Buffon that animals cannot be taught to speak. Rousseau, who examined the origin of language in detail, both in the *Discourse on the Origin of Inequality* (1755) and in the *Essay on the Origin of Languages* (1755), distinguishes between two kinds of language: natural language and artificial. Natural language is that which is exhibited by animals and also by natural man when he articulates the cry of nature. Artificial language is what man devised as he realized his perfectibility: it is more complex and requires abstract thought, something that animals do not have and never will.

In the *Essay on the Origin of Languages, Chapter 1*, Rousseau says that animals that live and work together, such as beavers, ants, and bees, have a natural language for communicating with each other. He emphasizes the point by declaring that he has no doubt about it. He believes that these animals possess a gestured language that speaks only to the eyes. These languages are natural and not acquired; these animals have these languages at birth, they all have them, and everywhere they all have the same language. However, these languages do not require abstract thought; animals, by virtue of their very limited intelligence, can never be taught to speak: "Animals have a structure more than adequate for this kind of communication, yet none of them has ever put it to this use. Here, it seems to me, is a most distinctive difference. Those among them that work and live together, Beavers, ants, bees, have some natural language for communicating with one another, I have no doubt about it. There is even reason to believe that the language of Beavers and that of ants is gestural and speaks only to the eyes. Be that as it may, precisely because these various languages are natural, they are not acquired; the animals that speak them have them at birth, they all have them, and everywhere they have the same one: they do not change languages, nor do they make any progress whatsoever in them. Conventional language belongs to man alone. This is why man makes progress in good as well as in evil, and why animals do not. This single distinction seems to be far-reaching: they say that it can be explained by the differenc in organs. I should be curious to see this explanation."[50] Hence, Rousseau stands in opposition to the materialists, who hold that language is merely a matter of physical organization. On the contrary, it requires abstract thought, which belongs to man alone. We see Buffon's influence here.

In the *Discourse on the Origin of Inequality*, Rousseau surmises that natural man also had language, the cry of nature: "The first language of mankind, the most universal and vivid, in a word the only language man needed, before he had occasion to exert his eloquence to persuade assembled multitudes, was the simple cry of nature. But as this was excited only by a sort of instinct on urgent occasions, to implore assistance in case of danger, or relief in case of suffering, it could be of little use in the ordinary course of life, in which more moderate feelings prevail. When the ideas of men began to expand and multiply, and closer communication took place among them, they strove to invent more numerous signs and a more copious language. They multiplied the inflections of the voice, and added gestures, which are in their own nature more expressive, and depend less for their meaning on a prior determination."[51]

He goes on to demonstrate that language is an artifice created by man when he started to form societies. David Gauthier, in *Rousseau: The Sentiment of Existence*, explains that Rousseau hypothesized that people began to speak when they became aware of others and were concerned with what others thought of them. When humans went to the nearby well to draw water and met and congregated with one another, they began to desire to gain the esteem of others and so, they created signs and sounds to accomplish this. They created language to get something. Natural man was self-sufficient and the sentiment of existence was enough to make his content. When man met others at the well, he turned his attention outward, to the other, and created signs and language to gain the love and/or assistance of others. Rousseau hypothesizes that that is why *aimez-moi* and *aidez-moi* sound similar. Both phrases arise from concern with gaining something from the other and wish fulfillment.

Rousseau explains that the man's first words had a broader signification than they did later on, as they were ignorant of what we recognize today to be the constituent parts of language; hence every single word signified a whole proposition. Later they began to distinguish subject and attribute, noun and verb. If one tree was called "A," another was called "B" because primitives thought that the two trees are not the same, and it required time for them to realize that two trees might have things in common and that they might both deserve the same word to identify them. Then man learned to arrange things by common and specific properties, which meant that he had to understand the distinguishing properties of objects. The need arose for observation and definition.

Rousseau explains that it is these complexities of language and the sophistication of thought that is required to develop language that illustrates why animals can never be taught to speak, not even apes, as smart as they appear to be: "Add to this, that general ideas cannot be introduced into the mind without the assistance of words, nor can the understanding seize them except by means of propositions. This is one of the reasons why animals cannot form such ideas, or ever acquire that capacity for self-improvement which depends on them. When a monkey goes from one nut to another, are we to conceive that he entertains any general idea of that kind of fruit, and compares its archetype with the two individual nuts? Assuredly he does not; but the sight of one of these nuts recalls to his memory the sensations which he received from the other, and his eyes, being modified after a certain manner, give information to the palate of the modification it is about to receive. Every general idea is purely intellectual; if the imagination meddles with it ever so little, the idea immediately becomes particular."[52]

Here we see that Rousseau is arguing that monkeys go from one nut to another because the sensations that their palate receives from the first are relegated to memory and then, when the sight of a second nut appears, their memory calls up the sensation and they are drawn to the second nut. This falls under the notion of instinct and absence of free will that Rousseau discusses elsewhere. He argues that man has free will and animals are governed by instinct. Instinct dictates that animals always behave in the same way in a set of circumstances. Rousseau gives the example of a pigeon starving next to a bowl of meat and a cat, atop a bowl of fruit or grain; conversely, man observes that different animals eat different foods and that consequently, a variety of foods are edible, and so he varies his diet according to what is available at the time. Hence, man associates ideas, while animals do not.

Rousseau proves that apes could never learn language. He declares that general ideas can enter the mind only with the help of words and animals are incapable of forming such ideas, nor ever acquire the perfectibility that depends on them. Hence, animals, which are not perfectible, will never acquire the ability to connect ideas that is the foundation of language. Humans learn and grow over a lifetime; conversely, animals are all that they will ever be after the first few months of life. It is evident that animals lack perfectibility and the implication is that if they are capable of developing language, they would have done so by now. We see the influence that Buffon had on Rousseau, who identified free will and perfectibility as two absolutes that differentiate man from all other creatures.

Diderot, on the other hand, had no doubt that apes were the link between man and other animals on the great chain of beings. He declared as much in the *Elements of Physiology* (1774–1780): "The intermediary between man and other animals is the monkey."[53] This statement is a reiteration of the views that he had expressed in his 1769 trilogy.

In 1769 Diderot explored and developed the notion that not only might apes be taught to speak, but that animals might even be crossbred with humans to create useful hybrids that would behave as servants and make man's life easier. He demonstrates that not only is the plenitude and continuity of the great chain of beings infinite, but that man can engineer at will what nature does randomly. There is no limit to what man can do with biology. Furthermore, as all matter is conscious at every level of organization, animals, too, have aversion, desire, memory, and intelligence.

In the *Conversation between d'Alembert and Diderot*, the character d'Alembert declares that consciousness is based on the memory of one's own actions. Without memory, there would be no "I" because we would experience our existence only at the moment of receiving an impression and there would be no connected string of events. Our lives would be a broken sequence of isolated sensations, we would live a schizophrenic existence of unrelated incidents and live from moment to moment, from birth until death.

The character Diderot maintains that if a conscious being who had the organization needed for memory links together the impressions he receives, and constructs from this a story, that of his life, and so acquires consciousness of self, then he is able to deny, affirm, conclude, and think. Diderot goes on to explain how memory works. He compares the fibers of our organs to conscious, vibrating strings. A vibrating string goes on vibrating long after I has been plucked. Furthermore, it has the ability to make other strings vibrate, and so memory is called up. Man is not the only animal capable of remembering: so is a finch and a nightingale, two examples of conscious beings that sing and remember tunes. Hence, birds have memory, learn tunes, remember them, and warble them, having learned them.

Diderot goes on to show that finches and nightingales are not the only animals that have consciousness: so do chicks. Diderot paints a very moving portrait of a tiny chick breaking out of its shell, living its first moments in the world. Anyone who has ever witnessed the sight of a chick slowly and relentlessly working very hard to break the shell of its egg with its beak and emerging, trying very hard to stand up on its wobbly legs that are unaccustomed to supporting its weight, repeatedly falling, and then finally succeeding in standing up, is filled with awe, amazement, and great empathy for the tiny creature who has worked so hard to enter into what we know to be a hostile world. Diderot captures the beauty, the mystery, and the great pathos of the event: "This creature stirs, moves about, makes a noise-I can hear it cheeping through the shell-it takes on a downy covering, it can see. The weight of its wagging head keeps on banging the beak against the inner wall of its prison. Now the wall is breached and the bird emerges, walks, flies, feels pain, runs away, comes back again, complains, suffers, loves, desires, enjoys, it experiences all your affections and does all the things you do. And will you maintain, with Descartes, that it is an imitating machine pure and simple? Why, even little children will laugh at you, and philosophers will answer that if it is machine you are one too!"[54]

Diderot makes it clear that the newborn chick has consciousness, life, memory, passions, thought. He goes on to reiterate that not just the chick, but every animal has joy, sorrow, hunger, thirst, anger, admiration, and fright.

In the *Sequel to the Conversation*, Diderot blurs the boundaries between species: "They claim to have seen in the Archduke's farmyard an abominable rabbit which acted as cock to a score of shameless hens who seemed quite willing to put up with it, and they will add that they have been shown chickens covered with fur which were the fruit of this bestiality. Of course they were laughed at."[55] While this passage is humorous, it serves as an entrée, one of two, to the real dish. The next entrée to the real dish is the suggestion of creating hybrids of goats and humans to be a fleet-footed race that could act as servants. This specter immediately raises a theological issue: would goat-men have souls that require forgiveness and redemption by a Savior? Julie responds, "That would make a rare hullabaloo at the Sorbonne."[56] And yet, this passage is still not the climax of the story: it is merely an intermediary step leading up to what Diderot is really driving at: teaching orangutans to speak. The character Bordeu asks, "Have you noticed in the Zoo, in a glass cage, the orang-outang that looks like St John preaching in the wilderness?" Julie replies, "Yes, I have seen him." Bordeu continues, "One day Cardinal Polignac said to him: 'If you will speak, old chap, I will baptize you.'"[57]

It is highly significant that Diderot saved the dialogue about the orangutan last. This passage marks not only the end of the *Sequel to the Conversation*, but to the entire trilogy, as well. This is the climax, the philosophical statement, the main point of the whole biological treatise. He blurs the boundaries between man and ape on the chain of beings. He suggests that man does not have a unique place in Creation far removed from all other creatures. He is a transformist and is declaring that one species has metamorphosed from the other. The salient points that Diderot makes with this joke, which is the capstone not only of the trilogy, but the statement that embodies his findings after all of his years as chief editor of the *Encyclopedia*, his attendance at autopsies and surgical operations, his perusal of the scientific journals of the time, are as follows:

- Man does not have a special place at the head of all nature; there is not an insurmountable chasm between man and the other animals.
- We can turn an ape into a man or a man into an ape by modifying prenatal fibers in the embryo, education, environment, climate, food, and applying the fourth dimension, time.
- Once the first ape has learned to speak, Cartesian duality will be laid to rest forever.
- Man can willfully engineer his future and accomplish what nature does by random chance.
- Perhaps there are multiple species of man, some more advanced than others.

Years later he would reiterate his view in one terse statement in the *Elements of Physiology* (1774–1780): "The intermediary between man and other animals is the monkey."[58]

In the next chapter we will see that an even more challenging issue arose from man's voyages around the world: the enigma of the origin of multiple races. How the vast diversity of man's physical characteristics could have derived from a single ancestral pair was problematic. Some naturalists thought that all of humanity arose from a single ancestral pair, while others did not.

Furthermore, the diversity of races entailed more than just biological questions: the transatlantic slave trade was an evil against which the *philosophes* battled throughout their lives. The *philosophes* often used their biological treatises as an opportunity to exercise their masterful skills in composing vociferous manifestos against slavery and racial injustice; conversely, they also used their political manifestos to promulgate bold new biological premises. They voiced outrage at the way that Europeans invaded the lands of indigenous peoples, merchandized slaves, and brutalized and killed natives for crimes as trivial as stealing a worthless trinket. They declaimed slavery as being contrary to religion, ethics, and natural law.

The *philosophes* also used the races discovered in Africa, Asia, the New World, and the Pacific islands, as a foil to caustically criticize the evils of European civilization, not the least of which were slavery and the inequality inherent in distinct social classes. They continually reiterated that slavery does not exist in far away lands such as Tahiti. At one point, in the middle of a political and social manifesto (the *Supplement to the Voyage of Bougainville*), Diderot presciently suggested that some races are older than others. With astonishing acuity, he maintained that the Oriental race is younger than the Caucasian race! He implied that Tahitians belonged to a younger race that was newer on the planet and that had not yet developed the vices that had arisen in the Caucasian race: "The Tahitian is close to the origins of the world and the European near its old age. The gulf between us is greater than that separating the new-born child from the decrepit dotard."[59] We will see that Diderot, in the *Supplement to the Voyage of Bougainville*, would utilize race as a platform to denounce European imperialism, slavery, murder, cruelty, and a lot more.

Chapter 10

Race

If you think that you have the authority to oppress me because you are stronger and more clever than me, then do not complain when my forceful arm rips open your breast to find your heart...[1]

—Diderot, *History of Two Indias* (1770–1780)

THE GENEALOGY OF THE CONCEPT OF RACE

Eighteenth-century naturalists recognized that by classifying similar entities, they could acquire a better knowledge of them: they could readily observe similarities in structures, quickly retrieve information about relationships that exist, analyze the information, and draw conclusions. Classification was used to represent what was known and to generate a new cycle of experimentation, comparisons, and theorizing.

As travelers explored the world and became aware of the great diversity of humankind, scientists and philosophers tried to compile the information into a meaningful classification. Whenever they compared races, they identified and compared physical characteristics, perceived intelligence, personality, customs, climate, and geographical location.

In 1684 the French medical doctor and traveler, François Bernier, classified humanity according to four or five races in an anonymous work entitled, *A New Division of the Earth, according to the Different Races of Men who Inhabit It*.[2] He used the term "race" [*race*] and "species" [*espèce*] interchangeably. Bernier opens by recalling that in the past, geographers had divided the earth according to countries or regions. He proposes something new: to divide the earth according to the physical characteristics of the people indigenous to each region; he also declares that certain distinguishing physical characteristics are undeniable. His system of racial classification is based on skin color, facial profile, and the texture and color of hair. Observing these physical traits, he distinguished among "four or five races of people, whose differences are so obvious, that by right these should be used as a basis for a new division of the world."

Bernier distinguishes among the white color (generally indigenous to Europe and parts of Africa and Asia), and the black color (most of Africa), which he presciently points out is not caused by the sun because if Africans are transported to a cold country, their children are still born black. Bernier hypothesizes that the cause must be sought in the "seed" or in the "blood." This was highly intuitive, especially since he articulated this a century before Maupertuis would posit the inheritance of parental elements and conversely, Buffon would go down the road of acquired characteristics, argue the effect of climate, and assert that if blacks moved to a cold climate, after many generations, their progeny would turn white. Bernier also stated that a third race was the group residing in the Orient, and a fourth race was the Lapp, comprised of "little stunted creatures with thick legs," who are "very ugly."

Bernier was not sure whether the number of races could be pegged at four or five. He opens by stating that there are four or five, but, after enumerating Europeans, Africans, Orientals and Lapps, he says that the olive color and facial features of Native Americans are not significantly different from those of Europeans so as to cause them to be categorized as a race distinct from that of the European. He points out that among Europeans, stature, facial profile, and hair color vary, as they do in other parts of the world. As an example he cites the Hottentots of the Cape of Good Hope, who seem to be of a different species from the rest of Africa. The reader draws the conclusion that Bernier was uncertain about Native Americans and Hottentots because he mentioned them last, apart from the previous groups, and that since he opened by stating that there are four or five races, that he was leaning toward classifying Native Americans as part of the European group, and Hottentots, as a distinct race of their own.

In 1753 Buffon, in the *History of the Ass*, declared that individuals belong to the same species if they can produce fertile offspring together. He articulated the definition of species thus: "...we can also unite into one species two successions of individuals who reproduce by mixing"[3] and "Species being thus confined to a constant succession of individuals endowed with the power of reproduction..."[4] Conversely, Buffon declared that individuals belong to different species if they cannot produce fertile offspring together: "...we can draw a line of separation between two species, that is, between two successions of individuals who reproduce, but cannot mix..."[5] and "Every species, every succession of individuals, who reproduce and cannot mix, shall be considered and treated separately; and we shall employ no other families, genera, orders, and classes, that what are exhibited by Nature herself."[6]

Buffon went on to observe that all the different races of man belong to the same species because they can produce fertile offspring together: "But these differences in colour and dimensions prevent not the Negro and White, the Laplander and Patagonian, the giant and dwarf, from mixing together and producing fertile individuals; and, consequently, these men, so different in appearance, are all of one species, because this uniform reproduction is the very circumstance which constitutes distinct species."[7]

In 1755 Diderot, in his unsigned article, "Species (*Natural History*)" [*Espèce (Histoire naturelle)*], in the *Encyclopedia,* cites verbatim Buffon's definition of species from the *History of the Ass,* using quotation marks and a reference: "...it is the constant succession and uninterreupted renewal of these individuals that constitute it...we can always draw a line of separation between two *species*, that is to say, between two successions of individuals that reproduce and cannot mix, as we can also combine into a single *species* two successions of individuals who reproduce themselves while mixing...*Species,* therefore, being nothing other than a constant succession of similar individuals & who reproducing themselves...M. de Buffon, *Nat. Hist., Gen. & Part. &c. v. 4, p. 784 & thereafter.*"[8] It must be noted, however, that while Buffon's definition of species appears in the *Histoire naturelle* in 4:384–86, Diderot erroneously refers to p. 784. Diderot also cites Buffon profusely in his articles, "Animal" [*Animal*] (1751), "Man (*Natural History*)" [*Homme (Histoire naturelle)*] (1765), and "Human Species (*Natural History*)" [*Humaine espèce (Histoire naturelle)*] (1765).

In 1758 Linnaeus classified man into seven races in the *Systema naturæ.* Robert Bernasconi identifies the salient points in Linnaeus' classification thus: "So one finds in the tenth edition of 1758, after the feral or wild man, the following classes: *Homo americanus,* who was allegedly obstinate, content, free, and governed by habit; *Homo europæus*, who was allegedly gentle, very acute, inventive and governed by customs or religious observances (*ritus*); *Homo asiaticus*, who was allegedly severe, haughty, covetous, and governed by opinions; and *Homo africanus*, who was allegedly crafty, indolent, negligent and governed by caprice."[9] Linnaeus identified feral or wild man as *Homo ferus*, who is four-footed, unable to speak, and covered with hair [*tetrapus, mutus, hirsutus*].[10] He also listed monsters [*Monstrosus*] and Troglodytes [*Troglodytes*] as races.[11]

In 1765 Diderot, in his unsigned article, "Human Species (*Natural History*)," in the *Encyclopedia*, affirms the unity of the human race. After describing distinguishing physical characteristics and customs of different races around the globe, he concludes that there is only one human race, comprised of individuals who are more or less tanned: "From what precedes it follows that in the entire new continent that we have just traversed, there is only one and the same race of men, more or less tanned. Americans have the same origin. Europeans have the same origin. From north to south we see the same varieties in both hemispheres. Therefore, everything goes to prove that humankind is not comprised of essentially different species. The difference between whites and browns arises from food, morals, customs, climate; that between browns and blacks has the same cause. Therefore, originally there was only one race of men, which, having multiplied and spread across the surface of the earth, gave rise in the long run to all the varieties that we have just mentioned..."[12]

In 1766 Buffon, in the *Degeneration of Animals*, explores ways that man can tamper with race and undo what nature has done. For example, because he believed in acquired characteristics, he thought that with time, white people who have been transported from the north to the Equator would become brown or black and vice-versa. The process could be accelerated if they abandoned the food from their native climate and ate only food from the country to which they were relocated.

Buffon proposes that another way to change skin color is to interbreed the races. He surmises that if "man were forced to abandon those climates which he had invaded, and to return to his native country, he would, in the progress of time, resume his original features, his primitive stature, and his natural colour. But the mixture of races would produce this effect much sooner. A white male with a black female, or a black male with a white female, equally produce a mulatto, whose colour is brown, or a mixture of black and white. This mulatto intermixing with a white, produced a second mulatto less brown than the former; and, if the second mulatto unites with a white, the third will leave only a slight shade of brown, which will entirely vanish in future generations. Hence, by this mixture, 150, or 200 years, are sufficient to bleach the skin of a Negro. But, to produce the same effect by the influence of climate alone, many centuries would perhaps be necessary."[13]

Buffon also thought that a race of men or a species of animals could degenerate due to climate, geography, and food. Taking animals out of their lands of origin, domesticating them, working them too hard, and giving them food not indigenous to their lands of origin, caused the degeneration of their species. Domestication causes degeneration because it prevents animals from exercising and eating food that is suitable for them in the wilderness, which is their natural habitat. Their weakened state, which is acquired, is passed on their offspring through inheritance.

Buffon thought that man was originally white and that his land of origin is Europe; as he traversed the planet and became exposed to extreme weather conditions, his skin color, facial features, and height, underwent changes. He regarded these changes as evidence of degeneration. However, despite these differences, Buffon reiterated that all the different races of man constitute one species: "And, after many ages had elapsed; after he had traversed whole continents, and intermixed with races already degenerated by the influence of different climates; after he was habituated to the scorching hears of the South, and the frozen regions of the North; the changes he underwent became so great and so conspicuous, as to give room for suspecting, that the Negro, the Laplander, and the White, were really different species, if, on the one hand, we were not certain, that only one man was originally created, and, on the other, that the White, the Laplander, and the Negro, are capable of uniting, and of propagating the great and undivided family of the human kind."[14]

RACE IN EIGHTEENTH-CENTURY THOUGHT

Many *philosophes* believed that all of mankind originated from a single ancestor or ancestral pair. In the nineteenth century the term "monogenesis" would be coined to express this idea. Christians were monogenecists and they pointed to the Genesis account in the Bible to uphold the notion that all humans are of a single origin; they argued the authority of the Word of God. Furthermore, atheist materialists such as Diderot and Maupertuis also argued monogenesis: they hypothesized that random errors in the generative process could explain the origin of all races, species, even kingdoms from a single prototype. Conversely, there were thinkers who hypothesized the origination of mankind from a number of different ancestors. In the nineteenth century, this concept would be called polygenesis. Voltaire was a proponent of the independent origin of the different races.

Although the eighteenth-century dictionary did not yet have the terms "monogenesis" and "polygenesis" to identify these ideas, the *philosophes* had devised the concepts and argued over them. For example, in the *Treatise on Metaphysics* (1734), Voltaire makes a polygenecist statement. He declares that the multiplicity of races could not have posibly have metamorphosed from a single ancestral pair: "Therefore, it seems to me that I am quite justified in believing that it is with men as with trees; that pear trees, pines, oaks, and apricot trees do not come from the same tree, and that bearded white men, woolly haired black men, yellow men with manes, and beardless men do not come from the same man."[15] Voltaire thought that it was absurd to think that all men, who have greatly diverse physical characteristics, could be traced back to the same mother and father. Because Voltaire denied that there have been any physical changes since Creation, he was forced to deny the common origin of all men. It is important to note, however, that although polygenecists believed that all humans are not of the same origin, they were not necessarily racists or in favor of slavery. For example, Voltaire, who believed that different races are different species of man, devoted his life to opposing slavery and injustice.

On the other hand, Maupertuis was a monogenecist and in 1745, in *Physical Venus*, he hypothesized that the first humans were all of the same color, and that after many generations, errors in the generative process caused the appearance of all the different hues of skin color that exist today. The second part of *Physical Venus* had been published the previous year under the title, *Dissertation on the Origin of Blacks*. In this work Maupertuis points out the inadequacies of the preformation theory (its inability to explain hybrids or monsters), promotes the theory of epigenesis (because it can explain the chance origin of the different colors of man, as well as hybrids and

Race

monsters), and hypothesizes that random errors in the generative process can explain the vast diversity that we see in nature.

Maupertuis begins by taking a stand against his contemporaries who adhere to preformation. Preformation cannot explain the origin of the different colors of man, hybrids, monsters, or why inherited traits often skip a generation: "Therefore, I apologize to modern physicists if I cannot accept the systems that they have so ingeniously conceived. For I am not among those who believe that we can advance physics by adhering to a system despite some phenomenon that is clearly incompatible with it; & who, having noticed some aspect that will inevitably lead to the downfall of the edifice, nevertheless finish building it, & inhabit it with as much assurance, as if it were the most solid in the world."[16]

Next Maupertuis points out that the laws of attraction (vital laws and vital forces), hypothesized by chemists on a small scale and astronomers on a grand scale, may play a critical role in generation, as well. He points out that "the most famous chemists today admit Attraction, and extend its function farther than did the astronomers. Why should this force, if its exists in nature, not be involved in the formation of animal bodies?"[17] Hence, he begins by positing that the seeds circulating throughout the bloodstream of the father and mother, that would one day form the offspring, have certain affinities or attraction to each other and to other biological elements. This is not unlike gravity or chemical attraction. These affinities or attractions must be the basis of biological theory.

By the time that Maupertuis concludes *Part 1* of *Physical Venus*, he has shown the flaws in preformation theory and he has explained that there must exist affinities or attraction that biological elements have for one another. This serves as the basis for *Part 2*, which is entitled, *Dissertation on the Origin of Blacks*.

Before Maupertuis begins his examination of the origin of the black race, he has something to say about racism and passionately so. He begins *Part 2, Chapter 1* by unleashing a scathing criticism of the white European's attitude towards blacks. The invective states, "If the first white men who saw blacks had found them in the forests, perhaps they would not have accorded them the name of men. However, those that were found in large cities, governed by wise queens, who made the arts and sciences flourish, who built the temple of Jupiter Ammon in times when nearly all other peoples were barbarians, these Blacks may well have not been able to regard Whites as their brothers."[18] Maupertuis dispenses with the stereotype that blacks live only in jungles by pointing out that "those that are found in large cities" are responsible for landmark achievements. One such feat is an architectural one, namely, the construction of the Temple of Jupiter Ammon. Actually, there are two temples of Jupiter Ammon that are noteworthy: one is in Karnok, Luxor (Egypt) and it owes its huge size to 1,300 years of construction. It was started by the 12th Dynasty's Sesostris I and was continually built through the Ptolemic period. The other Temple of Jupiter Ammon is at Siwa in the Libyan desert and its inscriptions date from 4th century BC. In this passage Maupertuis is showing the great subjectivity inherent in ethnocentrism and he is demonstrating contempt for that of the white European.

In the enumeration, "governed by wise queens," "made the arts and sciences flourish," and "built the temple of Jupiter Ammon," Maupertuis is hyperbolizing the intellectual and cultural achievements of blacks. He further hyperbolizes their accomplishments by undercutting those of the white European: "when nearly all other people were barbarians" and "not be able to regard whites as their brothers."

It is significant that Maupertuis puts an asterisk after "who were governed by wise queens," and in a footnote, refers the reader to Diodorus of Sicily. This reference to Diodorus substantiates his commentary on the achievements of African peoples. Diodorus Siculus was a Greek historian residing in Argyrium in Sicily circa 80–20 BC. In *Book 3* of his 40 volume *History*, Diodorus chronicles the achievements and customs of the Ethiopians, Libyans and Atlantians, their great sense of honor and loyalty to their kings, and their adaptiveness and ability to make the most of available natural resources . For example, *Book 3, Chapter 2* says that the Ethiopians were not immigrants from abroad, but that they were natives of the land and that they bear the name of "autochthones" (sprung from the soil itself); this indicates that this race originated in Africa and that it was not an immigrant white race turned black, as the *philosophes* (even Maupertuis himself, further on in *Physical Venus*) would argue is probable; the Ethiopians enjoyed a state of freedom and peace with one another. *Chapter 4* discusses the Ethiopians' system of writing, hieroglyphics. In *Chapter 5* Ethiopian priests select a king from the noblest men from their own number. *Chapter 7* says that Ethiopians' loyalty to their king is so great that if the king has been maimed in some part of his body, all of his companions maim themselves in the same part of their body because it would be a disgraceful thing if the king had been maimed in his leg and his friends should be of sound limb; it is customary for the comrades of kings to commit suicide when the king dies, out of loyalty; for this reason conspiracies against the king are rare because all his friends are equally concerned both for his safety and their own. *Chapters 12–14* describe the mining of gold. *Chapters 15–20* and *22* describe a people called the "fish-eaters," how they harvest fish, prepare it for consumption, and also, leave their dead on the shore so that the tide would carry the bodies out to the fish, and so, be part of a cycle that continues for all eternity. *Chapter 21* describes the "turtle-eaters," how they catch huge turtles, eat them, and use their oversize shells as boats. *Chapter 23* discusses the methods of survival of the "root-eaters"; *Chapter 24*, the "wood-eaters"; *Chapter 25*, hunters who hunt because their land is not suitable for agriculture; *Chapter 26*, "elephant-fighters," who hunt and eat elephants; *Chapter 28*, "bird-eaters"; *Chapter 29*, "locust-eaters"; *Chapters 53–55* describe the Amazons, a race in Libya ruled by women who practiced the art of war and were required to serve in the army for a fixed period; the queen of the Amazons, Myrina, with an army of 30,000 foot-soldiers and 3,000 cavalry, defeated the Atlantians, Syria, Taurus, and some islands, such as Lesbos. Hence, Maupertuis' invitation to the reader to peruse the *History* of Diodorus is an invitation to learn that white Europe does not have a monopoly on intellect, achievement, creativity, or adaptation when faced with survival. Furthermore, it is a feminist statement attesting to the "rule of wise queens" and the valiant Amazons chronicled in Diodorus, *Book 3, Chapters 53–55*, who surpass the achievements and valor of European women.

After opening with a caustic commentary on whites and a footnote on Diodorus, Maupertuis devotes the rest of *Part 2, Chapter 1*, to enumerating the various races that inhabit the earth and to describing their diverse physical characteristics. Then, in *Chapter 2*, he sets out to explain how all of humanity arose from a single pair of parents, and how errors in the generative process could explain the great diversity that we see today.

Maupertuis begins *Chapter 2* by declaring, "All these peoples that we have just covered, such diverse men, have they come from the same mother? We cannot doubt it. What remains for us to examine, is how, from a single individual, so many different races could have arisen. I will venture some speculation."[19] He goes on to set forth the theory of epigenesis, or the unfolding of successive accretions, which could explain the origin of different colors of men, hybrids, and monsters. In epigenesis, the offspring is a unique new individual, rather than a copy of either parent. Maupertuis attributes heredity to errors that occur in the arrangement of patterns of parental elements supplied by either parent; once an error has occurred, it is repeated in subsequent generations, but not necessarily in every generation without skipping one or more.

Maupertuis believed in metamorphoses that would one day be called "acquired characteristics": the eighteenth century believed that every part of the parents' bodies contributed seeds that would eventually form that part of the embryo's body from which they came. That is why he believed that prolonged exposure to the sun would create black offspring. If the parents' skin turned black from the sun, and that skin contained seeds that would one day go on to form the skin of the offspring, then acquired characteristics can be passed on from generation to generation.

Maupertuis was intrigued by the fact that traits disappear in offspring and then reappear generations later. Observing that birth anomalies often skip a generation, he hypothesized that some of the elements circulating throughout the parents' bodies that find their way to the offspring must come from their parents (the offspring's grandparents). Parental elements are passed along from generation to generation, but do not always become manifest. They seem to have disappeared, only to reappear in some future generation. They may not be seen, but they are passed along, nevertheless: "These varieties, if we could trace them, perhaps would have their origin in some unknown ancestor. They perpetuate themselves in repeated generations of individuals that have them, and disappear from generations of individuals that do not have them. But, what is perhaps even more astonishing, is that after an interruption of these varieties, to see them reappear, to see the offspring who resemble neither his father, nor his mother, born with the traits of his ancestor."[20]

Nature contains the basis of all varieties, but it is random chance or, conversely, the art of men, that set them in motion. With careful breeding or "art" humans could engineer the appearance of species from the distant past: "Nature contains the basis of all these varieties: but chance or art set them in motion. It is thus that those whose industry is applied to satisfying the taste of the curious, are, so to speak, the creators of new species. We see the appearance of species of dogs, pigeons, canaries, that did not previously exist in nature. They had at first only been fortuitous individuals; art and repeated generations have made species of them."[21]

In *Part 2, Chapter 6*, Maupertuis observes that black children are born from white parents and white children are born from black parents. However, he finds it significant that black children are rarely born from white parents and that white children are more often born from black parents, such as the albino Negro that had been exhibited in the Parisian salons. Unfortunately, he erroneously concludes that the first race must have been the white race and that the black race is derived from a hereditary error in white parents: "Therefore, it seems to be proven that if blacks are born from white parents, these births are incomparably more rare than the births of white children from black parents. Perhaps that would suffice to lead us to think that white was the color of the first men, & that it is only by some accident that black became an inherited color in the large families that populate the Torrid Zone; among which however the primitive color has not been so perfectly eradicated that it does not sometimes reappear."[22]

It is to Maupertuis' great credit that he presciently attributed the origination of race to random chance. He employs the language of probability and games of chance throughout *Part 2*: chance [*le hasard*], accident [*accident*], probability [*la probabilité*], rare [*rares*], reappear sometimes [*reparoisse quelquefois*], and fortuitous combinations [*combinaisons fortuites*]. All varieties that we see in nature (races, monsters, hybrids) perpetuate themselves because a chance or random error in the combinations of parental elements repeats itself: "Nature contains the bases of all these varieties: but it is chance or art that set them going [*La Nature contient le fonds de toutes ces variétés: mais le hasard ou l'art les mettent en œuvre.*]; "He considers this whiteness as a disease of the skin; according to him it is an accident, but an accident that perpetuates itself and exists for several generations" [*Il regarde cette blancheur comme une maladie de la peau; c'est selon lui un accident, mais un accident qui se perpetue & qui subsiste pendant plusieurs genérations.*]; "For one considers this whiteness as a disease, or such accident" [*Car qu'on prenne cette blancheur pour une maladie, ou pour tel accident*]; "Now to explain al these phenomena: the production of accidental varieties..." [*Pour expliquer maintenant tous ces Phénomenes: la production des variétés accidentelles...*]; "Chance, or the scarcity of family traits will sometimes be of other combinations..." [*Le hasard, ou la disette des traits de famille seront quelquefois d'autres assemblages...*]; "These productions are at first only accidental..." [*Ces productions ne sont d'abord qu'accidentelles...*]; "...that a new accident would be necessary to reproduce the original species." [*...qu'il faudroit un nouveau hasard pour réproduire l'espece originaire.*]; "Now if these marvels sometimes occur, the probability that they would sooner occur among children of the lower classes than among children of upper classes is immense..." [*Or si ces Prodiges arrivoient quelquefois, la probabilité qu'ils arrivoient plutôt parmi les enfans du peuple que parmi les enfans des grands, est immense...*]; "...if these Phenomena are not even more rares..." [*...si ces Phénomenes ne sont pas même fort rares...*]; "...these births are incomparably more rare..." [*...ces naissances sont incomparablement plus rares...*]; "...and that it is only by random chance that black became an inherited color..." [*...que ce n'est que par quelque accident que le noir est devenu une couleur héréditaire...*]; and "...that is does not sometimes reappear." [*...qu'elle ne reparoisse quelquefois.*]; "...are only fortuitous combinations..." [*...ne sont que des combinaisons fortuites...*].

It is also to Maupertuis' credit that he used mathematics to calculate the statistical probability that albinism would occur in a given population. In *Part 2, Chapter 6*, he observes that more black children are born from white parents among the lower classes than among the upper classes; then he calculates that for every black child born from a white nobleman, there would have to be 1,000 black births among the general populace. During his scientific career he would go on to calculate the statistical probability that polydigitism would occur in a given population.

After Voltaire read Maupertuis' work, he decided to reiterate his belief that the different races of men are different species in *Relation touchant un maure blanc amené d'Afrique à Paris en 1744* (1744). This piece was written after the albino Negro had been exhibited in Paris in 1743 and as a response to Maupertuis' *Dissertation physique à l'occasion du nègre blanc* (1744). Voltaire's comment on the phenomenon of the white Negro was, "Here at last is a new wealth of nature, a species that no more resembles ours than spaniels do greyhounds."[23] Voltaire thought that albino Negroes belong to a species different from that of the other blacks in Africa and from that of white Europeans; he uses the term "species" eleven times in the piece to hyperbolize his thesis that there are multiple species of men. However, despite his polygencist beliefs, Voltaire is not a racist and it is significant that he finishes his work by declaring the equal worth of all men: "...if we think that we are worth much more than they, we are very mistaken.[24]

In 1753 Maupertuis, in the *Essay on the Formation of Organized Bodies*, reiterates the role of random errors on creation. However, now he expands his thought to explain not just the appearance of new races, hybrids, and monsters, but new species and even kingdoms, as well: "Could one not explain by that how from two single individuals, the multiplication of the most dissimilar species could have resulted? They would owe their first origin only to some chance productions in which the elementary parts would not have retained the order that they had in the father and mother animals: each degree of error would have made a new species; and with repeated deviations there would arise the infinite diversity of animals that we see today, which perhaps will increase more with time, but to which the course of centuries will perhaps bring only imperceptible developments.[25] Here Maupertuis presciently hypothesizes that elementary parts would not have retained the order that they had in the father and mother animals. He also originates and emphasized the importance of the arrangement or pattern [*ordre*] of parental elements.

THE PHILOSOPHES' WAR ON RACISM

Robert Bernasconi, in the first chapter of his book, *Race*, traces the genealogy of racism in European thought: "One need only think of the purity of blood statutes of fifteenth-century Spain that were used against the *conversos*, Jews who had converted to Christianity but who were still not accepted. Then there were the debates in sixteenth-century Spain when the opponents of Bartolomé de Las Casas justified the mistreatment of Native Americans on the grounds that they were not human. One can also look at the Atlantic trade in African slaves that began in the sixteenth century and was already a large operation in the seventeenth century. It was possible for the Spanish or the English to exploit Jews, Native Americans, and Africans, as Jews, Native Americans, and Africans, without having the concept of race let alone being able to appeal to a rigorous system of racial classification. We have no difficulty identifying these as cases of racism, but they were not sustained by a scientific concept of race."[26]

The Black Code [*Code Noir*], an edict of the King of France dated March 1685, contains 60 articles concerning laws governing slaves in the French West Indies. Article 1 begins by having the King's officers evict all Jews who have taken residence in the French West Indies.[27] Article 2 ordains that all slaves be baptized and instructed in the Roman Catholic religion.[28] Article 11 prohibits priests from presiding over the marriages of slaves who do not have the consent of their owners.[29] Article 12 specifies that children born from slave marriages will be slaves themselves and will be the property of the wife's owner and not the husband's, if the wife and husband have different owners.[30] Article 33 states that a slave who strikes his owner or the spouse or children of his owner, causing a contusion or release of blood, will be punished by death.[31] Article 35 states that certain thefts (ie: horses, asses, cows) will be punished by afflictive punishment, even death, if the case requires it.[32] Article 44 says, "We declare slaves to be movable goods" [*les biens meubles*].[33]

In response to the injustices of the times, Buffon, Diderot, and Maupertuis, declared the unity of the human race. They pointed out that the skin color of man is an infinite continuum that very gradually changes as one travels away from the Equator towards the poles; there is one human race and man is more or less tanned than his brothers. What disturbed the *philosophes* deeply was man's inhumanity to man that was rationalized by racial or religious differences. The *philosophes* defended the integrity of the "other"; they were abolitionists and passionately so. We have seen that in 1745, Maupertuis, in what was essentially a biological treatise on heredity and epigenesis, took the opportunity to boldly point out that the Africans who lived in large cities accomplished much to lay the building blocks of civilization and that women were "wise queens" and, via the footnote to Diodorus, fearless warriors and conquerors.

The *philosophes* employed various rhetorical strategies to denounce slavery: these included satire and irony and conversely, blunt, outright condemnations of slavery as being contrary to natural law. Montesquieu, in the *Persian Letters* (1721), overtly denounces slavery as being contrary to natural law, from the First Eunuch's *Letter 9* to Roxane's suicide letter (*Letter 161*). It is significant that the laws of nature are mentioned in both of these letters. In *Letter 9*, the First Eunuch reveals the tragedy of the wasted life of one who has been mutilated so that he can tend to the sultan's seraglio. Every day he lives in misery in a private hell. He writes to his friend, Ibbi, that he, Ibbi, is fortunate to be able to travel with his owner and see provinces and kingdoms; the First Eunuch, on the other hand, is "shut inside this dreadful prison," a eunuch surrounded by beautiful women, whose life is analogous to the torments of Tantalus; everything that has been done to him, and that he has agreed to, in order to avoid more strenuous work, is contrary to natural law. Consequently, his existence is a wasted life lived against nature.

In Roxane's suicide letter (*Letter 161*), she also proclaims that slavery is contrary to the laws of nature: "How could you have thought me credulous enough to imagine that I was in the world only in order to worship your caprices? that while you allowed yourself everything, you had the right to thwart all my desires? No: I may have lived in servitude, but I have always been free. I have amended your laws according to the laws of nature, and my mind has always remained independent."[34]

In the *Spirit of Laws* (1748), *Books 15, 16, and 17*, Montesquieu again addresses the practice of slavery. Although he posits that a hot climate creates the tendency towards slavery, he holds that climate is not deterministic or absolute: man has the power to change laws and overthrow tyrannical governments and institutions when he is ready. Montesquieu maintains that slavery is intrinsically evil and that it should be avoided.

In *Book 15, Chapter 1* Montesquieu declares that slavery is, by its own nature, bad. It is bad both for the slave and for the slave master: for the slave because he can do nothing through the motive of virtue; for the slave owner because unlimited authority corrupts him (absolute power corrupts absolutely); he becomes fierce, hasty, severe, angry, voluptuous, and cruel.

In *Book 15, Chapter 3* Montesquieu surmises that slavery must originate from the scorn that one nation has for another's customs. As an example he cites the fact that the Spaniards who traveled to the Americas were so astonished that the natives there ate crabs, snails, grasshoppers and locusts, that they smoked, and trimmed their beards in a different manner than they did, they justified making them slaves. Montesquieu concludes that knowledge and experience makes man gentle; reason humanizes him; prejudice eradicates gentleness and compassion. Hence, slavery is founded on ignorance and the prejudice that arises from the absence of knowledge and experience.

In *Book 15, Chapter 4* Montesquieu boldly asserts that religion gives rise to slavery: white people think that they have the divine right to own slaves if they can convert them to Christianity. This is a reference to the Black Code, Article 2, which says that all slaves will be baptized in the Roman Catholic faith. Montesquieu demonstrates that Article 2 is a total corruption and reversal of the Great Commission, in which the resurrected Christ instructs his disciples to spread the faith to all the world (Matt 28:19–20, Mark 16:15–18, Luke 24:47, Acts 1:8, and John 20:21–23).

It is in *Book 15, Chapter 5*, entitled, *On the Slavery of Negroes* [*De l'esclavage des nègres*] where Montesquieu demonstrates his mastery in employing irony and sarcasm to hyperbolize an abolitionist statement. He begins the chapter by announcing his intention to play the advocate and argue in favor of slavery in America. What follows is a trail of one ironic statement after the next. First, because the Americans killed all the American Indians, they need to import slaves to clear the land and to keep the price of sugar low. Next he conducts a comical tirade against the physical appearance of blacks. The point is obvious: esthetics are subjective and highly personal and should not be used as a slide rule to judge intelligence or whether a person has a soul.

Next Montesquieu sarcastically argues that blacks must lack common sense because they prefer to own a glass necklace rather than one of gold, which is so highly valued among nations having a police. It is parenthetically understood that Europe, which prides itself in being a civilized continent, has so many thieves that law enforcement has risen to the ranks of a profession.

Then comes the strongest statement in the chapter, the climax of his sarcastic exposition: it is impossible for whites to assume that blacks are men because if they did, they (whites) would have to assume that they are not Christians. This is a very powerful statement because it evokes all of the injustice and crimes perpetrated by cruel, tyrannical slave masters. The only way that whites can rationalize their atrocities and still regard themselves as Christians, that is, as souls that have been saved for Christ for all eternity, is if they regard those they oppress as animals. Hence, Montesquieu gathers momentum in this chapter, only to deliver the final blow at the end, the capstone of a sharpened, polemical tool.

Voltaire, too, in a short tale, *The Travels of Scarmentado* (1753), shows that it is ignorance of the "other" that causes men to commit atrocities. The phenomenon is universal: no matter where in the world our protagonist, Scarmentado, travels, people are always prejudiced-against foreigners and also against their own kind who happen to do things differently. First, Scarmentado, born and raised in Crete, travels to Rome and must rapidly make his escape from a monsignor who is on the verge of "placing him in the category of his minions." Voltaire leaves it to the reader's lurid imagination to deduce what this could possibly mean. Throughout his travels, our protagonist learns that most men are oblivious to the fact that the "other," too, may have a will of his own and aspirations to fulfil. It is significant that while in Rome, Scarmentado witnesses processions, exorcisms, and robberies: all three of these are related to the notion of free will or rather, the relegation of it to a more powerful entity. Scarmentado must flee to Holland, where a man is being beheaded for his religious doctrine; he is imprisoned in Spain after he jokes about the Grand Inquisitor's throne; in Constantinople he has misfortunes due to the differences among the Eastern Orthodox, Roman Catholic, and Moslem faiths; in Persia, the preference for white mutton or black mutton is taken so seriously, that he flees the country before getting involved in that dispute; in China he is seized and bound by clerics who believe that he is a spy of the Pope; in Delphi, which is under Ottoman rule at the time, he is nearly beheaded because a fellow-lodger declares that European sovereigns govern without killing their fathers or brothers, or cutting of the heads of their subjects; his companion is beheaded.

It is significant that Scarmentado's last misadventure is in Africa: by saving Africa for last, Voltaire is making a statement against slavery all over the world, but especially against the African slave trade and the Black Code. In Africa, Scarmentado is captured by Africans and made a slave because of his facial characteristics and the color of his skin. The black corsair (pirate) aboard the vessel that captures him explains that because whites buy blacks in public markets on the coast of Guinea like beasts of burden, turn them into slaves, and submit them to hard labor under the pain of beatings, Africans, in turn, will capture and enslave whites, whenever and wherever possible. However, the black slave owners are merciful to Scarmentado because after a year of labor, they permit him to buy his freedom and return home. Hence, Voltaire makes an abolitionist statement via an itinerary of countries where ignorance prevails over reason, and fear and contempt for the "other" gives rise to murder, torture, imprisonment, and slavery. He reiterates Montesquieu's view that slavery is born of prejudice against the customs of other nations and extends it to show that all injustice is born of ignorance and fear.

In 1756 Voltaire employed his sharp wit and satirical pen again to condemn slavery. In the *Essay on Morals [Essai sur les mœurs], Chapter 152*, he points out the injustice of forcing human beings to toil under inhumane working conditions and shortening their lifespan so that others might comfortably live a life of decadence: "One hundred thousand slaves, black or mulatto, work in sugar mills, indigo and cocoa plantations, sacrificing their lives to gratify our newly acquired appetites for sugar, cocoa, coffee, and tobacco, needs unknown to our ancestors. We are going to purchase these Negroes from the Guinea Coast, the Gold Coast, the Ivory Coast. Thirty years ago a good Negro was purchased for 50 pounds; that is a bit less than five times the price of a fat ox…We tell them that they are men like us…and then they are made to work like beasts of burden; they are poorly fed: if they want to escape, one of their legs is broken, and they are made to turn the sugar mills by hand, since they have been given a wooden leg. After that we dare to talk about the rights of people!"[35]

The encyclopedist Louis de Jaucourt also speaks out vociferously against slavery in the articles, "Slavery" [*Esclavage*] (1755) and "Traité des negres" [*Slave Trade*] (1765). In the article "Slavery," Jaucourt begins by defining slavery as the establishment of a right based on force, a right that gives one man ownership of another to the point that he is the absolute master of his life, his possessions, and his liberty. Jaucourt borrows heavily from Montesquieu's treatment of slavery (*Spirit of Laws, Books 15, 16, and 17*) and declares that he will not stop to praise the soundness of Montesquieu's principles because he has nothing to add to his glory.

Jaucourt begins his article by bolding declaring that all men are born free; in the beginning, they had only one name, one condition. Humanity was not classified according to race or social class. As his authority he cites Plutarch, who said that in the times of Saturn and Rhea there were neither masters, nor slaves; nature made everyone equal. This equality did not last long and servitude was introduced, little by little; it was based on mutual agreement due to necessity: the multiplication of the human race caused men to leave the simplicity of the first centuries and look for ways to add to their conveniences and acquire superfluous possessions.

Jaucourt uses the terms "slavery" and "servitude" interchangeably. However, he distinguishes between real and personal servitude. He borrows the definitions of real slavery and personal slavery from *Spirit of Laws, Book 15, Chapter 9*, entitled, "Several Kinds of Slavery": real servitude annexes the slave to the land; personal servitude concerns domestic services and relates more to the master's person.

Jaucourt traces the genealogy of slave laws from biblical to modern times. Moses handed down laws to protect servants from their owners and to give them certain rights. Moses defined the term "slavery" and ordained that it would not last longer than the jubilee year for foreigners and for Jews, six years (Lev. 25:39). Furthermore, the purpose of the Sabbath was to provide rest for servants and slaves (Ex 22 and 23, Deut 16). He goes on to point out that it is very strange that civil law would relax natural law. He recommends that civil laws regarding slavery should heed the words of Saint Paul, who said "Masters, give unto your servants that which is just and equal; knowing that ye also have a Master in heaven" (Col 4:1). Jaucourt continues to follow the history of slavery laws through the Greeks, the Romans, and the Middle Ages. By the 15the century, slavery was abolished in Europe, except for Poland, Hungary, Bohemia and lower Germany. However, Jaucourt points out the irony in the fact that at the time when slavery was being abolished in Europe, Christian powers made conquests abroad and bought and sold slaves, forgetting the principles of nature and Christianity that render all men equal.

After a four page exposition on slavery, Jaucourt states his conclusions: "Therefore, it is to go directly against the right of people and against nature, to believe that the Christian religion gives to those who profess it, the right to reduce to servitude those who do not profess it, in order to facilitate its propagation. It is, then, this manner of thinking that encouraged the destroyers of America to commit their crimes; & it is not the only time that religion has been used against its own maxim, that teach us that the quality of the other extends across the whole universe."[36]

In the article "Slave Trade" [*Traité des nègres*] (1765), Jaucourt denounces slavery once more as being a business practice contrary to religion and natural law. He opens by defining slavery as intrinsically evil: "slave trade (African commerce). It is the purchase of blacks by Europeans from the coasts of Africa for use as slaves in their colonies. The purchase of blacks in order to reduce them to slavery is commercial trafficking violates religion, ethics, natural laws and all the rights of human nature."[37] He employs "purchase" [*achat*] twice and "commercial trafficking" [*négoce*] and sets the business practice of slavery in opposition to religion {*religion*}, morality [*morale*], and natural law [*lois naturelles*]. He goes on to point out that the Africans' leaders abroad do not have the natural right or authority to sell the citizens that live in their jurisdictions. He lays the blame not only on buyers and merchants of slaves, but also on tribal leaders who step outside the boundaries of their natural authority to sell their fellow citizens as commodities in a marektplace: "Kings, princes and magistrates do not own their subjects, and therefore do not have the right to take away their liberty and sell them as slaves."[38] By the same token, no one has the right to purchase a slave: "On the other hand, no man has the right either to purchase them, or to become their master. Man and his liberty are not for sale; they cannot be sold, bought or paid for at any price."[39]

Jaucourt surmises that if the slave trade were abolished in the colonies, they would suffer economic ruin, but only for a period of time, and then there would arise a new system that would be superior to that of slavery. If the colonies set the slaves free, society would adjust, more people would emigrate abroad, and the arts and sciences would flourish there.

Jean-Jacques Rousseau, in *Julie, or the New Heloïse* [*Julie, ou La Nouvelle Héloïse*] (1761), denounces slavery by painting a tragic portrait of human suffering with a few deft strokes of his pen. The following passage is stirring and typical of Rousseau's emotive style: "I have seen those vast unfortunate regions that only seem to be destined to cover the earth with herds of slaves. From their sordid sight I have averted my eyes with disdain, horror and pity; and, seeing one fourth of my fellow humans changed into beasts for the service of others, I have grieved to be a man."[40]

"Unfortunate" anthropomorphoses "regions"; people are unfortunate, not property. "Unfortunate" is hyperbolized by the preceding "vast"; it sets the somber mood for the tragic scene that will follow. "Destined" is fatalistic and hyperbolizes the sense of tragedy by removing the element of choice, free will. The regions of land are destined, locked into a tragic picture and the reader does not yet fully understand why the scene is tragic. "To cover the earth with slaves" is a surprise: suddenly the reader visualizes multitudes of human beings. These people are metaphorized as animals: "herds" reduces them to a large gathering of animals, faceless, inhuman Cartesian machines who are thought not to think or fell; they are robbed of their humanity. The animalization of the slaves, implied by "herds" is hyperbolized by the beginning of the next sentence: "sordid sight" connotes the unpleasant, ugly, hideous. These slaves, reduced to the level of animals by "herds," are monsters, deformed beings, creatures that cannot be called men. The speaker is so repulsed by this hideous sight of people who have lost their identity as human beings, he is forced to avert his eyes: "I turned my eyes" is involuntary, the emotional response is so strong, it overwhelms him and he must look away. The feelings that fill his soul are "disdain, sorrow and pity." These three nouns show that the slaves have not totally lost their humanity: they are still human enough that an onlooker can empathize with them and react to them. If they had lost all of their humanity, they would not elicit an emotional response. It is precisely because they have the bodies of humans, the facial expressions which undoubtedly reflect exhaustion, pain, suffering, that their humanity remains visible despite the fact that mistreatment has marred their appearance. Rousseau contends that something that violates nature is taking place: one fourth of humanity is serving the other three fourths; one fourth of "my fellow humans" [*mes semblables*], people like the speaker himself, are living wasted lives so that the other three fourths can live in decadence, enjoy products that they did not need before, increase their conveniences and profits. Rousseau says that "my kind is changed into animals for the service of others." The conclusion: "I grieve being a man." The speaker grieves for many reasons. First and foremost, he grieves because he identifies with the victims. His empathy causes him to momentarily set his own identity aside and experience the anguish, exhaustion, muscular discomfort, anger, sadness of the unfortunate workers he is observing. Secondly, he grieves how far humanity has metamorphosed from natural man, how it gave up its freedoms by mutual consent, how perverse it has become. Thirdly, he grieves because the institution of slavery is so entrenched, that he, as one man alone, is powerless to put an end to it. He can write about it, but he must live with his rage. Fourthly, he may feel guilty and a sense of shared responsibility with the *bourgeois* living in big cities, if he enjoys coffee, sugar, cocoa, or uses products containing indigo. Fifthly, if he has been raised as a Christian or has read Christ's words in the Bible, he is ashamed and outraged at what other Christians are doing in violation of the Golden Rule.

In this brief passage, Rousseau combines his masterful skill as a fluid, seductive novelist with his political agenda. The landscape that he paints is truly an iconic representation of his belief system that "man is born free, but that everywhere he is in chains." In a few lines, the reader visualizes the landscape covered with wretched slaves as if he were standing on a hill gazing at the panorama below. He is provided with an iconic representation of men who have left the simple virtues of living in the countryside to be exploited by decadent civilization. Rousseau's literary style is flowing and seductive. His long, sinuous, elegant sentences, like ocean waves, carry the reader's emotions where he dictates. The panorama he paints stirs a deep, involuntary emotional response from the reader.

Rousseau's objectives are to attack the institutions of slavery and also of private property. He believed that the notions of slavery and private property did not exist in natural man: it was not until men left the woods to join civilization that they developed the notions of "mine" and "yours." The "vast regions" are "unfortunate" now that man has become civilized; originally, when they belonged to everyone, they were not "unfortunate." Slavery is contrary to natural law, which dictates the self-determination of the individual. Natural man was free; civilized men buy and sell themselves by mutual consent to gain advantages.

Rousseau, in *The Social Contract* [*Du Contrat social*] (1762), Book 1, Chapter 4, entitled, "Slavery" [*De l'esclavage*], again exercises his masterful skill as a polemicist to boldly declare that slavery is an abomination that must end; is it morally wrong, illegitimate, always, everywhere. Rousseau begins, "Since no man has a natural authority over his fellow, and force creates no right," slavery exists by convention.[41] However, no one has the right to agree to it by convention; either to sell himself into slavery, or to sell another, either his children or the people in his village, or to buy an individual. Rousseau contends, "To say that a man gives himself gratuitously, is to say what is absurd and inconceivable; such an act is null and illegitimate, from the mere fact that he who does it is out of his mind. To say the same of a whole people is to suppose a people of madmen; and madness creates no right....To renounce liberty is to renounce being a man, to surrender the rights of humanity and even its duties...such a renunciation is incompatible with man's nature; to remove all liberty from his will is to remove all morality from his acts."[42] Rousseau concludes the chapter by declaring that slavery is against natural law: "So, from whatever aspect we regard the question, the right of slavery is null and void, not only as being illegitimate, but also because it is absurd and meaningless. The words *slave* and *right* contradict each other, and are mutually exclusive. It will always be equally foolish for a man to say to a man or to a people: 'I make with you a convention wholly at your expense and wholly to my advantage; I shall keep it as long as I like, and you will keep it as long as I like.'"[43]

Diderot also staunchly defended the self-determination of the individual by contributing a considerable amount of abolitionist material to the abbé Raynal's *Philosophical and Political History of the Settlements and Trade of the Europeans in the East and West Indies* [*Histoire philosophique et politique des établissemens et du commerce des européens dans les deux Indes*] (1770–1780). This work is more commonly known as Raynal's *History of Two Indias* [*Histoire des deux Indes*]. Since its publication, manuscripts have been found in the Fonds Vandeul that have been published under the abbé Raynal's name, but that have actually been written by Diderot.

Diderot contributed material to all three editions of Raynal's *History*: the first edition of 1770, the second edition of 1774, and the third edition of 1780. An excellent anthology of Diderot's political works that includes his contributions to Raynal's *History* that we recommend is Denis Diderot, *Œuvres: Politique*, edited by Laurent Versini (Paris: Robert Laffont, 1995). The fragments that Diderot authored are also documented by Michèle Duchet in *Diderot et l'Histoire des deux Indes ou l'Ecriture fragmentaire* (Paris: A.-G. Nizet, 1978).

In Raynal's *History, Book 11, Chapter 24*, entitled, *Slavery and Liberty*, Diderot defines freedom thus: "Freedom is owning yourself."[44] He also defines natural freedom: "Natural freedom is the right that nature has given every man to make what use he likes of himself, at his will."[45] The chapter also includes a dialogue between two interlocutors, one who advances arguments in defense of slavery, and the other, against. When the former defends the divine right of white people to own slaves, the latter, the abolitionist, provides an ardent response: "Men or demons, whoever you are, will you dare to justify attempts on my independence by the right of the strongest? What! The person who wants to make me a slave is not at all guilty; he is making use of his rights? Where are these rights? Who has given them such a sacred character that they can silence my own? Nature has given me the right to protect myself; it has certainly not given you the right to attack me. If you think that you have the authority to oppress me because you are stronger and more clever than me, then do not complain when my forceful arm rips open your breast to find your heart…"[46]

Diderot goes on to attack the argumentation provided by the racist interlocutor. The white supremacist states: "But Negroes are a species of man born for slavery. They are limited, deceitful, nasty; they admit among themselves the superiority of our intelligence, and almost recognize the justice of our empire."[47] Diderot, speaking through the abolitionist, has this to say to him: "Negroes are limited because slavery crushes the energy of the soul. They are nasty: but not enough to you. They are deceitful because one does not owe truth to tyrants. They acknowledge the superiority of our minds because we have perpetuated their ignorance; the justice of our empire because we have abused their weakness. In the impossibility of maintaining our superiority by force, criminal politics has had recourse only to trickery. You have almost succeeded in persuading them that they were a unique species born for humiliation and dependence, for work and punishment. You have neglected nothing to degrade these wretched ones, and then you reproach them for being low."[48]

The *philosophe* Condorcet also spoke out vociferously against slavery. His work, *Reflections on the Slavery of Negroes* [*Réflexions sur l'esclavage des nègres*] (1781), written under the pseudonym Joachim Schwartz, is a 99 page manifesto against slavery. He wrote this piece to refute a justification of slavery that had been published in Paris the year before. Condorcet's book systematically challenges all justifications for slavery, including the notion that slavery is needed for the economic survival of the colonies. For example, in *Chapter 2*, he debunks the myth that Africans are fortunate to be sold into slavery because, in their native lands, they are criminals condemned to death or prisoners of war awaiting execution. In an ironic statement he extrapolates that by this reasoning, the slave trade becomes a humanitarian act. This is how Condorcet employs reasoning and logic to debunk that myth that slaves are condemned criminals: first, the belief is not only unproven, it is improbable. Africans did not behead all of their prisoners before Europeans came along to buy them. Judging from the slave trade, they must have executed not only married women, but unmarried girls, which has never be reported anywhere, by any people. Furthermore, selling criminals into slavery is not legitimate law. One of the conditions that the punishment be just is that it be determined by law, and as to its duration and its form. The law can never dictate that a man be a slave of another because the punishment is then contingent upon the whims of the master, and it is necessarily undetermined. Condorcet concludes, "Moreover, it is as absurd as it is atrocious to dare to posit that most of the unfortunate ones are criminals. Are you afraid of not having enough scorn for them, of not treating them with enough harshness? And how do you suppose that there exists a country where so many crimes are committed, and yet where such exact justice is executed?"[49]

THE USE OF RACE AS A FOIL FOR EUROPEAN VICES

A favorite polemical technique of the *philosophes* was to write pieces elaborating on the virtues and vices of dark skinned peoples in order to provide a foil for the evils of European civilization. An example of this is Diderot's *Supplement to the Voyage of Bougainville*, which contains several dialogues and a monologue inspired by Bougainville's 1771 non-fictional narrative of his circumnavigation of the earth.[50] Diderot uses the work as a platform to denounce racism, criticize European vices, and hyperbolize the polarities that exist between nature and European culture. *Part 1* is a dialogue between two interlocutors, "A" and "B," who discuss Bougainville's voyage. "B" begins by hypothesizing the common origin of all life: perhaps all of the land surface on the earth was once joined together and it later broke up into fragments that drifted apart. "A" asks, "How does he explain the presence of certain animals on islands separated from every continent by a vast expanse of sea? Who could have transported wolves, foxes, dogs, deer and snakes there?"[51] "B" replies, "Who knows the early history of our earth? How many great tracts of land, now isolated, were once joined?"[52] "B" reiterates this by pointing out that there are men who live on a remote tiny island northeast of New Zealand: "In considering its position on the globe, wouldn't anyone ask how it was that men came to be there? What form of communication once linked them to the rest of their species?"[53]

Here Diderot is affirming not only the unity of the human race and of each species, but also the common origin of all living things: perhaps originally there was only one land mass on which life emerged. Hence, early in the dialogue he provides a scientific reason to oppose racism and slavery: all men may have a common origin because groups of them exist on remote islands. The entirety of the dialogue will be devoted to observing principles that are true in all societies and that provide more evidence as to the commonalities that exist among all men.

Diderot examines how a burgeoning population of men on the tiny remote island northeast of New Zealand might have been able to survive with a finite supply of natural resources. He surmises that the population must have been deliberately reduced through cannibalism, murder, and the use of abortion and sexual mutilation as birth control. He wonders whether these evils are derived from the need for survival. Here Diderot seeks to explain the origin of practices and rituals that horrified Europeans. The character "A" hypothesizes that it is the need for survival that gives rise to murder and barbaric rituals: "...gives rise to so many bizarre customs, at once cruel and necessary, for which the justification is lost in the mists of antiquity, leaving philosophers at their wits' end to explain them. It appears to be a fairly universal rule that supernatural and divinely inspired practices grow stronger and more durable with time, eventually becoming transformed into civil and national laws, while civil and national institutions become consecrated and degenerate into supernatural and divine precepts."[54] Here Diderot endeavors to explain the origin of cruel rituals practiced in Africa and Asia, but in addition, he is, by analogy, taking a swipe at Europeans, as well: his principles explain the fact that European civil law, too, is derived from religious law (Mosaic law), the divine right of kings, burning people at the stake, and torturing them on the rack by pulling their limbs apart, are examples that are implied here. If the origins of the beliefs and rituals of indigenous peoples have disappeared "in the night of time," so have those of the Europeans.

Diderot surprises the reader by providing a stunning exposé of how the Catholic Church mistreated dark skinned people in Paraguay. The Jesuits' oppression of the Paraguayans is startling and graphic. He begins by identifying the oppressors by name before enumerating their crimes. When "A" asks, "Wasn't Bougainville in Paraguay at the very moment the Jesuits were expelled?" he is announcing that the subject of the conversation will be the order of the Society of Jesus.[55] This announcement is necessary because otherwise the reader could easily mistake the oppressors in the lines that follow as rich, white plantation owners who are imperialistic and exploitative, but certainly, he would not assume that they are members of the clergy.

Diderot hyperbolizes the Jesuits' cruelty to their slaves through the use of metaphor, simile, and a long enumeration of atrocious acts: "...these cruel sons of Sparta in their black habits mistreated their Indian slaves no less than the Lacedemonians abused their helots, condemning them to incessant work, slaking their own thirst with their sweat, leaving them no right of property, brutalising them by the force of superstition, demanding the deepest reverance, striding among them, whip in hand, lashing out against everyone, of any age or sex."[56]

"B" metaphorizes Jesuits as "these cruel sons of Sparta in their black habits" and compares them, through simile, to the Lacedmonians [*comme les Lacédemoniens*]; he metamphorizes the Paraguayans as "Indian slaves" and compares them to the Helots. The Helots are a metaphor for a people that has been conquered and then treated in the most inhumane manner in the world. Helots were Peloponnesian Greeks that were enslaved under Spartan rule and treated with cruelty. They lived in their master's household, but were owned by the state; unlike ordinary slaves, their master could not declare them free. Montesquieu advises that the slaves of Lacedemonia could expect no justice for either insult or injury. They were slaves not only of citizens, but of the public; they belonged to everyone and not just to one master.[57] They were the backbone of Lacedemonian agriculture, functioned as domestic slaves, and during wartime, were used in heavy infantry, light infantry, and as rowers in ships. According to Plutarch, Spartan officials (Ephors) declared war on the Helots so that citizens could kill them without reprisal.[58] Young Spartans (Krypteia), armed only with their daggers and a few provisions, went into the countryside by day, and at night, killed Helots and took their food. This kept the Helots in line and it provided an opportunity for Spartan adolescents to prove their bravery. During the Peloponnesian War, the Helots were promised their freedom if they did well in battle, but after 2,000 of them were freed, they were later assassinated.

Diderot uses the verb "condemn" [*condamner*] to identify what the Jesuits did to the slaves. "Condemn" is what the law does to the worse criminals and connotes punishment that is deserved for crimes committed; here it is ironic because the Jesuits are not justices and the indigenous people of Paraguay are not criminals. Then Diderot provides a long enumeration of the atrocities that the Jesuits have inflicted on their slaves. Each of the acts that he enumerates hyperbolizes "condemn": "incessant work," "slaking their own thirst with their sweat," "leaving them no right of property," "brutalising them by the force of superstition," "demanding the deepest reverence," "striding among them, whip in hand," and "lashing out against everyone, of any age of sex."

Here Diderot oxymorically pairs the concepts of barbarism and civilization in order to hyperbolize civilized man's inhumanity to man. As a stark contrast to this, there immediately follows an example of how native Patagonians treat white people who disembark on their shores: "They're fine people, strong and energetic, who spring at you with embraces, crying out 'Chaoua.'"[59] Here we have an essay in contrasts: Paraguayans vs. Patagonians, slavery vs. freedom, the evils of civilization vs. the rights of natural law, despair vs. strong and energetic, "black habits" versus dark skin, "cruel sons of Sparta" versus "they're fine people," "lashing out against everyone, of any age or sex" versus "spring at you with embraces."

Having shown that civilization is not so civilized, Diderot reiterates Rousseau's view that natural man was neither good nor evil, but became evil when he joined society, and became progressively more evil as society matured: "The Tahitian is close to the origins of the world and the European near its old age. The gulf between us is greater than that separating the newborn child from the decrepit dotard."[60] This is more than a statement as to the innocence of natural man. Diderot is likening the Asian Pacific race to infancy and the Caucasian race to senility. The former has just begun its lifespan, the latter is approaching the completion of its own. Here Diderot is making a highly prescient statement with stunning accuracy: not all races have been on earth the same amount of time. He surmises that the yellow race is younger than the white race, that it metamorphosed later. This conclusion is not only evident from his language, "The Tahitian is close to the origins," "the European near its old age," "new-born child" and "decrepit dotard": it is in line with Diderot's view that all life metamorphosed from a single prototype. Since Diderot believed in the metamorphosis of species, his statement can be understood to mean that whites have been here longer than yellows and that this explains the Europeans' advanced moral turpitude.

In summation, eighteenth-century naturalists classified plants, animals and races of humans because they recognized that classifying similar things facilitates the processes of understanding and drawing conclusions. They classified the various races according to physical characteristics, perceived intelligence, personality, customs, climate, and geographical location. Buffon devised the definition of species as a group of individuals who can produce a fertile offspring together; he recognized that all of the different races of man belong to the same species because they meet that definition. Maupertuis and Diderot advanced the theory that all the races can be traced back to a single ancestral pair: through epigenesis, the offspring is a unique and new entity, different from either of its parents. Accidents in the patterns or arrangements of parental elements causes new characteristics to arise in the offspring that are transmitted from generation to generation. In this way, we can explain how various races derived from a single ancestral pair.

The *philosophes*, whose intellectual curiosity caused them to examine the similarities and differences between the races, always concluded by affirming the unity of the human race: all men have the same wants and needs; they are merely more or less tanned than their brothers. The *philosophes* were polemicists who defended the self-determination of the individual as a basic tenet of natural law. They wrote prolifically in favor of abolishing the practice of slavery. Montesquieu, Diderot, Voltaire, Jaucourt, Condorect, Rousseau, and many others, each using his own unique signature style, whether it was satire and irony, or a flowing, pre-romantic, emotive, seductive style, or blunt, outright declamations, repeatedly condemned the Black Code and the practice of slavery as antithetical to religion, ethics, and natural law.

Conclusion

The intermediary between man and other animals is the monkey.[1]
—Denis Diderot, *Elements of Physiology* (1774–1780)

Eighteenth-century thought regarding the origin of man was a mosaic comprised of diverse opinions, many brilliantly prescient, and as varied and colorful as the naturalists and *philosophes* who held them. There were the Creationists, who held that all species left the hands of the Creator in perfect condition and that no new species have arisen since Creation (Buffon and Voltaire). There were also the atheist materialists, such as Maupertuis and Diderot, who embraced random creation propelled by the motive and conscious properties of atoms. On the other hand, there were the panspermists, such as Maillet and La Mettrie, who held that preexistent seeds fertilized the earth, sky and sea. Finally, there was Rousseau, who posited anthropological (intraspecies) change, but not biological (interspecies) transformism. Rousseau creatively applied Buffon's theory of the physical degeneration of species to the dissolution of man's morality; he borrowed many of Buffon's observations regarding the physical bodies of creatures, and ingeniously followed a parallel route, applying them to hypothesize a psychic and moral dissolution that occurred during man's anthropological (intraspecies) metamorphosis from his natural state to his civilized.

Maillet was stunningly prescient, for although he wrote c. 1700, he proposed measuring the rate of sea level decline to date the earth. Furthermore, his character, Telliamed, says that fish developed wings and fins that helped them to walk on the ocean floor and later on land. He also tells the tale of a Dutch cabin boy who fell overboard and reappeared years later as a merman with scales and a fish tail. Hence, Maillet's great contributions to the eighteenth century were transformism, the idea that life originated in the sea, and the notion that the physical characteristics of living beings adapt to changes in their environment.

Montesquieu also surmised that man must have metamorphosed from animals in the distant past. In the *Persian Letters* (1721), he describes the animal-like ancestors of man, those Troglodytes of former times, who were deformed, hairy like bears, and who hissed. He derives his imagery of man's monstrous ancestors from Aristotle, Herodotus, Pomponius Mela, and Pliny the Elder. This notion was promulgated further by Lucretius, who declared that in the beginning nature created many monsters and only those without significant self-contradictions survived.

In the *Spirit of Laws* (1748), Montesquieu elaborates further on the impermanence of nature and the flux that it delivers. He hypothesizes that a relationship exists between climate and human physiology and also between climate and government. Climate, geography, topography, and soil all affect man's body and temperament and contribute to an array of tendencies that a nation has. However, climate is not absolutely determinative: man has the power to progress beyond the influence of climate and improve his life through legislation, when his intellect and the society in which he lives have matured to the point where he is ready to fight for self-determination. However, Montesquieu warns us that civilizations rise and fall, that all societies have a finite lifespan, and that systems of government are cyclical..

La Mettrie, for his part, made many significant contributions to the notion of the metamorphosis of man through the random motive property of matter. He declared that physical matter is the only reality through which all phenomena can be explained. He dispensed with the notion of the immortal soul and declared that consciousness is solely contingent upon the functioning of the brain, central nervous system, and the five senses. Consciousness is influenced by food, age, learning, inheritance, climate, and the environment. All life is the result of random chance and random molecular collisions: flux+time=dispersion of chaos. Hence, he held that the motive property of atoms, not God, created everything. Atoms are continually colliding and eventually form every viable organized being that exists. He rejected final causes and attributed all events to the random flux of nature.

Among Buffon's contributions to biology was his observation "all that can be, is {*tout ce qui peut être, est*}. He observed that nature produces every imaginable variation in each species. Furthermore, he observed that physical characteristics are shared among species. Hence, he developed the two dimensional, linear chain of beings into a three dimensional cone. The materialists Maupertuis and Diderot added the fourth dimension, time, and posited the metamorphosis of species over great length of time.

Maupertuis posited that inherited errors explain the vast diversity that the flux of nature delivers. His study of polydactyly in a Berlinese family caused him to arrive at some landmark conclusions in biology. He identified birth anomalies as "traits" and noted that they were carried by either parent. He also noted that birth defects often skip a generation or two and reappear further down the family tree. In addition, he was able to calculate the statistical probability that polydactyly would recur in a given population. His great influence on Diderot's thought was his hypothesis that errors in the arrangement of parental elements are inherited and could explain how the vast variety of living beings that we see today many have developed from a single prototype.

Maupertuis' other great contribution to Diderot was an emergent consciousness: when particles unite, each particle loses its consciousness of self and acquires the consciousness of the larger body to which it belongs. It loses its individual memory and consciousness and acquires that of the whole. This explains why man is conscious of his existence and of the presence of others, but not of every molecule that constitutes his body.

Diderot seized upon this notion, cited Maupertuis in *Thoughts on the Interpretation of Nature* (1753), and adopted his two main contributions to biology, namely, 1) that inherited errors in the arrangement of parental elements could explain how all living beings developed from a single prototype and 2) emergent consciousness (that all matter is conscious and that when particles combine, they lose their memory and consciousness of self and acquire the consciousness of the larger body that they form).

Diderot's originality lies in the fact that he viewed species as mutable, not static, and that he posited the appearance, lifespan and extinction of species over time. He surmised that microscopic animalcules, species, star systems, and perhaps the universe itself, randomly come into existence via the motive and conscious properties of atoms, exist for a time, and then fall out of existence. Diderot's transformism rested on a fulcrum of three pivotal points: probability theory, the motive property of atoms, and the conscious property of matter.

Rousseau, on the other hand, was not a transformist. He embraced anthropological (intraspecies) metamorphosis and sociological change, but not biological (interspecies) transformism, as Diderot and the materialists did. Rousseau's legacy is that he posited anthropological change in a new and creative way: he applied Buffon's theory of the physical degeneration of species to the dissolution of man's morality. Rousseau borrowed many of Buffon's observations regarding the physical bodies of creatures, and ingeniously followed a parallel route, applying them to hypothesize a psychic and moral dissolution that occurred during man's anthropological (intraspecies) metamorphosis from his natural state to his civilized. Rousseau's "natural man" was neither good, nor evil, but a *tabula rasa*, on which his experiences would imprint. Hence, all of the vices that exist in society today, most notably war, slavery, theft, the notion of honor, pride, and greed, were unknown to natural man, and are artifices created by civilized man and consented to by him

Rousseau noted the great similarities between man and the great apes, who appeared to be conscious, thinking, intelligent animals. Rousseau suggested widening the umbrella of the human species to include the great apes and accepted the notion that perhaps there are multiple varieties of the human species. This was a political statement, not a biological one: he opposed racism, slavery and the subjugation of people living in extremely primitive conditions in Africa and Asia. He declared the human dignity and basic human rights of all men, even apes, who if experimentation could show had perfectibility and could be taught language, deserved the compassion and esteem accorded to all men. Therefore, his proposal to include apes in the human species was not a statement as to biological transformism, which he denied, but rather, a scathing criticism of civilized Europe in which men consent to inequality and to valuing some people more than others according to the artifices of social class.

On the other hand, Voltaire, the deist, saw contemporary science as a threat to God. The motive property of atoms, determinative of all Creation, could be held to replace God's will. If particles in motion always existed, there is no need for a Prime Mover. The discovery of fossils and the placement of marine shells on mountaintops and far from water indicated that the earth was much older than had been previously thought; the great size of the fossil bones discovered in Siberia and the New World indicated that species do metamorphose over time and that perhaps, Diderot might be right after all.

Conclusion

The debate over whether apes can be taught a language challenged man's unique place on the chain of beings: if the great apes could be taught to speak, a gap would be filled on the great chain between man and the apes. There would no longer be an insurmountable chasm between man and all other animals. Talking apes would also pose innumerable problems for Cartesians: if a soul impart the intellect necessary for language, do talking apes have a soul that requires forgiveness from Original Sin? How many species of animals, especially those that work together and communicate (beavers, ants, bees), have souls? The great apes also raised the possibility of multiple varieties of the human race or multiple species of men.

The multiplicity of races posed more than intriguing biological questions: there was the pressing issue of the transatlantic slave trade that needed to be abolished. The *philosophes* declared that all slavery violates the basic tenets of natural law. All men come into the world with nothing and leave with nothing. Hence, nature, itself, indicates that they are all equal. Furthermore, all people have the natural right to freedom and self-determination. The *philosophes*' polemics against slavery culminated in the Declaration of the Rights of Man and of the Citizen [*La Déclaration des droits de l'homme et du citoyen*] on August 26, 1789. The first article states, "Men are born and remain free and equal in rights. Social distinctions may be based only on considerations of the common good." Undeniable rights are set forth not only for French citizens, but for all man without exception. While this document did not abolish slavery in the French colonies, it served as a precursor to subsequent human rights instruments. On February 4, 1794, the National Assembly voted to end slavery in all French colonies. However, slavery was reinstated by Napoleon in 1802 and not banned for good until 1848.

Notes

INTRODUCTION

1. Aristotle, *History of Animals*, edited by D. M. Balme (Cambridge, MA: Harvard University Press, 1991), 7.1.588b–5, pp. 61, 63.
2. "Chain," *Oxford English Dictionary Online*, 4a, http://dictionary.oed.com (Apr. 12, 2006).
3. Arthur O. Lovejoy, *The Great Chain of Being: A Study of the History of an Idea* (Cambridge, MA: Harvard University Press, 2001), 55.
4. Aristotle, *Metaphysics: Books I-IX*, translated by Hugh Tredennick (Cambridge, MA: Harvard University Press, 1933), 2.2.994a–10, pp. 87, 89.
5. Ibid., 2.2.994b, p. 91.
6. Aristotle, *Metaphysics: Books X-XIV*, translated by Hugh Tredennick (Cambridge, MA: Harvard University Press, 1935), 12.6.1071b15, p. 141.
7. Arthur O. Lovejoy, *The Great Chain of Being: A Study of the History of an Idea* (Cambridge, MA: Harvard University Press, 2001), 56. Also, Aristotle, *History of Animals*, edited by D. M. Balme (Cambridge, MA: Harvard University Press, 1991), book 7.1.588b, pp. 61, 63.
8. Aristotle, *Parts of Animals*, translated by A. L. Peck (Cambridge, MA: Harvard University Press, 1937), 4.13.697b, pp. 427, 429. Aristotle discusses the shared forms and functions of various species of animals. See also Aristotle, *History of Animals*, edited by D. M. Balme (Cambridge, MA: Harvard University Press, 1991), 2.8–9, pp. 103, 105, 107. Discussed by Arthur O. Lovejoy, *The Great Chain of Being: A Study of the History of an Idea* (Cambridge, MA: Harvard University Press, 2001), 57.
9. This is my translation. Unless otherwise noted, all translations are my own. "Parcourant ensuite successivement & par ordre les différens objets qui composent l'Univers, & se mettant à la tête de tous les êtres créez, il verra avec étonnement qu'on peut descendre par des degrés presqu'insensibles, de la créature la plus parfaite jusqu'à la matière la plus informe, de l'animal le mieux organisé jusqu'au minéral le plus brut…" Georges-Louis Leclerc de Buffon, *Premier discours* (1749) in *Histoire naturelle, générale et particulière*, 15 volumes (Paris: Imprimerie royale, 1749–1767), 1:12.

10. "...on voit clairement qu'il est impossible de donner un système général, une méthode parfaite, non seulement pour l'Histoire Naturelle entière, mais même pour une seule de ses branches; car pour faire un système, un arrangement, en un mot une méthode générale, il faut que tout y soit compris; il faut diviser ce tout en différentes classes, partager ces classes en genres, sous-diviser ces genres en espèces, & tout cela suivant un ordre dans lequel il entre nécessairement de l'arbitraire. Mais la Nature marche par des gradations inconnues, & par conséquent elle ne peut pas se prêter totalement à ces divisions, puisqu'elle passe d'une espèce à une autre espèce, & souvent d'un genre à un autre genre, par des nuances imperceptibles; de sorte qu'il se trouve un grand nombre d'espèces moyennes & d'objets mi-partis qu'on ne sçait où placer, & qui dérangent nécessairement le projet du système général: cette vérité est trop importante pour que je ne l'appuie pas de tout ce qui peut la rendre claire & évidente." Ibid., 1:13.
11. Georges-Louis Leclerc de Buffon, *The Seals, Walrus, and Manati* (1765) in *Natural History, General and Particular*, translated by William Smellie, 9 vols., 2nd edition (London: W. Strahan and T. Cadell, 1785), 7:328–29. "Assemblons pour un instant tous les animaux quadrupèdes, faisons-en un groupe, ou plutôt formons-en une troupe dont les intervalles et les rangs représentent à peu près la proximité ou l'éloignement qui se trouve entre chaque espèce; plaçons au centre les genres les plus nombreux, et sur les flancs, sur les ailes ceux qui le sont le moins; resserrons-les tous dans le plus petit espace, afin de les mieux voir; et nous trouverons qu'il n'est pas possible d'arrondir cette enceinte: Que quoique tous les animaux quadrupèdes tiennent entr'eux de plus près qu'ils ne tiennent aux autres êtres, il s'en trouve néanmoins en grand nombre qui font des pointes au dehors, et semblent s'élancer pour atteindre à d'autres classes de la Nature; les singes tendent à s'approcher de l'homme et s'en approchent en effet de très-près; les chauve-souris sont les singes des oiseaux qu'elles imitent par leur vol; les porc-épics. Les hérissons par les tuyaux dont ils sont couverts, semblent nous indiquer que les plumes pourroient appartenir à d'autres qu'aux oiseaux; les tatous par leur test écailleux s'approchent de la tortue et des crustacées; les castors par les écailles de leur queue ressemblent aux poissons; les fourmillers par leur espèce de bec ou de trompe sans dents et par leur longue langue, nous rappellent encore les oiseaux; enfin les Phoques, les Morses et les Lamantins font un petit corps à part qui forme la pointe la plus saillante pour arriver aux cétacées." Georges-Louis Leclerc de Buffon, *Les Phoques, les morses et les lamantins* (1765) in *Histoire naturelle, générale et particulière*, 15 volumes (Paris: Imprimerie royale, 1749–1767), 13:330–31.
12. "Avez-vous vu au Jardin du Roi, sous une cage de verre, un orang-outang qui a l'air d'un saint Jean qui prêche au désert?...Le cardinal de Polignac lui disait un jour: 'Parle, et je te baptise.'" Denis Diderot, *Suite de l'Entretien* in *Œuvres complètes de Diderot*, edited by Jean Assézat and Maurice Tourneux (Paris: Garnier, 1875–1877), 2:190.
13. Mary Efrosini Gregory, *Diderot and the Metamorphosis of Species* (New York and London: Routledge, 2007), 19-51.
14. Arthur O. Lovejoy, *The Great Chain of Being: A Study of the History of an Idea* (Cambridge, MA: Harvard University Press, 2001).
15. Jacques Roger, *The Life Sciences in Eighteenth-Century French Thought* (Stanford: Stanford University Press, 1997), 420–25.

16. Francis Haber, "Fossils and the Idea of a Process of time in Natural History" in *Forerunners of Darwin: 1745–1859*, edited by Bentley Glass, *et al* (Baltimore: The John Hopkins Press, 1959), 222–61.
17. Arthur O. Lovejoy, "Buffon and the Problem of Species" in *Forerunners of Darwin: 1745–1859*, edited by Bentley Glass, *et al* (Baltimore: The John Hopkins Press, 1959), 84–113.
18. Bentley Glass, "Maupertuis, Pioneer of Genetics and Evolution" in *Forerunners of Darwin: 1745–1859*, edited by Bentley Glass, *et al* (Baltimore: The John Hopkins Press, 1959), 51–83.
19. Lester Crocker, "Diderot and Eighteenth-Century French Transformism" in *Forerunners of Darwin: 1745–1859*, edited by Bentley Glass, *et al* (Baltimore: The John Hopkins Press, 1959), 114–43.
20. Jacques Roger, *The Life Sciences in Eighteenth-Century French Thought* (Stanford: Stanford University Press, 1997), 420–25.
21. Otis E. Fellows and Stephen F. Milliken, *Buffon* (New York: Twayne Publishers, Inc., 1972), 109.
22. John H. Eddy, Jr., "Buffon, Organic Change, and the Races of Man" (Ph. D diss., University of Oklahoma, 1977).
23. John H. Eddy, Jr. "Buffon, Organic Alterations, and Man." *Studies in the History of Biology* 7 (1984): 1–45.
24. Paul Lawrence Farber, "Buffon's Concept of Species" (Ph. D diss., Indiana University, 1970).
25. Paul Lawrence Farber, "Buffon and the Concept of Species." *Journal of the History of Biology* 5, no. 2 (Fall 1972): 259–84.
26. Phillip R. Sloan, "The Idea of Racial Degeneracy in Buffon's *Histoire naturelle.*" *Racism in the Eighteenth Century*. Volume 3 of *Studies in Eighteenth-Century Culture*. Cleveland: Case Western Reserve University Press, 1973, 293–321.
27. Michèle Duchet, "L'anthropologie de Diderot," *Anthropologie et histoire au siècle des Lumières: Buffon, Voltaire, Rousseau, Helvétius, Diderot* (Paris: Maspéro, 1971), 407–75.
28. Jean Ehrard, "Diderot, l'Encyclopédie, et l'Histoire et théorie de la Terre," *Buffon 88* (Paris: Vrin, 1988), 135–42.
29. Arthur O. Lovejoy, "Buffon and the Problem of Species" in *Forerunners of Darwin, 1745–1859*, edited by Bentley Glass *et al.* (Baltimore: The Johns Hopkins Press, 1959), 84–113.
30. Jacques Roger. *Buffon: A Life in Natural History* (Ithaca: Cornell University Press, 1997).
31. Jacques Roger, "Diderot et Buffon en 1749," *Diderot Studies* 4 (1963): 221–36.
32. Jacques Roger, *The Life Sciences in Eighteenth-Century French Thought* (Stanford: Stanford University Press, 1997).
33. Aram Vartanian, "Buffon et Diderot," *Buffon 88* (Paris: Vrin, 1988), 119–33.
34. Jean E. Perkins, "Diderot and La Mettrie," *Studies on Voltaire and the Eighteenth Century* 10 (1959): 49–100.
35. Ann Thomson, "La Mettrie et Diderot," http://www.sigu7.jussieu.fr/diderot/travaux/revseance2.htm (Jan. 24, 2006).
36. Ann Thomson, "L'unité matérielle de l'homme chez La Mettrie et Diderot," *Colloque International Diderot* (1985): 61–68.

37. Aram Vartanian, "La Mettrie and Diderot Revisited: An Intertextual Encounter," *Diderot Studies* 21 (1983): 155–97.
38. Aram Vartanian, "Trembley's Polyp, La Mettrie, and Eighteenth-Century French Materialism," *Journal of the History of Ideas* 2 (1950): 259–86.
39. Marx W. Wartofsky, "Diderot and the Development of Materialist Monism," *Diderot Studies* 2 (1952): 279–329.
40. Otis Fellows, "Buffon and Rousseau: Aspects of a Relationship," *PMLA* 75, no. 3 (June 1960): 184–96.
41. Francis Moran, "Of Pongos and Men: Orangs-Outang in Rousseau's *Discourse on Inequality*," *Review of Politics* 57, no. 4 (Autumn 1995): 641–64.
42. Leonard Sorenson, "Natural Inequality and Rousseau's Political Philosophy in his *Discourse on Inequality*," *Western Political Quarterly* 43, no. 4 (December 1990): 763–88.

CHAPTER ONE

1. Benoît de Maillet, *Telliamed: Or, Discourses between an Indian Philosopher, and a French Missionary, on the Diminution of the Sea, the Formation of the Earth, the Origin of Men and Animals, and other Curious Subjects, relating to Natural History and Philosophy* (London: Jacob Loyseau, 1750), 224. "Les petits ailerons qu'ils avoient sous le ventre, & qui, comme leurs nageoires, leur avoient aidé à se promener dans la mer, devinrent des pieds, & leur servirent à marcher sur la terre." Benoît de Maillet, *Telliamed ou Entretiens d'un philosophe indien avec un missionnaire françois sur la diminution de la mer, la formation de la terre, l'origine de l'homme, &c., Chapter 6*, "Sixième journée. De l'origine de l'homme & des animaux, & de la propagation des espèces par les semences," 2 vols. (Amsterdam: L'Honoré et fils, 1748), 2:140.
2. *Biographie universelle ancienne et moderne*, ed. Michaud, 45 vols. (Paris: Ch. Delagrave et Cie, 1870–1873), 26:125.
3. James Lawrence Powell, *Mysteries of Terra Firma: The Age and Evolution of the Earth*. http://www.powells.com/biblio?show=HARDCOVER:SALE:068487282X: 10.98&page=excerpt#page (September 11, 2006).
4. Ibid.
5. Ibid.
6. Ibid.
7. Ibid.
8. Benoît de Maillet, *Telliamed: Or, Discourses between an Indian Philosopher, and a French Missionary, on the Diminution of the Sea, the Formation of the Earth, the Origin of Men and Animals, and other Curious Subjects, relating to Natural History and Philosophy*, Chapter 1, "First Day. Proofs of the Diminution of the Sea" (London: Jacob Loyseau, 1750), 4. "Une observation que mon Ayeul avoit faite…dans sa jeunesse, ainsi qu'il l'assura à mon père, que dans le plus grand calme la mer restoit toujours supérieure au rocher, & le couvroit de ses eaux. Cependant 22 ans avant sa mort la superficie de ce rocher parut à sec, ou pour me servir de vos termes, commença à veiller." Benoît de Maillet, *Telliamed ou*

Entretiens d'un philosophe indien avec un missionnaire françois sur la diminution de la mer, la formation de la terre, l'origine de l'homme, &c., Chapter 1, "Première journée. Preuves de la Diminution de la Mer," 2 vols. (Amsterdam: L'Honoré et fils, 1748), 1:5–6.

9. Ibid. "…on y rencontroit, comme dans ces derniers, des coquillages de mer colés & inserés à leur superficie. Vingt sortes de pétrifications qui n'avoient entr'elles aucune ressemblance, s'offroient à ses yeux…Le principe d'une si grande variété dans les terrains, jointe aux lits divers en épaisseur & en substance, ainsi qu'en couleur, dont la plupart de ces carrières étoient composées, embarassoient étrangement sa raison." Ibid., 1:6–7.

10. "Or de l'estimation que je viens de faire de la diminution des eaux de la mer, c'est-à-dire, d'environ un pied dans l'espace de trois siècles, & de trois pieds quatre pouces en mille ans…" Ibid., 1:204.

11. Benoît de Maillet, *Telliamed: Or, Discourses between an Indian Philosopher, and a French Missionary, on the Diminution of the Sea, the Formation of the Earth, the Origin of Men and Animals, and other Curious Subjects, relating to Natural History and Philosophy* (London: Jacob Loyseau, 1750), 218. "En effet les herbes, les plantes, les racines, les bleds, les arbres, & tout ce que la terre produit & nourrit de cette espèce, n'est-il pas sorti de la mer, N'est-il pas du moins naturel de le penser, sur la certitude que toutes nos terres habitables sont originairement sorties de ses eaux?" Benoît de Maillet, *Telliamed ou Entretiens d'un philosophe indien avec un missionnaire françois sur la diminution de la mer, la formation de la terre, l'origine de l'homme, &c.*, Chapter 6, "Sixième journée. De l'origine de l'homme & des animaux, & de la propagation des espèces par les semences," 2 vols. (Amsterdam: L'Honoré et fils, 1748), 2:131.

12. Ibid., 219. "Pour venir à présent à ce qui regarde l'origine des animaux terrestres, plus je remarque qu'il n'y en a aucun marchant volant, ou rampant, dont la mer ne renferme des espèces semblables, ou approchantes, & dont le passage d'un de ces élémens à l'autre ne soit possible, probable, même soutenu d'un grand nombre d'exemples." Ibid., 2:133.

13. Ibid., 220. "Or la ressemblance de figure, même d'inclination, qui se remarque entre certains poissons & quelques animaux terrestres, est non seulement digne d'attention; il est même surprenant que personne, que je scache, n'ait travaillé jusqu'ici à approfondir les raisons de cette conformité." Ibid., 2:134–35.

14. Ibid., 222. "Il y a dans la mer des poissons de presque toutes les figures des animaux terrestres, même des oiseaux. Elle renferme des plantes & des fleurs, & quelques fruits: l'ortie, la rose, l'œillet, le melon, le raisin y trouvent leurs semblables." Ibid., 2:138.

15. Ibid., 223–24. "Car il peut arriver, comme nous sçavons qu'en effet il arrive assez souvent, que les poissons aîlés & volans chassant ou étant chassés dans la mer, emportés du désir de la proie ou de la crainte de la mort, ou bien poussés peut-être à quelques pas du rivage par les vagues qu'excitoit une tempête, soient tombés dans des roseaux ou dans des herbages, d'où ensuite il ne leur fut pas possible de reprendre vers la mer l'effort qui les en avoit tirés, & qu'en cet état ils ayent contracté une plus grande faculté de voler. Alors leurs nageoires n'étant plus baignées des eaux de la mer, se sendirent & se déjetterent par la sécheresse. Tandis qu'ils trouverent dans les roseaux & les herbages dans lesquels ils étoient tombés, quelques alimens pour se soutenir, les toyaux de leurs nageoires séparés les uns des

autres se prolongerent, & se revêtirent de barbes; ou pour parler plus juste, les membranes qui auparavant les avoient tenus collés les uns aux autres, se métamorphoserent. La barbe formée de ces pellicules déjettées s'allongea elle-même; la peau de ces animaux se revêtit insensiblement d'un duvet de la même couleur dont elle étoit peinte, & ce duvet grandit. Les petits ailerons qu'ils avoient sous le ventre, & qui, comme leurs nageoires, leur avoient aidé à se promener dans la mer, devinrent des pieds, & leur servirent à marcher sur la terre. Il se fit encore d'autres petits changemens dans leur figure. Le bec & le col des uns s'alongerent; ceux des autres se racourcirent: il en fut de même du reste du corps. Cependant la conformité de la première figure subsists dans le total; & elle est, & sera toujours aisé de reconnoître." Ibid., 2:139–40.

16. Ibid., 225. "Quant aux animaux à quatre pied, nous ne trouvons pas seulement dans la mer des espèces de leur figure & de leur mêmes inclinations, vivant dans le sein des flots des mêmes alimens, dont ils se nourrissent sur la terre: nous avons encore cent exemples de ces espèces vivant également dans l'air & dans les eaux. Les Singes marins n'ont-ils pas toute la figure des singes de terre? Il y en a de même de plusieurs espèces." Ibid., 2:143.

17. Ibid., 226. "Le lion, le cheval, le bœuf, le cochon, le loup, le chameau, le chat, le chien, la chèvre, le mouton, ont de même leur semblable dans la mer." Ibid., 2:144.

18. Ibid., 228. "C'est ainsi sans doute, que tous les animaux terrestres ont passé du séjour des eaux à la respiration de l'air, & ont contracté la faculté de mugir, de hurler, d'aboyer & de se faire entendre, qu'ils n'avoient point dans la mer, ou qu'ils n'avoient du moins que fort imparfaitement." Ibid., 2:148.

19. Ibid., 230–31. "J'ai lû dans vos histoires, qu'...un Officier d'une des villes du Delta, ou de la basse Egypte, se promenant sur le soir avec quelques uns de ses amis sur les bords du Nil, ils apperçurent assez proche du rivage un homme marin suivi de sa femelle..." Ibid., 2:151.

20. Ibid., 245–56. "Rien n'est plus commun, que ces hommes sauvages. En 1702...ces Hollandois se saisirent dans une descente de deux animaux mâles, qu'ils amenerent à Batavia, & qu'ils nommerent dans la Langue du pays (Orans-outans,) c'est-à-dire, hommes silvains. Ils avoient toute la forme humaine, & marchoient comme nous sur deux pieds. Leurs jambes & leurs bras étoient très déliés, & revêtus de poil...Ces Orans-outans avoient les ongles des doigts des pieds fort longs, & un peu crochus..." Ibid., 2:172–73.

21. Ibid., 246. "Pour revenir aux diverses espèces d'hommes, ceux qui ont des queues peuvent ils être les fils de ceux qui n'en ont point? Comme les singes à queue ne descendent certainement point de ceux qui sont sans queue, ne seroit-il pas naturel de penser de même, que les hommes qui naissent avec des queues sont d'une espèce diverse de ceux qui n'en ont jamais eu?" Ibid., 2:174.

22. "Panspermia," *Oxford English Dictionary Online*, http://www.oed.com (September 18, 2006).

23. Titus Lucretius Carus, *De rerum natura*, translated by W.H.D. Rouse and revised by Martin Ferguson Smith (Cambridge, MA: Harvard University Press, 2002), 5.793–94, p. 441. "nam neque de cælo cecidisse animalia possunt nec terrestria de salsis exisse lacunis." Titus Lucretius Carus, *De Rerum natura*, translated by W.H.D. Rouse and revised by Martin Ferguson Smith (Cambridge, MA: Harvard University Press, 2002), 5.793–94, p. 440. Cited in Benoît de Maillet, *Telliamed ou Entretiens d'un philosophe indien avec un missionnaire françois sur la diminution de la mer,*

Notes 259

       la formation de la terre, l'origine de l'homme, &c., Chapter 6, "Sixième journée. De l'origine de l'homme & des animaux, & de la propagation des espèces par les semences," 2 vols. (Amsterdam: L'Honoré et fils, 1748), 2:131note(a).

24. Ibid., 5.791–92, p. 441. "...inde, loci mortalia sæcla creavit multa modis multis varia ratione coorta." Ibid., 5.791–92, p. 440.
25. Jacques Roger, *The Life Sciences in Eighteenth-Century French Thought* (Stanford: Stanford University Press, 1997), 423.
26. "Tous les Animaux, & l'homme par conséquent, qu'aucun Sage ne s'avisa jamais de soustraire à leur catégorie, seroient-ils véritablement fils de la Terre, comme la Fable le dit des Géants? La mer couvrant peut-être originairement la surface de nôtre Globe, n'auroit-elle point été elle-même le berçeau flottant de tous les Etres éternellement enfermés dans son sein? C'est le système de l'auteur de *Telliamed*, qui revient à peu près à celui de Lucrèce; car toujours faudroit-il que le mer, absorbée par les pores de la Terre, consumée peu à peu par la chaleur du Soleil & le laps infini des temps, eût été formée, en se retirant, de laisser l'œuf humain, comme elle fait quelque fois le poisson, à sec sur le rivage. Moyennant quoi, sans autre incubation que celle du Soleil, l'homme & tout autre animal seroient sortis de leur coque, comme certains éclosent encore aujourd'hui dans les païs chauds, & comme sont aussi les Poulets dans un fumier chaud par l'art des Physiciens." Julien Offray de la Mettrie, *Le système d'Epicure* in *Œuvres philosophiques* (London: Jean Nourse, 1751), 341–42.
27. John Turberville Needham, *Observations upon the Generation, Composition, and Decomposition of Animal and Vegetable Substances. Communicated in a Letter to Martin Folkes, Esq; President of the Royal Society* (London, 1749), 39.
28. Julien Offray de La Mettrie, *Man a Machine* in *Man a Machine, including Frederick the Great's "Eulogy" on La Mettrie and Extracts from La Mettrie's "The Natural History of the Soul,"* notes by Gertrude Carman Bussey (Chicago: The Open Court Publishing Company, 1912), 125. "Nous ne connoissons point la Nature: Des causes cachées dans son sein pourroient avoir tout produit. Voiez à votre tour le Polype de Trembley! Ne contient il pas en soi les causes qui donnent lieu à sa régénération?" Julien Offray de La Mettrie, *L'Homme-machine* in *Œuvres philosophiques* (London: Jean Nourse, 1751), 51.
29. Jacques Roger, *The Life Sciences in Eighteenth-Century French Thought* (Stanford: Stanford University Press, 1997), 397.
30. "...qui m'apprit que les montagnes et les hommes sont produits par les eaux de la mer. Il y eut d'abord de beaux hommes marins qui ensuite devinrent amphibies. Leur belle queue fourchue se changea en cuisses et en jambes. J'étais encore tout plein des Métamorphoses d'Ovide, et d'un livre où il était démontré que la race des hommes était bâtarde d'une race de babouns: j'aimais autant descendre d'un poisson que d'un singe." François-Marie Arouet de Voltaire, *L'Homme aux 40 écus* in *Œuvres complètes de Voltaire avec préfaces, avertissements, notes, etc.*, 70 vols., edited by M. Beuchot (Paris: Lefèvre, 1829–1840), 34:43.
31. Otis E. Fellows and Stephen F. Milliken, *Buffon* (New York: Twayne Publishers, Inc., 1972), 109.
32. Peter J. Bowler, *Evolution: The History of an Idea*, 3rd ed. (Berkeley: University of California Press, 2003), 34–35.
33. Ibid., 72.
34. Ibid., 72–73.

35. Ibid., 73.
36. Jacques Roger, *The Life Sciences in Eighteenth-Century French Thought* (Stanford: Stanford University Press, 1997), 420.
37. Ibid., 423.
38. Ibid., 424.
39. Ibid.

CHAPTER TWO

1. Charles-Louis de Secondat de Montesquieu, *Persian Letters, Letter 11* (London: Penguin Books Ltd, 2004), 53. "...ces anciens Troglodites...ressembloient plus à des bêtes qu'à des hommes." Charles-Louis de Secondat de Montesquieu, *Lettres persanes, Lettre 11* (Cologne: Pierre Marteau, 1754), 23–24.
2. Ibid. "Il y avoit en Arabie un petit peuple, appellé Troglodite, qui descendoit de ces anciens Troglodites, qui, si nous en croyons les historiens, ressembloient plus à des bêtes qu'à des hommes. Ceux-ci n'étoient point si contrefaits, ils n'étoient point velus comme des ours, ils ne siffloient point, ils avoient deux yeux: mais ils étoient si méchans & si féroces, qu'il n'y avoit parmi eux aucun principe d'équité, ni de justice." Ibid.
3. Aristotle, *History of Animals*, edited by D. M. Balme (Cambridge, MA: Harvard University Press, 1991), 8.12.597a–5, pp. 131, 133.
4. Herodotus, *The Persian Wars*, translated by A. D. Godley (Cambridge, MA: Harvard University Press, 1921), 4.183, p. 387.
5. Montesquieu, *Lettres persanes*, edited and annotated by Paul Vernière (Paris: Garnier, 1960), 28n3.
6. Pomponius Mela, *Geography/De situ orbis A.D. 43*, translated by Paul Berry (Lewiston, NY: The Edwin Mellen Press, 1997), 11. "tum primos ab oriente Garamantas, post Augilas et Troglodytas, et ultimos ad occasum Atlantas audimus. intra, si credere libet, vix iam homines magisque semiferi Aegypanes et Blemyes et Gamphasantes et Satyri sine tectis ac sedibus passim vagi habent potius terras quam habitant." Pomponius Mela, *De chorographia*, introduced and annotated by Piergiorgio Parroni (Rome: Edizioni di Storia e Letteratura, 1984), 1.4.23, p. 115.
7. Ibid., 21. "Troglodytæ, nullarum opum domini strident magis quam loquuntur, specus subeunt alunturque serpentibus...Blemyis capita absunt, vultus in pectore est." Ibid., 1.8.44, 48, p. 119.
8. "Le géographe disait: "Troglodyta, nullarum opum domini, strident magis quam loquuntur"; Hérodote leur faisser pousser des cris aigus comme des chauvres-souris...(Histoires, IV, 183)." Montesquieu, *Lettres persanes*, edited and annotated by Paul Vernière (Paris: Garnier, 1960), 28n3.
9. "Montesquieu a lu Pomponius Mela. Celui-ci écrit: *Troglodytae, nullarum opum domini, strident magis quam loquuntur*, et ce mot explique celui de Montesquieu: 'ils ne siffloient point.' Lorsqu'il ajoute: 'ils avoient deux yeux,' il pense au même passage de Pomponius Mela qui disait, sur un peuple voisin des Troglodytes: *Blemmyis capita absunt; vultus in pectore est*, ou peut-être à Pline l'Ancien:

Blemmyis traduntur capita abesse, ore et oculis pectore affixis." Montesquieu, *Lettres persanes*, annotated by Antoine Adam (Geneva: Librairie Droz, 1954), 36n1.
10. Titus Lucretius Carus, *De rerum natura*, 5.837–50, 855–56.
11. Montesquieu, *Persian Letters, Letter 12* (London: Penguin Books Ltd, 2004), 56. "…ils avoient de l'humanité; ils connoissoient la justice; ils aimoient la vertu…" Montesquieu, *Lettres persanes, Lettre 12* (Cologne: Pierre Marteau, 1754), 28.
12. Ibid., 57–58. "Le soir, lorsque les troupeaux quittoient les prairies, & que les bœufs fatigués avoient ramené la charrue, ils s'assembloient…ils décrivoient ensuite les délices de la vie champêtre, & le bonheur d'une condition toujours parée de l'innocence…Dans ce pays heureux, la cupidité étoit étrangère: ils se faisoient des présens, où celui qui donnoit, croyoit toujours avoir l'avantage: le peuple Troglodite se regardoit comme une seule famille: les troupeaux étoient presque toujours confondus; la seule peine qu'on s'épargnoit ordinairement, c'étoit de les partager." Ibid., 31.
13. "Le souvenir de Fénelon demeure ici constant. C'est la Bétique du *Télémaque* (Wetstein, 1719, livre VIII, p. 170), avec tous les thèmes fabuleux de l'âge d'or. Les Troglodytes, vertueux comme le peuple de Bétique, sont des bergers: 'L'innocence des mœurs, la bonne foi, l'obéissance et l'horreur du vice habitent dans cette heureuse terre.' 'Tous les biens sont communs; les fruits des arbres, le lait des troupeaux sont des richesses si abondantes que des peuples si sobres et si modérés n'ont pas besoin de les partager.' Communisme et fraternité: 'Ils s'aiment tous d'un amour fraternel que rien ne trouble.' Même idéal arcadien qui exclut l'argent, le commerce, la vie urbaine, les conquêtes et la guerre. Le style même de Montesquieu imite l'onction fénelonienne…La religion n'est pas révélée aux Troglodytes. Elle est naturelle et dérive spontanément de cœurs vertueux." Montesquieu, *Lettres persanes*, edited and annotated by Paul Vernière (Paris: Garnier, 1960), 31n1–32, 33n1.
14. Otis E. Fellows and Norman L. Torrey, *The Age of Enlightenment*, 2nd edition (New York: Appleton-Century Crofts, 1971), 105n5.
15. J. Robert Loy, *Montesquieu* (New York: Twayne Publishers, Inc., 1968), 24.
16. Ibid., 42–43.
17. Ibid., 43.
18. Ibid., 54.
19. The causality between geographic latitude and tendencies towards virtue or vice is examined in Jean-Patrice Courtois, "Le Physique et le moral dans la théorie du climat chez Montesquieu" in *Le Travail des Lumières pour Georges Benrekassa*, edited by Caroline Jacot Grapa, *et al.* (Paris: Honoré Champion, 2002), 139–56; Christopher S. Jones, "Politicizing Travel and Climatizing Philosophy: Watsuji, Montesquieu and the European Tour," *Japan Forum* 14, no. 1 (March 1, 2002), 41–62; James W. Pennebaker, *et al.*, "Stereotypes of Emotional Expressiveness of Northerners and Southerners: A Cross-Cultural Test of Montesquieu's Hypotheses," *Journal of Personality and Social Psychology* 70, no. 2 (February 1996), 372–80; Judith N. Shklar, "Virtue in a Bad Climate; Good Men and Good Citizens in Montesquieu's *L'Esprit des lois*" in *Enlightenment Studies in Honour of Lester G. Crocker*, edited by Alfred J. Bingham and Virgil W. Topazio (Oxford: The Voltaire Foundation at the Taylor Institution, 1979), 315–28.

20. "...il s'arme d'un microscope pour donner à ce lieu commun de la sagesse des nations un fondement scientifique solide..." Montesquieu, *De l'esprit des lois*, introduced and annotated by J. Ehrard (Paris: Editions Sociales, 1969), 149.
21. Montesquieu, *The Spirit of the Laws*, translated and edited by Anne M. Cohler, Basia Carolyn Miller, and Harold Samuel Stone (Cambridge: Cambridge University Press, 1989), 231. "...le caractère de l'esprit, & les passions du cœur, soient extrêmement différentes dans les divers climats..." Montesquieu, *De l'esprit des loix*, in 2 volumes (Edinburgh: G. Hamilton and J. Balfour, 1750), 1:317.
22. Ibid. "L'air froid resserre les extrêmités des fibres extérieures de notre corps; cela augmente leur ressort, & favorise le retour du sang des extrêmités vers le cœur. Il diminue la longueur de ces mêmes fibres; il augmente donc encore par-la leur force. L'air chaud, au-contraire, relâche les extrêmités des fibres & les allonge; il diminue donc leur force & leur ressort." Ibid., 1:317–18.
23. Ibid., 231–32. "On a donc plus de vigueur dans les climats froids. L'action du cœur & la réaction des extrêmités des fibres s'y sont mieux, les liqueurs sont mieux en équilibre, le sang est plus déterminé vers le cœur, & réciproquement le cœur a plus de puissance." Ibid., 1:318.
24. Ibid., 232. "Cette force plus grande doit produire bien des effets, par exemple, plus de confiance en soi-même, c'est-à-dire, plus de courage; plus de connoissance de sa supériorité, c'est-à-dire, moins de desir de la vengeance; plus d'opinion de sa sureté, c'est-à-dire, plus de franchise, moins de soupçons, de politique, & de ruses. Enfin, cela doit faire des caractères bien différens." Ibid.
25. Ibid. "Approchez des païs du Midi, vous croirez vous éloigner de la morale même; des passions plus vives multiplieront les crimes; chacun cherchera à prendre sur les autres, tous les avantages qui peuvent favoriser ces mêmes passions." Ibid.
26. Ibid., 234. "Dans les païs du Nord, une machine saine & bien constituée, mais lourde, trouve ses plaisirs dans tout ce qui peut remettre les esprits en mouvement, la chasse, les voyages, la guerrre, le vin." Ibid., 1:321.
27. "Engin, instrument propre à faire mouvoir, à tirer, lever, traisner, lancer quelque chose. *Grande machine. machine admirable, merveilleuse, nouvelle machine, machine fort ingenieuse. machine de guerre. machine de ballet. machine qui lançoit de gros carreaux de pierre, qui decochoit cent traits à la fois. machine pour tirer de l'eau. machine à lever des pierres sur le haut d'un bastiment. machine hydraulique,* ou pour les eaux. *Inventer une machine. faire joüer une machine. cette machine joüe bien, va bien. l'effet d'une machine. les pieces, les ressorts d'une machine.*" "Machine," *Dictionnaire de l'Académie française* (Paris: Baptiste Coignard, 1694), 1.
28. "On appelle aussi, *Machine,* Certain assemblahge de ressorts dont le mouvement & l'effet se termine en luy-mesme. *L'horloge est une belle machine. les automates sont des machines fort ingenieuses.*" Ibid.
29. "On dit fig. *Que l'homme est un machine admirable.* Les Anciens Poëtes appelloient l'Univers, *La machine ronde.*" Ibid.
30. Julien Offray de La Mettrie, *Man a Machine* in *Man a Machine, including Frederick the Great's "Eulogy" on La Mettrie and Extracts from La Mettrie's "The Natural History of the Soul,"* notes by Gertrude Carman Bussey (Chicago: The Open Court Publishing Company, 1912), 148. "Concluons donc hardiment que l'Homme est une Machine; & qu'il n'y a dans tout l'Univers qu'une seule substance diversement

Notes 263

modifiée." Julien Offray de La Mettrie, *L'Homme-machine* in *Œuvres philosophiques* (London: Jean Nourse, 1751), 79.

31. Montesquieu, *The Spirit of the Laws*, translated and edited by Anne M. Cohler, Basia Carolyn Miller, and Harold Samuel Stone (Cambridge: Cambridge University Press, 1989), 242. "…chez les Anglois elle est l'effet d'une maladie, elle tient à l'état physique de la machine, & est indépendante de toute autre cause." Montesquieu, *De l'esprit des loix*, in 2 volumes (Edinburgh: G. Hamilton and J. Balfour, 1750), 1:332.
32. Ibid. "Il y a apparence que c'est un défaut de filtration du suc nerveux; la machine dont les forces motrices se trouvent à tout moment sans action, est lasse d'elle-même; l'ame ne sent point de douleur, mais une certaine difficulté de l'existence." Ibid.
33. "…loin d'établir un déterminisme unilatéral, il croit que le législateur peut et doit combattre les 'vices du climat' (chap. V à IX)." Montesquieu, *De l'esprit des lois*, introduced and annotated by J. Ehrard (Paris: Editions Sociales, 1969), 149.
34. C. P. Courtney, "Montesquieu and Natural Law," in *Montesquieu's Science of Politics: Essays on The Spirit of Laws*, edited by David W. Carrithers, Michael A. Mosher, and Paul A. Rahe (Lanham, MD: Rowman and Littlefield Publishers, Inc., 2001), 57.
35. Ibid., 58.
36. Ibid, 59.
37. Montesquieu, *The Spirit of Laws*, Book 19, Chapter 14.
38. "Selon Jean Deprun, dans l'introduction à son grand ouvrage sur *La Philosophie de l'inquiétude en France au XVIIIe siècle*…les hommes…voyant le Dieu paternel et créateur s'éloigner d'eux de plus en plus, ils se sont sentis abandonnés; chassés du centre de l'univers dont ils n'étaient plus la fin, ils ont été immergés dans le flux changeant des phénomènes et dans les enchaînements inépuisables de causes et d'effets." Henri Coulet, "Diderot et le problème du changement," *Recherches sur Diderot et sur l'Encyclopédie* 2 (April 1987): 59.
39. Ibid., 60.
40. Robert Shackleton, *Montesquieu: A Critical Biography* (Oxford: Oxford University Press, 1961), 312–13.
41. Ibid., 313.

CHAPTER THREE

1. Julien Offray de La Mettrie, *Man a Machine* in *Man a Machine, including Frederick the Great's "Eulogy" on La Mettrie and Extracts from La Mettrie's "The Natural History of the Soul,"* notes by Gertrude Carman Bussey (Chicago: The Open Court Publishing Company, 1912), 103. "…alors il ne se croioit pas Roi, n'étoit distingué du Singe et des autres Animaux…" Julien Offray de La Mettrie, *L'Homme-machine* in *Œuvres philosophiques* (London: Jean Nourse, 1751), 28.
2. Jacques Roger, *The Life Sciences in Eighteenth-Century French Literature* (Stanford: Stanford University Press, 1997), 398.

3. "Celui ou celle qui n'admet que la matière." "Matérialiste," *Dictionnaire de L'Académie française* (Paris: Brunet, 1762), 105.
4. "Opinion de ceux qui n'admettent point d'autre substance que la matière." "Matérialsme," *Dictionnaire de L'Académie française* (1762), 105.
5. "Materialism," *Webster's Third New International Dictionary of the English Language Unabridged* (Springfield: Merriam-Webster Inc., 1993), 1392.
6. René Descartes, *Letter to the Marquess of Newcastle, 23 November 1646* in *The Philosophical Writings of Descartes*, translated by John Cottingham, *et al.* (Cambridge: Cambridge University Press, 1991), 3:303. "…que notre corps n'est pas seulement une machine qui se remue de soi-même…" René Descartes, *Lettre au Marquis of Newcastle, le 23 novembre* in *Œuvres et Lettres*, edited by André Bridoux (Paris: Gallimard, 1953), 1255.
7. René Descartes, *Treatise on Man* in *The Philosophical Writings of Descartes*, translated by John Cottingham, *et al.* (Cambridge: Cambridge University Press, 1991), 1:99. "Nous voyons des horloges, des fontaines artificielles, des moulins, et autres semblables machines, qui n'étant faites que par des hommes, ne laissent pas d'avoir la force de se mouvoir d'elles-mêmes en plusieurs diverses façons." René Descartes, *Traité de l'Homme* in *Œuvres et Lettres*, edited by André Bridoux (Paris: Gallimard, 1953), 807.
8. The full sentence reads, "Thus every movement we make without any contribution from our will-as often happens when we breathe, walk, eat and, indeed, when we perform any action which is common to us and the beasts-depends solely on the arrangement of our limbs and on the route which the spirits, produced by the heat of the heart, follow naturally in the brain, nerves and muscles. This occurs in the same way as the movement of a watch is produced merely by the strength of its spring and the configuration of its wheels." René Descartes, *The Passions of the Soul, Article 16* in *The Philosophical Writings of Descartes*, translated by John Cottingham, *et al.* (Cambridge: Cambridge University Press, 1991), 1:335. "…tous les mouvements que nous faisons sans que notre volonté y contribue (comme il arrive souvent que nous respirons, que nous marchons, que nous mangeons, et enfin que nous faisons toutes les actions qui nous sont communes avec les bêtes) ne dépendent que de la conformation de nos membres et du cours que les esprits, excités par la chaleur du cœur, suivent naturellement dans le cerveau, dans les nerfs et dans les muscles, en même façon que le mouvement d'une montre est produit par la seule force de son ressort et la figure de ses roues." René Descartes, *Les Passions de l'âme, Article 16* in *Œuvres et Lettres*, edited by André Bridoux (Paris: Gallimard, 1953), 704.
9. Julien Offray de La Mettrie, *Treatise on the Soul* in *Machine Man and Other Writings*, translated and edited by Ann Thomson (Cambridge: Cambridge University Press, 1996), 43. "Ce n'est ni Aristote, ni Platon, ni Descartes, ni Malebranche, qui vous apprendront ce que c'est votre Ame. En vain vous vous tourmentez pour connoître la nature, n'en déplaise à votre vanité & à votre indocilité, il faut que vous vous soumettiez à l'ignorance & à la foi." Julien Offray de La Mettrie, *Traité de l'âme* in *Œuvres philosophiques* (London: Jean Nourse, 1751), 85.
10. Ernst Cassirer, *The Philosophy of the Enlightenment*, translated by Fritz C.A. Koelln and James P. Pettegrove (Princeton: Princeton University Press, 1979), 66–67.

11. Julien Offray de La Mettrie, *Treatise on the Soul* in *Machine Man and Other Writings*, translated and edited by Ann Thomson (Cambridge: Cambridge University Press, 1996), 45. "J'ouvre les yeux, & je ne vois autour de moi que matière, ou qu'étendue." Julien Offray de La Mettrie, *Traité de l'âme* in *Œuvres philosophiques* (London: Jean Nourse, 1751), 89.
12. "Extension," *Oxford English Dictionary Online*, http://www.oed.com (July 3, 2006).
13. Julien Offray de La Mettrie, *Treatise on the Soul* in *Machine Man and Other Writings*, translated and edited by Ann Thomson (Cambridge: Cambridge University Press, 1996), 50. "…du langage affectif, tel que les plaintes, les cris, les caresses, la fuite, les soupirs, le chant, & en un mot toutes les expressions de la douleur, de la tristesse, de l'aversion, de la craine, de l'audace, de la soumission, de la colère…" Julien Offray de La Mettrie, *Traité de l'âme* in *Œuvres philosophiques* (London: Jean Nourse, 1751), 98.
14. Ibid. "…car il n'est ici question que de la similitude des organes des sens, lesquels, à quelques modifications près, sont absolument les mêmes, & accusent évidemment les mêmes usages." Ibid., 99.
15. Ibid., 55. "Beaucoup d'expériences nous ont fait connoître que c'est effectivement dans le cerveau, que l'Ame est affectée des sensations propres à l'animal: car lorsque cette partie est considérablement blessée, l'animal n'a plus ni sentiment, ni discernement, ni connoissance…" Ibid., 109.
16. Ibid., 56. "…d'où comme de son trône, elle régit toutes les parties du corps." Ibid., 109.
17. Ibid., 58. "puisque les seuls nerfs moteurs portent à l'Ame l'idée des mouvemens…chaque nerf est propre à faire naître différentes sensations…" Ibid., 112.
18. Ibid. "Prenez un œil de bœuf, dépouillez-le adroitement de la sclérotique & de la choroïde; mettez où étoit la première de ces membranes, un papier dont la concavité s'ajuste parfaitement avec la convexité de l'œil. Présentez ensuite quelque corps que ce soit devant le trou de la pupille, vous verrez très-distinctement au fond de l'œil l'image de ce corps." Ibid., 113.
19. Ibid., 61. "Les idées de grandeur, de dureté, &c. ne sont déterminées que par nos organes. Avec d'autres sens, nous aurions des idées différentes des mêmes attributs, comme avec d'autres idées nous penserions autrement que nous ne pensons de tout ce qu'on appelle ouvrage de génie, ou de sentiment…D'ailleurs les sensations changent avec les organes; dans certaines jaunisses, tout paroît jaune. Changez avec le doigt l'axe de la vision, vous multiplierez les objets, vous en varierez à votre gré la situation & les attitudes. Les engelures, &c. font perdre l'usage du tact. Le plus petit embarras dans le canal d'Eustachi suffit pour rendre sourd." Ibid., 118–19.
20. Ibid., 67. "La cause de la mémoire est tout-à-fait mécanique, comme elle-même; elle paroît dépendre de ce que les impressions corporelles du cerveau, qui sont les traces d'idées qui se suivent, sont voisines; & que l'Ame ne peut faire la découverte d'une trace, ou d'une idée, sans rappeller les autres qui avoient coutume d'aller ensemble." Ibid., 129.
21. Ibid., 68. "L'imagination confond les diverses sensations incomplètes que la mémoire rappelle à l'Ame, & en forme des images, ou des tableaux, qui lui représentent des objets…différens des exactes sensations reçues autrefois par les sens." Ibid., 132.

22. Julien Offray de La Mettrie, *Man a Machine* in *Man a Machine, including Frederick the Great's "Eulogy" on La Mettrie and Extracts from La Mettrie's "The Natural History of the Soul,"* notes by Gertrude Carman Bussey (Chicago: The Open Court Publishing Company, 1912), 85-86. "Descartes, & tous les Cartésiens, parmi lesquels il y a long-tems qu'on a compté les Mallebranchistes, ont fait la même faute. Ils ont admis deux substances distinctes dans l'Homme, comme s'ils les avoient vuës & bien comptées." Julien Offray de La Mettrie, *L'Homme-machine* in *Œuvres philosophiques* (London: Jean Nourse, 1751), 10.
23. Aram Vartanian, *La Mettrie's L'Homme Machine: A Study in the Origins of an Idea; Critical Edition with an Introductory Monograph and Notes by Aram Vartanian* (Princeton: Princeton University Press, 1960), 13.
24. Julien Offray de La Mettrie, *Man a Machine* in *Man a Machine, including Frederick the Great's "Eulogy" on La Mettrie and Extracts from La Mettrie's "The Natural History of the Soul,"* notes by Gertrude Carman Bussey (Chicago: The Open Court Publishing Company, 1912), 88. "L'expérience & l'observation doivent donc seules nous guider ici. Elles se trouvent sans nombre dans les Fastes des Médecins, qui ont été Philosophes, & non dans les Philosophes, qui n'ont pas été Médecins. Ceux-ci ont parcouru, ont éclairé le Labyrinthe de l'Homme; ils nous ont seuls dévoilé ces ressorts cachés sous des envelopes, qui dérobent à nous yeux tant de merveilles." Julien Offray de La Mettrie, *L'Homme-machine* in *Œuvres philosophiques* (London: Jean Nourse, 1751), 13.
25. Ibid., 93. "Le corps humain est une Machine qui monte elle-même ses ressorts; vivante image du mouvement perpétuel. Les aliments entretiennent ce que la fièvre excite. Sans eux l'Ame languit, entre en fureur, & meurt abattuë." Ibid., 18.
26. Abraham Trembley, *Mémoires pour servir à l'histoire d'un genre de polypes d'eau douce, à bras en forme de cornes* (Paris: Durand, 1744). [*Memoirs for History of a Genus of Freshwater, Horn-shaped Polyps*"].
27. A discussion of the influence of Trembley's polyp on La Mettrie is found in Aram Vartanian, "Trembley's Polyp, La Mettrie, and Eighteenth-Century French Materialism," *Journal of the History of Ideas* 11(1950), 259–86.
28. Lester G. Crocker, "Diderot and Eighteenth-Century French Transformism," *Forerunners of Darwin: 1745–1859*, edited by Bentley Glass, *et al.*, (Baltimore: The Johns Hopkins Press, 1959), 117.
29. Ibid., 117n4. Crocker cites Julien Offray de La Mettrie, *L'Homme-machine*, edited by Maurice Solovine (Paris: Editions Bossard, 1921), 108–09.
30. Ibid., 117–18.
31. Ibid., 118.
32. Jacques Roger, *The Life Sciences in Eighteenth-Century French Thought*, edited by Keith R. Benson and translated by Robert Ellrich (Stanford: Stanford University Press, 1997), 396.
33. Ibid.
34. Ibid.
35. Aram Vartanian, *La Mettrie's L'Homme Machine: A Study in the Origins of an Idea; Critical Edition with an Introductory Monograph and Notes by Aram Vartanian* (Princeton: Princeton University Press, 1960), 21.
36. Vartanian's text reads, "…La Mettrie affirms that 'chaque petite fibre, au partie des corps organisés, se meut par un principe qui lui est propre, & dont l'action ne dépend point des nerfs.' With respect to the functional mode of the irritable

reaction, he observes further: 'Tel est le principe moteur des Corps entiers, ou des parties coupés en morceaux, qu'il produit des mouvemens non déréglés, comme on l'a cru, mais très réguliers.' The 'siège de cette force innée' is placed by him in the living tissues themselves." Ibid.
37. Ibid., 22.
38. Julien Offray de La Mettrie, *Man a Machine* in *Man a Machine, including Frederick the Great's "Eulogy" on La Mettrie and Extracts from La Mettrie's "The Natural History of the Soul,"* notes by Gertrude Carman Bussey (Chicago: The Open Court Publishing Company, 1912), 93. "Quelle puissance d'un Repas! La joie renaît dans un cœur triste; elle passe dans l'Ame des Convives qui l'expriment par d'aimables chansons..." Julien Offray de La Mettrie, *L'Homme-machine* in *Œuvres philosophiques* (London: Jean Nourse, 1751), 18.
39. Ibid., 94. "La viande crüe rend les animaux féroces; les Hommes le deviendront par la même nourriture." Ibid.
40. Ibid. "Il étoit homme à faire pendre l'innocent, comme le coupable." Ibid., 19.
41. Ibid., 95. "...tout dépend de la manière dont notre Machine est montée." Ibid.
42. Ibid. "A quels excès la faim cruelle peut nous porter! Plus de respect pour les entrailles auxquelles on doit, ou on a donné la vie; on les déchire à belles dents, on s'en fait d'horribles festins..." Ibid.
43. Ibid., 95. "...le plus foible est toujours la proie du plus fort." Ibid.
44. Ibid. "Il ne faut que des yeux pour voir l'Influence nécessaire de l'âge sur la Raison. L'âme suit les progrès du corps, comme ceux de l'Education." Ibid., 20.
45. Ibid., 96–97. "D'où cela vient-il, si ce n'est en partie, & de la nourriture qu'il prend, & de la semence de ses Pères..." Ibid., 21.
46. Bentley Glass, "Maupertuis, Pioneer of Genetics and Evolution" in *Forerunners of Darwin: 1745–1859*, edited by Bentley Glass, *et al* (Baltimore: The Johns Hopkins Press, 1959), 51–83.
47. Julien Offray de La Mettrie, *Man a Machine* in *Man a Machine, including Frederick the Great's "Eulogy" on La Mettrie and Extracts from La Mettrie's "The Natural History of the Soul,"* notes by Gertrude Carman Bussey (Chicago: The Open Court Publishing Company, 1912), 100. "Parmi les Animaux, les uns apprennent à parler & à chanter; ils retiennent des airs, & prennent tous les sons, aussi exactement qu'un Musicien." Julien Offray de La Mettrie, *L'Homme-machine* in *Œuvres philosophiques* (London: Jean Nourse, 1751), 25.
48. Ibid. "En un mot seroit-il absolument impossible d'apprendre une Langue à cet Animal? Je ne le crois pas." Ibid., 26.
49. Ibid. "...le grand Singe...nous ressemble si fort, que les Naturalistes l'ont apellé *Homme Sauvage*, ou Homme *des bois*." Ibid., 26.
50. Ibid., 101. "Pourquoi donc l'éducation des Singes seroit-elle impossible? Pourquoi ne pourroit-il enfin, à force de soins, imiter, à l'exemple des sourds, les mouvemens nécessaires pour prononcer? Je n'ose décider si les organes de la parole du Singe ne peuvent, quoiqu'on fasse, rien articuler; mais cette impossibilité absolüe me surprendroit, à cause de la grande Analogie du Singe & de l'Homme, & qu'il n'est point d'Animal connu jusqu'au présent, dont le dedans & le dehors lui ressemblent d'une manière si frappante." Ibid., 26–27.
51. Ibid., 103. "Des Animaux à l'Homme, la transition n'est pas violente; les vrais Philosophes en conviendront. Qu'étoit l'Homme, avant l'invention des Mots & la connoissance des Langues? Un Animal de son espèce, qui avec beaucoup moins

d'instinct naturel, que les autres, dont alors il ne se croioit pas Roi, n'étoit distingué du Singe & des autres Animaux, que comme le Singe l'est lui-même; je veux dire, par une physiognomie qui annonçoit plus de discernement. Réduit à la seule *connoissance intuitive* des Leibnitiens, il ne voioit que des Figures & des Couleurs, sans pouvoir rien distinguer entr'elles; vieux comme jeune, Enfant à tout âge, il bégaioit ses sensations & ses besoins, comme un chien affamé, ou ennuié du repos, demande à manger, ou à se promener.

Les Mots, les Langues, les Lois, les Sciences, les Beaux Arts sont venus; & par eux enfin le Diamant brut de notre esprit a été poli. On a dressé un Homme, comme un Animal..." Ibid., 28–29.

52. Ibid., 122. "Peut-être a-t-il été jetté au hasard sur un point de la surface de la Terre, sans qu'on puisse savoir ni comment, ni pourquoi; mais seulement qu'il doit vivre & mourir; semblable à ces champignons, qui paroissent d'un jour à l'autre, ou à ces fleurs qui bordent les fosses & couvrent les murailles." Ibid., 48.

53. Ibid., 135. "En faut-il davantage...pour prouver que l'Homme n'est qu'un Animal, ou un Assemblage de ressorts, qui tous se montent les uns par les autres, sans qu'on puisse dire par quel point du cercle humain la Nature a commencé? Ibid., 63.

54. Ibid., 140. "S'il a fallu plus d'instruments, plus de Rouages, plus de ressorts pour marquer les mouvemens des Planètes, que pour marquer les Heures, ou les répéter..." Ibid., 69.

55. Ibid., 143. "...ces êtres fiers & vains...ne sont au fond que des Animaux, & des Machines perpendiculairement rampantes." Ibid., 71.

56. William Harvey mentioned the rising point (*punctum saliens*) or when the heart first begins to beat in *Exercitatione de Generatione Animalium. Quibus accedunt Quædum de Partu: De Membranis ac Humoribus Uteri et de Conceptione* (London: Octavius Pulleyn, 1651), 149.

57. Julien Offray de La Mettrie, *Man a Machine* in *Man a Machine, including Frederick the Great's "Eulogy" on La Mettrie and Extracts from La Mettrie's "The Natural History of the Soul,"* notes by Gertrude Carman Bussey (Chicago: The Open Court Publishing Company, 1912), 145. "Telle est l'Uniformité de la Nature qu'on commence à sentir, & l'Analogie du règne Animal & Végétal, de l'Homme à la Plante. Peut-être même y a-t-il des Plantes Animales, c'est-à-dire qui en végétant, ou se battent comme les Polypes, ou font d'autres fonctions propres aux Animaux?" Julien Offray de La Mettrie, *L'Homme-machine* in *Œuvres philosophiques* (London: Jean Nourse, 1751), 74.

58. Ibid., 148. "Concluons donc hardiment que l'Homme est une Machine; & qu'il n'y a dans tout l'Univers qu'une seule substance diversement modifiée." Ibid., 79.

59. Julien Offray de La Mettrie, *Man a Plant* in *Man a Machine and Man a Plant*, translated by Richard A. Watson and Maya Rybalka, and introduced and annotated by Justin Leiber (Indianapolis: Hackett Publishing Company, Inc., 1994), 85. "Plus un Corps organisé a de besoins, plus la Nature lui a donné de moyens pour les satisfaire. Ces moyens sont les divers dégrés de cette Sagacité, connüe sous le nom d'Instinct dans les Animaux, & d'Ame dans l'Homme." Julien Offray de La Mettrie, *L'Homme-machine* in *Œuvres philosophiques* (London: Jean Nourse, 1751), 262.

60. "...les organes produisent les besoins, et réciproquement les besoins produisent les organes." Denis Diderot, *Rêve de d'Alembert* in *Œuvres complètes de Diderot*,

edited by Jean Assézat and Maurice Tourneux (Paris: Garnier, 1875–1877), 2:138–39.
61. Julien Offray de La Mettrie, *Man a Plant* in *Man a Machine and Man a Plant*, translated by Richard A. Watson and Maya Rybalka, and introduced and annotated by Justin Leiber (Indianapolis: Hackett Publishing Company, Inc., 1994), 85. "Moins un Corps organisé a de nécessités, moins il est dificile à nourrir & à élever, plus son partage d'Intelligence été mince." Julien Offray de La Mettrie, *L'Homme-plante* in *Œuvres philosophiques* (London: Jean Nourse, 1751), 262.
62. Ibid., 89. "...cette Echelle si impercepiblement graduée, qu'on voit la Nature exactement passer par tous les degrés, sans jamais sauter en quelque sorte un seul Echelon dans toutes ses productions diverses." Ibid., 267.
63. Lester G. Crocker, "Diderot and Eighteenth-Century French Transformism," *Forerunners of Darwin: 1745–1859*, edited by Bentley Glass, *et al.*, (Baltimore: The Johns Hopkins Press, 1959), 119.
64. Jacques Roger, *The Life Sciences in Eighteenth-Century French Thought*, edited by Keith R. Benson and translated by Robert Ellrich (Stanford: Stanford University Press, 1997), 397.
65. Julien Offray de La Mettrie, *Man a Plant* in *Man a Machine and Man a Plant*, translated by Richard A. Watson and Maya Rybalka, and introduced and annotated by Justin Leiber (Indianapolis: Hackett Publishing Company, Inc., 1994), 90. "Je sais que le Singe ressemble à l'Homme par bien d'autres choses que les Dents; l'Anatomie comparée en fait foi: quoiqu'elles ayent suffi à Linæus pour mettre l'Homme au rang des Quadrupèdes (à la tête à la vérité)." Julien Offray de La Mettrie, *L'Homme-plante* in *Œuvres philosophiques* (London: Jean Nourse, 1751), 270.
66. Ibid., 91. "...qui résulte visiblement de l'Organisation..." Ibid.
67. "...qu'elle ait ouvert son sein aux germes humains, déjà préparés, pour que ce superbe Animal, posées certaines loix, en pût éclore." Julien Offray de La Mettrie, *Le système d'Epicure* in *Œuvres philosophiques* (London: Jean Nourse, 1751), 334.
68. "...pourquoi la Terre, cette commune Mère & nourrice de tous les corps, auroit-elle refusé aux graines animales, ce qu'elle accorde aux végétaux les plus vils, les plus inutiles, les plus pernicieux?" Ibid.
69. "Les premières Générations ont dû être fort imparfaites. Ici l'Esophage aura manqué; là l'Estomac...les seuls animaux qui auront pû vivre, se conserver, & perpétuer leur espèce, auront été ceux qui se seront trouvés munis de toutes les Pièces nécessaires à la génération, & auxquels en un mot aucune partie essentielle n'aura manqué." Ibid., 335.
70. "La perfection n'a pas plus été l'ouvrage d'un jour pour la Nature, que pour l'Art." Ibid.
71. Denis Diderot, *Letter on the Blind* in *Diderot's Early Philosophical Works*, translated and edited by Margaret Jourdain (Chicago and London: The Open Court Publishing Company, 1916), 112. ""Voyez-moi bien, monsieur Holmes, je n'ai point d'yeux. Qu'avions-nous fait à Dieu, vous et moi, l'un pour avoir cet organe, l'autre pour en être privé?" Denis Diderot, *Lettre sur les aveugles* in *Œuvres complètes de Diderot*, edited by Jean Assézat et Maurice Tourneux (Paris: Garnier, 1875–1877), 1:310.
72. "Par quel infinité de combinaisons il a fallu que la nature ait passé, avant que d'arriver à celle-là seule de laquelle pouvoit résulter un Animal parfait!" Julien

Offray de La Mettrie, *Le système d'Epicure* in *Œuvres philosophiques* (London: Jean Nourse, 1751), 336–37.
73. "La Nature n'a plus songé de faire l'œil pour voir, que l'eau, pour servir de miroir à la simple Bergère." Ibid., 337.
74. *"Le hasard va souvent plus loin que la Prudence."* Ibid.
75. "Panspermia," *Oxford English Dictionary Online.* http://www.oed.com. July 19, 2006.
76. Jacques Roger, *The Life Sciences in Eighteenth-Century French Thought* (Stanford: Stanford University Press, 1997), 397.
77. Ibid., 398.
78. Ibid.
79. Ibid.
80. Ibid.
81. Lester G. Crocker, "Diderot and Eighteenth-Century French Transformism," *Forerunners of Darwin: 1745–1859*, edited by Bentley Glass, *et al.* (Baltimore: The Johns Hopkins Press, 1959), 125.
82. Aram Vartanian, "La Mettrie and Diderot Revisited: an Intertextual Encounter," *Diderot Studies* 21 (1983), 155–97.
83. Ibid., 155. Vartanian refers the reader to Jean E. Perkins, "Diderot and La Mettrie," *Studies on Voltaire and the Eighteenth Century* 10 (1959): 49–100; Leo Spitzer, "The Style of Diderot," in *Linguistics and Literary History* (Princeton: Princeton University Press, 1948), 188–89; Jean-Pierre Seguin, *Diderot, le discours et les choses: Essai de description du style d'un philosophe en 1750* (Paris: Klincksieck, 1978), 379–88; Jean Mayer's critical notes in Diderot's *Eléments de physiologie* (Paris: Marcel Didier, 1964); Paul Vernière's critical notes in Diderot's *Pensées sur l'interprétation de la nature* and *Le Rêve de d'Alembert* in Diderot's *Œuvres philosophiques* (Paris: Garnier, 1956).
84. Ibid., 157–58. "Non, Monsieur, je ne suis point l'auteur des *Pensées philosophiques.* Mes terres n'ont peut-être jamais porté de si beaux fruits." Julien Offray de La Mettrie,"Réponse à un libelle, inséré contre l'Auteur dans la *Bibliothèque Raisonnée* et dans les *Pensées Chrétiennes*" appended to the last volume of *Ouvrage de Pénélope; ou Machiavel en médecine* (Berlin, 1748–50), 3:360–61.
85. Ibid., 164.
86. Lester G. Crocker, "Diderot and Eighteenth-Century French Transformism," *Forerunners of Darwin: 1745–1859*, edited by Bentley Glass, *et al.* (Baltimore: The Johns Hopkins Press, 1959), 125.

CHAPTER FOUR

1. Georges-Louis Leclerc de Buffon, *The Hog, the Hog of Siam, and the Wild Boar* (1755) in *Natural History, General and Particular*, translated by William Smellie, 9 vols., 2nd edition (London: W. Strahan and T. Cadell, 1785), 3:504. "…il faut ne rien voir d'impossible, s'attendre à tout, et supposer que tout ce qui peut être, est. Les espèces ambigues, les productions irrégulières, les êtres anomaux cesseront dès-

lors de nous étonner..." Georges-Louis Leclerc de Buffon, *Le Cochon, le Cochon de Siam et le Sanglier* (1755) in *Histoire naturelle, générale et particulière*, 15 volumes (Paris: Imprimerie royale, 1749–1767), 5:102–03.
2. "Terme de Médecine. Traité sur le fœtus pendant son séjour dans la matrice." "Embryologie," *Dictionnaire de L'Académie française* (1762), 606.
3. Otis E. Fellows and Stephen F. Milliken, *Buffon* (New York: Twayne Publishers, Inc., 1972), 92.
4. Arthur O. Lovejoy, "Buffon and the Problem of Species," *Forerunners of Darwin: 1745–1859*, edited by Bentley Glass, *et al* (Baltimore: The John Hopkins Press, 1959), 94.
5. Georges-Louis Leclerc de Buffon, *Recapitulation* (1749) in *Natural History, General and Particular*, translated by William Smellie, 9 vols., 2nd edition (London: W. Strahan and T. Cadell, 1785), 2:346, 2:351–52. "Il existe donc une matière organique animée, universellement répandue dans toutes les substances animales ou végétales, qui sert également à leur nutrition, à leur développement et à leur reproduction...la reproduction ne se fait que par la même matière devenue surabondante au corps de l'animal ou du végétal; chaque partie du corps de l'un ou de l'autre renvoie les molécules organiques qu'elle ne peut plus admettre: ces molécules sont absolument analogues à chaque partie dont elles sont renvoyées, puisqu'elles étoient destinées à nourrir cette partie; dès-lors quand toutes les molécules renvoyées de tout le corps viennent à se rassembler, elles doivent former un petit corps semblable au premier, puisque chaque molécule est semblable à la partie dont elle a été renvoyée; c'est ainsi que se fait la reproduction dans toutes les espèces...Il n'y a donc point de germes préexistants, point de germes contenus à l'infini les uns dans les autres, mais il y a une matière organique toujours active, toujours prête à se mouler, à s'assimiler et à produire des êtres semblables à ceux qui la reçoivent: les espèces d'animaux ou de végétaux ne peuvent donc jamais s'épuiser d'elles-mêmes, tant qu'il subsistera des individus l'espèce sera toujours toute neuve, elle l'est autant aujourd'hui qu'elle l'étoit il y a trois mille ans..." Georges-Louis Leclerc de Buffon, *Récapitulation* (1749) in *Histoire naturelle, générale et particulière*, 15 volumes (Paris: Imprimerie royale, 1749–1767), 2:425–26. Arthur O. Lovejoy provides his own translation of this passage in his article, "Buffon and the Problem of Species," *Forerunners of Darwin: 1745–1859*, edited by Bentley Glass, *et al* (Baltimore: The John Hopkins Press, 1959), 94.
6. Otis E. Fellows and Stephen F. Milliken, *Buffon* (New York: Twayne Publishers, Inc., 1972), 93.
7. Ibid., 94.
8. Georges-Louis Leclerc de Buffon, *Analogies between Animals and Vegetables* in *History of Animals* (1749) in *Natural History, General and Particular*, translated by William Smellie, 9 vols., 2nd edition (London: W. Strahan and T. Cadell, 1785), 2:14. "...et qu'enfin le vivant et l'animé, au lieu d'être un degré métaphysique des êtres, est une propriété physique de la matière." Georges-Louis Leclerc de Buffon, *Comparaison des Animaux & des Végétaux* in *Histoire des animaux* (1749) in *Histoire naturelle, générale et particulière*, 15 volumes (Paris: Imprimerie royale, 1749–1767), 2:17.
9. Otis E. Fellows and Stephen F. Milliken, *Buffon* (New York: Twayne Publishers, Inc., 1972), 94–95.

10. Georges-Louis Leclerc de Buffon, *The Horse* (1753) in *Natural History, General and Particular*, translated by William Smellie, 9 vols., 2nd edition (London: W. Strahan and T. Cadell, 1785), 3:344–45. "Il y a dans la Nature un prototype général dans chaque espèce sur lequel chaque individu est modelé, mais qui semble, en se réalisant, s'altérer ou se perfectionner par les circonstances; en sorte que, relativement à de certaines qualités, il y a une variation bizarre en apparence dans la succession des individus, et en même temps une constance qui paroît admirable dans l'espèce entière: le premier animal, le premier cheval, par exemple, a été le modèle extérieur et le moule intérieur sur lequel tous les chevaux qui sont nés, tous ceux qui existent et tous ceux qui naîtront ont été formés; mais ce modèle, dont nous ne connoissons que les copies, a pû s'altérer ou se perfectionner en communiquant sa forme et se multipliant: l'empreinte originaire subsists en son entier dans chaque individu…" Georges-Louis Leclerc de Buffon, *Le cheval* (1753) in *Histoire naturelle, générale et particulière*, 15 volumes (Paris: Imprimerie royale, 1749–1767), 4:215–16.

11. Ibid., 3:345. "…mais quoiqu'il y en ait des millions, aucun de ces individus n'est cependant semblable en tout a un autre individu, ni par conséquent au modèle dont il porte l'empreinte: cette différence qui prouve combien la Nature est éloignée de rien faire d'absolu, et combien elle fait nuancer ses ouvrages, se trouve dans l'espèce humaine, dans celles de tous les animaux, de tous les végétaux, de tous les êtres en un mot qui se produisent…" Ibid., 4:216.

12. Georges-Louis Leclerc de Buffon, *The Lion* (1761) in *Natural History, General and Particular*, translated by William Smellie, 9 vols., 2nd edition (London: W. Strahan and T. Cadell, 1785), 5:65. "…le lion n'a jamais habité les régions du nord, le renne ne s'est jamais trouvé dans les contrées du midi, et il n'y a peut-être aucun animal dont l'espèce soit comme celle de l'homme généralement répandue sur toute la surface de la terre; chacun à son pays, sa patrie naturelle dans laquelle chacun est retenu par nécessité physique, chacun est fils de la terre qu'il habite, et c'est dans ce sens qu'on doit dire que tel ou tel animal est originaire de tel ou tel climat." Georges-Louis Leclerc de Buffon, *Le Lion* (1761) in *Histoire naturelle, générale et particulière*, 15 volumes (Paris: Imprimerie royale, 1749–1767), 9:2.

13. Georges-Louis Leclerc de Buffon, *The Camel and Dromedary* (1764) in *Natural History, General and Particular*, translated by William Smellie, 9 vols., 2nd edition (London: W. Strahan and T. Cadell, 1785), 6:123–25. "Il paroît être originaire d'Arabie; car non seulement c'est le pays où il est en plus grand nombre, mais c'est aussi celui auquel il est le plus conforme; l'Arabie est le pays du monde le plus aride, et où l'eau est la plus rare; le chameau est le plus sobre des animaux et peut passer plusieurs jours sans boire; le terrein est presque partout sec et sablonneux; le chameau a les pieds faits pour marcher dans les sables, et ne peut au contraire se soûtenir dans les terreins humides et glissans; l'herbe et les pâturges manquant à cette terre, le bœuf y manque aussi, et le chameau remplace cette bête de somme. On ne se trompe guère sur le pays naturel des animaux en le jugeant par ces rapports de conformité; leur vraie patrie est la terre à laquelle ils ressemblent, c'est-à-dire, à laquelle leur nature paroît s'être entièrement conformée: surtout lorsque cette même nature de l'animal ne se modifie point ailleurs et ne se prête pas à l'influence des autres climats. On a inutilement essayé de multiplier les chameaux en Espagne, on les a vainement transportés en Amérique, ils n'ont réussi ni dans l'un ni dans l'autre climat, et dans les grandes Indes on n'en trouve guère au delà de Surate et

d'Ormus." Georges-Louis Leclerc de Buffon, *Le Chameau et le dromadaire* (1764) in *Histoire naturelle, générale et particulière*, 15 volumes (Paris: Imprimerie royale, 1749–1767), 11:216–18.
14. Aristotle, *History of Animals: Books I-III*, translated by A.L. Peck (Cambridge, MA: Harvard University Press, 1965), 1.1.486a 15–25, pp. 3, 5.
15. Ibid., 1.1.486b 15–20, p. 7.
16. Georges-Louis Leclerc de Buffon, *The Ass* (1753) in *Natural History, General and Particular*, translated by William Smellie, 9 vols., 2nd edition (London: W. Strahan and T. Cadell, 1785), 3:402. "…si cette conformité constante et ce dessein suivi de l'homme aux quadrupèdes, des quadrupèdes aux cétacés, des cétacés aux oiseaux, des oiseaux aux reptiles, des reptiles aux poissons…ne semblent pas indiquer qu'en créant les animaux, l'Etre suprême n'a voulu employer qu'une idée, et la varier en même temps de toutes les manières possibles, afin que l'homme pût admirer également, et la magnificence de l'exécution, et la simplicité du dessein." Georges-Louis Leclerc de Buffon, *L'Asne* (1753) in *Histoire naturelle, générale et particulière*, 15 volumes (Paris: Imprimerie royale, 1749–1767), 4:381.
17. "Parcourant ensuite successivement & par ordre les différens objets qui composent l'Univers, & se mettant à la tête de tous les êtres créés, il verra avec étonnement qu'on peut descendre par des degrés presqu'insensibles, de la créature la plus parfaite jusqu'à la matière la plus informe, de l'animal le mieux organisé jusqu'au minéral le plus brut…" Georges-Louis Leclerc de Buffon, *Premier discours* (1749) in *Histoire naturelle, générale et particulière*, 15 volumes (Paris: Imprimerie royale, 1749–1767), 1:12.
18. "…on voit clairement qu'il est impossible de donner un système général, une méthode parfaite, non seulement pour l'Histoire Naturelle entière, mais même pour une seule de ses branches; car pour faire un système, un arrangement, en un mot une méthode générale, il faut que tout y soit compris; il faut diviser ce tout en différentes classes, partager ces classes en genres, sous-diviser ces genres en espèces, & tout cela suivant un ordre dans lequel il entre nécessairement de l'arbitraire. Mais la Nature marche par des gradations inconnues, & par conséquent elle ne peut pas se prêter totalement à ces divisions, puisqu'elle passe d'une espèce à une autre espèce, & souvent d'un genre à un autre genre, par des nuances imperceptibles; de sorte qu'il se trouve un grand nombre d'espèces moyennes & d'objets mi-partis qu'on ne sçait où placer, & qui dérangent nécessairement le projet du système général: cette vérité est trop importante pour que je ne l'appuie pas de tout ce qui peut la rendre claire & évidente." Ibid., 1:13.
19. Georges-Louis Leclerc de Buffon, *The Seals, Walrus, and Manati* (1765) in *Natural History, General and Particular*, translated by William Smellie, 9 vols., 2nd edition (London: W. Strahan and T. Cadell, 1785), 7:328–29. "Assemblons pour un instant tous les animaux quadrupèdes, faisons-en un groupe, ou plutôt formons-en une troupe dont les intervalles et les rangs représentent à peu près la proximité ou l'éloignement qui se trouve entre chaque espèce; plaçons au centre les genres les plus nombreux, et sur les flancs, sur les ailes ceux qui le sont le moins; resserrons-les tous dans le plus petit espace, afin de les mieux voir; et nous trouverons qu'il n'est pas possible d'arrondir cette enceinte: Que quoique tous les animaux quadrupèdes tiennent entr'eux de plus près qu'ils ne tiennent aux autres êtres, il s'en trouve néanmoins en grand nombre qui font des pointes au dehors, et semblent s'élancer pour atteindre à d'autres classes de la Nature; les singes tendent à

s'approcher de l'homme et s'en approchent en effet de très-près; les chauve-souris sont les singes des oiseaux qu'elles imitent par leur vol; les porc-épics. Les hérissons par les tuyaux dont ils sont couverts, semblent nous indiquer que les plumes pourroient appartenir à d'autres qu'aux oiseaux; les tatous par leur test écailleux s'approchent de la tortue et des crustacées; les castors par les écailles de leur queue ressemblent aux poissons; les fourmillers par leur espèce de bec ou de trompe sans dents et par leur longue langue, nous rappellent encore les oiseaux; enfin les Phoques, les Morses et les Lamantins font un petit corps à part qui forme la pointe la plus saillante pour arriver aux cétacées." Georges-Louis Leclerc de Buffon, *Les Phoques, les morses et les lamantins* (1765) in *Histoire naturelle, générale et particulière*, 15 volumes (Paris: Imprimerie royale, 1749–1767), 13:330–31.

20. Jacques Roger, *Buffon: A Life in Natural History* (Ithaca: Cornell University Press, 1997), 88. Roger cites "Premier discours," *Histoire, naturelle, générale et particulière* (Paris: Imprimerie royale, 1749–1767), 1:12.

21. Otis E. Fellows and Stephen F. Milliken, *Buffon* (New York: Twayne Publishers, Inc., 1972), 116.

22. "Quand on voit les métamorphoses successives de l'enveloppe du prototype, quel qu'il ait été, approcher un règne d'un autre règne par des degrés insensibles, et peupler les confins des deux règnes (s'il est permis de se servir du terme de *confins* où il n'y a aucune division réelle), et peupler, dis-je, les confins des deux règnes, d'êtres incertains, ambigus, dépouillés en grande partie des formes, des qualités et des fonctions de l'un, et revêtus des formes, des qualités, des fonctions de l'autre, qui ne se sentirait porté à croire qu'il n'y a jamais eu qu'un premier être prototype de tous les êtres? Mais, que cette conjecture philosophique soit admise avec le docteur Baumann, comme vraie, ou rejetée avec M. de Buffon comme fausse, on ne niera pas qu'il ne faille l'embrasser comme une hypothèse essentielle au progrès de la physique expérimentale…" Diderot, *De l'interprétation de la nature, Pensée 12*, in *Œuvres complètes de Diderot*, edited by Jean Assézat and Maurice Tourneux (Paris: Garnier, 1875–1877), 2:16.

23. Georges-Louis Leclerc de Buffon, *The Lion* (1761) in *Natural History, General and Particular*, translated by William Smellie, 9 vols., 2nd edition (London: W. Strahan and T. Cadell, 1785), 5:65. "…les différences mêmes des espèces semblent dépendre des différens climats; les unes ne peuvent se propager que dans les pays chauds, les autres ne peuvent subsister que dans des climats froids; le lion n'a jamais habité les régions du nord, le renne ne s'est jamais trouvé dans les contrées du midi, et il n'y a peut-être aucun animal dont l'espèce soit comme celle de l'homme généralement répandue sur toute la surface de la terre; chacun à son pays, sa patrie naturelle dans laquelle chacun est retenu par nécessité physique, chacun est fils de la terre qu'il habite, et c'est dans ce sens qu'on doit dire que tel ou tel animal est originaire de tel ou tel climat." Georges-Louis Leclerc de Buffon, *Le lion* (1761) in *Histoire naturelle, générale et particulière*, 15 volumes (Paris: Imprimerie royale, 1749–1767), 9:2.

24. Georges-Louis Leclerc de Buffon, *The Horse* (1753) in *Natural History, General and Particular*, translated by William Smellie, 9 vols., 2nd edition (London: W. Strahan and T. Cadell, 1785), 3:344–45. "…en sorte que pour avoir de bon grain, de belles fleurs, etc. il faut en échanger les graines et ne jamais les semer dans le même terrain qui les a produits; et de même, pour avoir de beaux chevaux, de bons

chiens, et. il faut donner aux femelles du pays des mâles étrangers, et réciproquement aux mâles du pays des femelles étrangères; sans cela les grains, les fleurs, les animaux dégénèrent, ou plutôt prennent une si forte teinture du climat, que la matière domine sur la forme et semble l'abâtardir: l'empreinte reste, mais défigurée par tous les traits qui ne lui sont pas essentials: en mêlant au contraire les races, et sur-tout en les renouvelant toujours par des races étrangères, la forme semble se perfectionner, et la Nature se relever et donner tout ce qu'elle peut produire de meilleur…on sait par expérience que des animaux ou des végétaux transplantés d'un climat lointain, souvent dégénèrent et quelquefois se perfectionnent en peu de temps, c'est-à-dire, en un très petit nombre de générations: il est aisé de concevoir que ce qui produit cet effet est la différence du climat et de la nourriture; l'influence de ces deux causes doit à la longue rendre ces animaux exempts ou susceptibles de certaines affections, de certaines maladies; leur tempérament doit changer peu à peu; le développement de la forme, qui dépend en partie de la nourriture et de la qualité des humeurs, doit donc changer aussi dans les générations…" Georges-Louis Leclerc de Buffon, *Le cheval* (1753) in *Histoire naturelle, générale et particulière*, 15 volumes (Paris: Imprimerie royale, 1749–1767), 4:216–17.

25. Georges-Louis Leclerc de Buffon, *Le mouflon et les autres brébis* (1764) in *Histoire naturelle, générale et particulière*, 15 volumes (Paris: Imprimerie royale, 1749–1767), 9:363–64.

26. Georges-Louis Leclerc de Buffon, *Le chameau et le dromadaire* (1764) in *Histoire naturelle, générale et particulière*, 15 volumes (Paris: Imprimerie royale, 1749–1767), 11:228–32.

27. Jacques Roger, *Buffon: A Life in Natural History* (Ithaca: Cornell University Press, 1997), 178.

28. Georges-Louis Leclerc de Buffon, *The Lion* (1761) in *Natural History, General and Particular*, translated by William Smellie, 9 vols., 2nd edition (London: W. Strahan and T. Cadell, 1785), 5:64–65. "Dans l'espèce humaine l'influence du climat ne se marque que par des variétés assez légères, parce que cette espèce est une, et qu'elle est très-distinctement séparée de toutes les autres espèces; l'homme, blanc en Europe, noir en Afrique, jaune en Asia, et rouge en Amérique, n'est que le même homme teint de la couleur du climat: comme il est fait pour régner sur la terre, que le globe entier est son domaine, il semble que sa nature se soit prêtée à toutes les situations; sous les feux du midi, dans les glaces du nord il vit, il multiplie, il se trouve partout si anciennement répandu, qu'il ne paroît affecter aucun climat particulier. Dans les animaux au contraire, l'influence du climat est plus forte…" Georges-Louis Leclerc de Buffon, *Le lion* (1761) in *Histoire naturelle, générale et particulière*, 15 volumes (Paris: Imprimerie royale, 1749–1767), 9:1–2.

29. Jacques Roger, *Buffon: A Life in Natural History* (Ithaca: Cornell University Press, 1997), 177. Roger cites from Georges-Louis Leclerc de Buffon, *Variétés* (1749) in *Histoire naturelle, générale et particulière*, 15 volumes (Paris: Imprimerie royale, 1749–1767), 3:483.

30. Ibid. Ibid., 3:446–47.

31. Georges-Louis Leclerc de Buffon, *On the Varieties of the Human Species* (1749) in *Natural History, General and Particular*, translated by William Smellie, 9 vols., 2nd edition (London: W. Strahan and T. Cadell, 1785), 3:206–07. "Tout concourt donc à prouver que le genre humain n'est pas composé d'espèces essentiellement

différentes entre elles, qu'au contraire il n'y a eu originairement qu'une seule espèce d'hommes, qui s'étant multipliée et répandue sur toutes la surface de la terre, a subi différens changemens par l'influence du climat, par la différence de la nourriture, par celle de la manière de vivre, par les maladies épidémiques, et aussi par le mélange varié à linfini des individus plus ou moins ressemblans; que d'abord ces altérations n'étoient pas si marquées, et ne produisoient que des variétés individuelles; qu'elles sont ensuite devenues variétés de l'espèce, parce qu'elles sont devenues plus générales, plus sensibles et plus constantes par l'action continuée de ces mêmes causes; qu'elles se sont perpétuées et qu'elles se perpétuent de génération en génération, comme les difformités ou les maladies des pères et mères passent à leurs enfans; et qu'enfin, comme elles n'ont été produites originairement que par le concours de causes extérieures et accidentelles, qu'elles n'ont été confirmées et rendues constantes que par le temps et l'action continuée de ces mêmes causes, il est très probable qu'elles disparoîtroient aussi peu à peu, et avec le temps, ou même qu'elles deviendroient différentes de ce qu'elles sont aujourd'hui, si ces mêmes causes ne subsistoient plus, ou si elles venoient à varier dans d'autres circonstances et par d'autres combinaisons." Georges-Louis Leclerc de Buffon, *Variétés* (1749) in *Histoire naturelle, générale et particulière*, 15 volumes (Paris: Imprimerie royale, 1749–1767), 3:529–30.

32. "De ce qui précede il suit que dans tout le nouveau continent que nous venons de parcourir, il n'y a qu'une seule & même race d'hommes, plus ou moins basanés. Les Américains sortent d'une même souche. Les Européens sortent d'une même souche. Du nord au midi on apperçoit les mêmes variétés dans l'un & l'autre hémisphere. Tout concourt donc à prouver que le genre *humain* n'est pas composé d'especes essentiellement différentes. La différence des blancs aux bruns vient de la nourriture, des moeurs, des usages, des climats; celle des bruns aux noir a la même cause. Il n'y a donc eu originairement qu'une seule race d'hommes, qui s'étant multipliée & répandue sur la surface de la terre, a donné à la longue toutes les variétés dont nous venons de faire mention; variétés qui disparoîtroient à la longue, si l'on pouvoit supposer que les peuples se déplaçassent tout-à-coup, & que les uns se trouvassent ou nécessairement ou volontairement assujettis aux mêmes causes qui ont agi sur ceux dont ils croient occuper les contrées." Denis Diderot, "Humaine espece (*Hist. nat.*)," *Encyclopédie, ou Dictionnaire raisonné des sciences, des arts et des métiers*, edited by Denis Diderot and Jean Le Rond d'Alembert (Paris: Briasson, David, Le Breton, Durant; Neuchâtel: S. Faulche, 1751–1765), 8:348.

33. Georges-Louis Leclerc de Buffon, *The History of Animals* (1749) in *Natural History, General and Particular*, translated by William Smellie, 9 vols., 2nd edition (London: W. Strahan and T. Cadell, 1785), 2:10. Georges-Louis Leclerc de Buffon, *L'histoire des animaux* (1749) in *Histoire naturelle, générale et particulière*, 15 volumes (Paris: Imprimerie royale, 1749–1767), 2:10–11.

34. Georges-Louis Leclerc de Buffon, *The Ass* (1753) in *Natural History, General and Particular*, translated by William Smellie, 9 vols., 2nd edition (London: W. Strahan and T. Cadell, 1785), 3:406–07. Georges-Louis Leclerc de Buffon, *L'Asne* (1753) in *Histoire naturelle, générale et particulière*, 15 volumes (Paris: Imprimerie royale, 1749–1767), 4:386–87.

35. Ibid., 3:411. Ibid., 4:391.

36. Georges-Louis Leclerc de Buffon, *The Goat* (1755) in *Natural History, General and Particular*, translated by William Smellie, 9 vols., 2nd edition (London: W. Strahan

Notes   277

and T. Cadell, 1785), 3:489–90. "…nous ignorons si le zèbre ne produiroit pas avec le cheval ou l'âne; si l'animal à large queue, auquel on a donné le nom de mouton de Barbarie, ne produiroit pas avec notre brebis; si le chamois n'est pas une chèvre sauvage; s'il ne formeroit pas avec nos chèvres quelque race intermédiaire; si les singes diffèrent réellement par les espèces, ou s'ils ne font, comme les chiens, qu'une seule et même espèce, mais variée par un grand nombre de races différentes; si le chien peut produire avec le renard et le loup; si le cerf produit avec la vache, la biche avec le dain, etc. Notre ignorance sur tous ces faits est, comme je l'ai dit, presque forcée, les expériences qui pourroient les décider demandant plus de temps, de soins et de dépense que la vie et la fortune d'un homme ordinaire ne peuvent le permettre. J'ai employé quelques années à faire des tentatives de cette espèce: j'en rendrai compte lorsque je parlerai des mulets; mais je conviendrai d'avance qu'elles ne m'ont fourni que peu de lumières, et que la plupart de ces épreuves ont été sans succès." Georges-Louis Leclerc de Buffon, *la Chèvre et la Chèvre d'Angora* (1755) in *Histoire naturelle, générale et particulière*, 15 volumes (Paris: Imprimerie royale, 1749–1767), 5:63. Arthur O. Lovejoy provides his own translation of Buffon's passage in his article, "Buffon and the Problem of Species," *Forerunners of Darwin: 1745–1859*, edited by Bentley Glass, *et al* (Baltimore: The John Hopkins Press, 1959), 95.

37. Arthur O. Lovejoy, *The Great Chain of Being* (Cambridge, MA; London: Harvard University Press, 2001), 230. Lovejoy cites Georges-Louis Leclerc de Buffon, *De la Nature, Seconde vue* (1765) in *Histoire naturelle, générale et particulière*, 15 volumes (Paris: Imprimerie royale, 1749–1767), 13:i.

38. "Ne pouroit-on pas expliquer par-là comment de deux seuls individus, la multiplication des especes les plus dissemblables auroit pû s'ensuivie?" Maupertuis, *Essai sur la formation des corps organisés* (Berlin, 1754), 40.

39. Georges-Louis Leclerc de Buffon, *The Ass* (1753) in *Natural History, General and Particular*, translated by William Smellie, 9 vols., 2nd edition (London: W. Strahan and T. Cadell, 1785), 3:400–01. "Si…nous choisissons un animal, ou même le corps de l'homme pour servir de base à nos connoissances, et y rapporter, par la voie de la comparaison, les autres êtres organisés, nous trouverons que…il existe en même temps un dessein primitif et général…le corps du cheval, par exemple, qui du premier coup d'œil paroît si différent du corps de l'homme, lorsqu'on vient à le comparer en détail et partie par partie, au lieu de surprendre par la différence, n'étonne plus que par la ressemblance singulière et presque complète qu'on y trouve…que l'on considère, comme l'a remarqué M. Daubenton, que le pied d'un cheval, en apparence si différent de la main de l'homme, est cependant composé des mêmes os, et que nous avons à l'extrémité de chacun de nos doigts, le même osselet en fer à cheval qui termine le pied de cet animal…" Georges-Louis Leclerc de Buffon, *L'Asne* (1753) in *Histoire naturelle, générale et particulière*, 15 volumes (Paris: Imprimerie royale, 1749–1767), 4:379–81. Lovejoy provides his own translation of Buffon's passage in his article, "Buffon and the Problem of Species," *Forerunners of Darwin: 1745-1859*, edited by Bentley Glass, *et al* (Baltimore: The John Hopkins Press, 1959), 96–97.

40. Ibid., 3:402-03. "…si l'on admet…que l'âne soit de la famille du cheval, et qu'il n'en diffère que parce qu'il a dégénéré, on pourra dire également que le singe est de la famille de l'homme, que c'est un homme dégénéré, que l'homme et le singe ont eu une origine commune comme le cheval et l'âne, que chaque famille, tant dans les

animaux que dans les végétaux, n'a eu qu'une seule souche, et même que tous les animaux sont venus d'un seul animal, qui, dans la succession des temps, a produit, en se perfectionnant et en dégénérant, toutes les races des autres animaux." Ibid., 4:382. Ibid., 97.
41. Ibid., 3:403–04. "Mais non, il est certain, par la révélation, que tous les animaux ont également participé à la grace de la création, que les deux premiers de chaque espèce et de toutes les espèces sont sortis tout formés des mains du Créateur, et l'on doit croire qu'ils étoient tels alors, à peu près, qu'ils nous sont aujourd'hui représentés par leurs descendans..." Ibid., 4:383. Ibid., 98.
42. Ibid., 3:404. "...depuis le temps d'Aristote jusqu'au nôtre, l'on n'a pas vû paroître d'espèces nouvelles, malgré le mouvement rapide qui entraîne, amoncelle ou dissipe les parties de la matière, malgré le nombre infini de combinaisons qui ont dû se faire pendant ces vingt siècles, malgré les accouplemens fortuits ou forcés des animaux d'espèces éloignées ou voisines, dont il n'a jamais résulté que des individus viciés et stériles, et qui n'ont pû faire souche pour de nouvelles générations." Ibid. Ibid., 98.
43. Ibid., 3:409–10. "...mais quel nombre immense et peut-être infini de combinaisons ne faudroit-il pas pour pouvoir seulement supposer que deux animaux, mâle et femelle, d'une certaine espèce, ont non seulement assez dégénéré pour n'être plus de cette espèce, c'est-à-dire, pour ne pouvoir plus produire avec ceux auxquels ils étoient semblables,mais encore dégénéré tous deux précisément au même point, et à ce point nécessaire pour ne pouvoir produire qu'ensemble!" Ibid., 4:389–90. Ibid., 98–99.
44. Ibid., 3:410–11. "...car si quelque espèce a été produite par la dégénération d'une autre, si l'espèce de l'âne vient de l'espèce du cheval, cela n'a pû se faire que successivement et par nuances, il y auroit eu entre le cheval et l'âne un grand nombre d'animaux intermédiaires...pourquoi ne verrions-nous pas aujourd'hui les représentans, les descendans de ces espèces intermédiaires? pourquoi n'en est-il demeuré que les deux extrêmes?" Ibid., 4:390–91. Ibid., 99.
45. Ibid., 3:410. "Quoiqu'on ne puisse donc pas démontrer que la production d'une espèce par la dégénération, soit une chose impossible à la Nature, le nombre des probabilités contraires est si énorme, que philosophiquement même on n'en peut guère douter..." Ibid., 4:390. Ibid., 99.
46. Arthur O. Lovejoy, "Buffon and the Problem of Species," *Forerunners of Darwin: 1745–1859*, edited by Bentley Glass, *et al* (Baltimore: The John Hopkins Press, 1959), 100.
47. Georges-Louis Leclerc de Buffon, *The Goat* (1755) in *Natural History, General and Particular*, translated by William Smellie, 9 vols., 2nd edition (London: W. Strahan and T. Cadell, 1785), 3:486. "Quoique les espèces dans les animaux soient toutes séparées par un intervalle que la Nature ne peut franchir, quelques-unes semblent se rapprocher par un si grand nombre de rapports, qu'il ne reste, pour ainsi dire, entre elles que l'espace nécessaire pour tirer la ligne de séparation...." Georges-Louis Leclerc de Buffon, *La Chèvre et la Chèvre d'Angora* (1755) in *Histoire naturelle, générale et particulière*, 15 volumes (Paris: Imprimerie royale, 1749–1767), 5:59.
48. Georges-Louis Leclerc de Buffon, *The Rat* (1758) in *Natural History, General and Particular*, translated by William Smellie, 9 vols., 2nd edition (London: W. Strahan and T. Cadell, 1785), 4:275–76. "Descendant par degrés du grand au petit, du fort au foible, nous trouverons que la Nature a sû tout compenser; qu'uniquement attentive à la conservation de chaque espèce, elle fait profusion d'individus, et se

soutient par le nombre dans toutes celles qu'elle a réduites au petit, ou qu'elle a laissées sans forces, sans armes et sans courage: et non seulement elle a voulu que ces espèces inférieures fussent en état de résister ou durer par le nombre; mais il semble qu'elle ait en même temps donné des supplémens à chacune, en multipliant les espèces voisines. Le rat, la souris...forment autant d'espèces distinctes et séparées..." Georges-Louis Leclerc de Buffon, *Le Rat* (1758) in *Histoire naturelle, générale et particulière*, 15 volumes (Paris: Imprimerie royale, 1749–1767), 7:278–79.

49. Georges-Louis Leclerc de Buffon, *Second View of Nature* (1765) in *Natural History, General and Particular*, translated by William Smellie, 9 vols., 2nd edition (London: W. Strahan and T. Cadell, 1785), 7:89–90. "Un individu, de quelque espèce qu'il soit, n'est rien dans l'Univers; cent individus, mille ne sont encore rien: les espèces sont les seuls êtres de la Nature; êtres perpétuels, aussi anciens, aussi permanens qu'elle; que pour mieux juger nous ne considérons plus comme une collection ou une suite d'individus semblables; mais comme un tout indépendant du nombre, indépendant du temps; un tout toujours vivant, toujours le même; un tout qui a été compté pour un dans les ouvrages de la création, et qui par conséquent ne fait qu'une unité dans la Nature...un jour, un siècle, un âge, toutes les portions du temps ne font pas partie de sa durée; le temps lui-même n'est relatif qu'aux individus, aux êtres dont l'existence est fugitive mais celle des espèces étant constante, leur permanence fait la durée, et leur différence le nombre. Comptons donc les espèces comme nous l'avons fait, donnons-leur à chacune un droit égal à la mense de la Nature; elles lui sont toues également chères, puisqu'à chacune elle a donné les moyens d'être, et de durer tout aussi longtemps qu'elle." Georges-Louis Leclerc de Buffon, *De la nature, Seconde vue* (1765) in *Histoire naturelle, générale et particulière*, 15 volumes (Paris: Imprimerie royale, 1749–1767), 13:i-ii.
50. Jacques Roger, *Buffon: A Life In Natural History* (Ithaca: Cornell University Press, 1997), 322–23.
51. Ibid., 323. Roger cites *L'Asne*, 4:383.
52. Jacques Roger, *The Life Sciences in Eighteenth-Century French Thought* (Stanford: Stanford University Press, 1997), 459, 672n214.
53. Ibid., 460, 672n216. Roger cites *Le Bœuf* (1753) in 4:470 and *Le Chien* (1755) in 5:210–17.
54. Diderot, *De l'Interprétation de la nature, Pensée 50*, in *Œuvres complètes de Diderot*, edited by Jean Assézat and Maurice Tourneux (Paris: Garnier, 1875–1877), 2:46, and Pierre-Louis Moreau de Maupertuis, *Essai sur la formation des corps organisés* (Berlin, 1754), 14, 18–19.
55. Ibid., 2:47. Ibid., 48-51.

CHAPTER FIVE

1. "...chaque degré d'erreur auroit fait une nouvelle espèce; & à force d'écarts répétés seroit venue la diversité infinie des animaux que nous voyons aujourd'hui..." Pierre-Louis de Maupertuis, *Essai sur la formation des corps organisés* (Berlin, 1754), 40–41.

2. Pierre-Louis Moreau de Maupertuis, *Vénus physique* (1745), second and third of unnumbered prefatory pages.
3. William Harvey, *Exercitatione de Generatione Animalium. Quibus accedunt Quædum de Partu: De Membranis ac Humoribus Uteri et de Conceptione* (London: Octavius Pulleyn, 1651), 148, and the English translation, *Anatomical Exercitations concerning the Generation of Living Creatures: To which are added Particular Discourses, of Births, and of Conceptions*, translated by Martin Llewellyn (London: Octavian Pullen, 1653), 272.
4. "Qu'un homme noir épouse une femme blanche, il semble que les deux couleurs soient mêlées; l'enfant naît olivâtre, & est mi-parti avec les traits de la mère & ceux du père." Pierre-Louis Mordeau de Maupertuis, *Vénus physique* (1745), 75.
5. "Voyage de Wafer, description de l'isthme de l'Amérique." Ibid., 125n.
6. Lionel Wafer, *A New Voyage and Description of the Isthmus of America* (London: James Kanpton, 1699), 133–36.
7. "Dans cet Isthme qui sépare la mer du Nord de la mer pacifique, on dit qu'on trouve des hommes plus blancs que tous ceux que nous connoissons: leurs cheveux seroient pris pour de la laine la plus blanche; leurs yeux trop foibles pour la lumiere du jour, ne s'ouvrent que dans l'obscurité de la nuit. Ils sont dans le genre des hommes ce que sont parmi les oiseaux, les chauve-souris & les hiboux. Quand l'astre du jour a disparu, & laisse la nature dans le deuil & dans le silence; quand tous les autres habitans de la terre accablés de leurs travaux, ou fatigués de leurs plaisirs, se livrent au sommeil; le Darien s'éveille, loue ses Dieux, se réjouit de l'absence d'une lumiere insupportable, & vient remplir le vide de la nature. Il écoute les cris de la chouette avec autant de plaisir que le berger de nos contrées entend le chant de l'alouette, lorsqu'à la premiere Aube, hors de la vue de l'épervier, elle semble aller chercher dans la nue le jour qui n'est pas encore sur la terre: elle marquee par le battement de ses ailes, la cadence de ses ramages; elle s'éleve & se perd dans la nue, on ne la voit plus, qu'on l'entend encore: ses sons qui n'ont plus rien de distinct, inspirent la tendresse & la rêverie; ce moment réunit la tranquillité de la nuit avec les plaisirs du jour. Le Soleil paroît: il vient rapporter sur la terre le mouvement & la vie, marquer les heures, & destine les différens travaux des hommes. Les Dariens n'ont pas attendu ce moment: ils sont déja tous retirés." Pierre-Louis Moreau de Maupertuis, *Vénus physique* (1745), 125–27.
8. "La nature contient le fonds de toutes ces variétés: mais le hasard ou l'art les mettent en œuvre. C'est ainsi que ceux dont l'industrie s'applique à satisfaire le goût des curieux, sont, pour ainsi dire, créateurs d'espèces nouvelles. Nous voyons paroître des races de chiens, de pigeons, de serins qui n'étoient point auparavant dans la nature. Ce n'ont été d'abord que des individus fortuits, l'art & les générations répétées en ont fait des espèces. Le fameux Lyonnès crée tous les ans quelqu'espece nouvelle, & détruit celle qui n'est plus à la mode. Il corrige les formes, & varie les couleurs: il a inventé les especes de l'*Arlequin*, du *Mopse*, &c." Ibid., 140–41.
9. "L'*arlequin* est une variété du petit Danois; mais au lieu que les Danois sont presque d'une seule couleur, les arlequins sont mouchetés, les uns blancs & noirs, les autres blancs & cannelés, les autres d'autre couleur." "Chien" *Encyclopédie, ou Dictionnaire raisonné des sciences, des arts et des métiers*, edited by Denis Diderot and Jean Le Rond d'Alembert (Paris: Briasson, David, Le Breton, Durant; Neuchâtel: S. Faulche, 1751–1765), 3:329.

10. "Pourquoi cet art se borne-t-il aux animaux? Pourquoi ces sultans blasés dans des serrails qui ne renferment que des femmes de toutes les espèces connues, ne se font-ils pas faire des espèces nouvelles?" Pierre-Louis Moreau de Maupertuis, *Vénus physique* (1745), 141.
11. "Un Roi du nord est parvenu à élever & embellir sa nation." Ibid., 142.
12. "Pour expliquer maintenant tous ces Phénomenes: la production des variétés accidentelles; la succession de ces variétés d'une génération à l'autre; & enfin l'établissement ou la destruction des espèces: voici ce me semble ce qu'il faudroit supposer." Ibid., 153–54.
13. "Le hasard, ou la disette des traits de famille feront quelquefois d'autres assemblages: & l'on verra naître de parens noirs un enfant blanc; ou peut-être même un noir, de parens blancs..." Ibid., 156.
14. "Si vous étiez assez généreux pour m'envoyer votre *Cosmologie*, je vous jurerais bien, par Newton et par vous, de n'en pas tirer de copie, et de vous la renvoyer après l'avoir lue." François-Marie Arouet de Voltaire, *Lettre à Maupertuis*, August 10, 1741, in *Œuvres de Voltaire avec préfaces, avertissements, notes, etc.*, 70 vols., edited by M. Beuchot (Paris: Lefèvre, 1829–1840), 54:390. Voltaire's letter is cited by Pierre Brunet, *Maupertuis* (Paris: A. Blanchard, 1929), 1:128; Jacques Roger, *The Life Sciences in Eighteenth-Century French Thought* (Stanford: Stanford University Press, 1997), 379 and 652n56; and *Maupertuis: Le savant et le philosophe; Présentation et extraits*, edited by Emile Callot (Paris: Marcel Rivière et Cie, 1964), 107.
15. "...et, en vérité, un homme qui a le malheur d'avoir lu la Cosmologie de Christian Wolf a besoin de la vôtre pour se dépiquer." Ibid. Cited by Jacques Roger, *The Life Sciences in Eighteenth-Century French Thought* (Stanford: Stanford University Press, 1997), 652n56.
16. Jacques Roger, *The Life Sciences in Eighteenth-Century French Thought* (Stanford: Stanford University Press, 1997), 652n56.
17. "Teleology," *Webster's Third New International Dictionary of the English Language Unabridged* (Springfield: Merriam-Webster Inc., 1993), 2350.
18. "Outre que pour expliquer une machine, on ne sçauroit mieux faire, que de proposer son but, & de montrer comment toutes les pièces y servent." Gottfried Wilhelm Leibniz, *Suite de la réponse aux réflexions sur les conséquences de quelques endroits de la philosophie de Descartes* (1697) in *Opera omnia*, 6 vols, edited by Louis Dutens (Geneva: De Tournes, 1768), 2:252.
19. "Si Dieu est Auteur des choses, & s'il est souverainement sage, on ne sçauroit assez bien raisonner de la structure de l'Univers, sans y faire entrer les vues de sa sagesse, comme on ne sçauroit assez bien raisonner sur un bâtiment, sans entrer dans les fins de l'Architecte." Ibid., 251.
20. "Mechanism," *Webster's Third New International Dictionary of the English Language Unabridged* (Springfield: Merriam-Webster Inc., 1993), 1401.
21. Jacques Roger, *The Life Sciences in Eighteenth-Century French Thought* (Stanford: Stanford University Press, 1997), 380.
22. "Mais ne pourrait-on pas dire que dans la combinaison fortuite des productions de la Nature, comme il n'y avait que celles où se trouvaient certains rapports de convenance, qui pussent subsister, il n'est pas merveilleux que cette convenance se trouve dans toutes les espèces qui actuellement existent? Le hasard, dirait-on, avait produit une multitude innombrable d'individus; un petit nombre se trouvait construit

de manière que les parties de l'animal pouvaient satisfaire à ses besoins; dans un autre infiniment plus grand, il n'y avait ni convenance, ni ordre: tous ces derniers ont péri; des animaux sans bouche ne pouvaient pas se perpétuer: les seuls qui soient restés sont ceux où se trouvaient l'ordre et la convenance; et ces espèces, que nous voyons aujourd'hui, ne sont que la plus petite partie de ce qu'un destin aveugle avait produit." Pierre-Louis Moreau de Maupertuis, *Essai de cosmologie* (1750), cited in *Maupertuis: Le savant et le philosophe; Présentation et extraits*, edited by Emile Callot (Paris: Marcel Rivière et Cie, 1964), 112- 13.

23. "Rapport, conformité. *ces choses là n'ont point de convenance l'une avec l'autre. quelle convenance y a-t-il entre des choses si différentes? pour bien discourir des choses, il en faut observer les convenances & les différences.*" "Convenance," *Dictionnaire de L'Académie française* (1694), 625.
24. "Convenance, ressemblance, conformité. *La langue italienne a un grand rapport a la langue latine. il y a un grand rapport d'humeurs entre ces deux hommes. le visage de cet homme a un grand rapport à celuy de l'autre. ce que vous dites n'a aucun rapport à ce que vous disiez hier.*" "Rapport," *Dictionnaire de L'Académie française* (1694), 281.
25. "Rapport qu'il y a entre les choses qui sont conformes. *Conformité d'inclinations. Conformité de sentiments. Conformité d'humeurs. Conformité d'esprit. Conformité d'Arrêts, de Traités.*" "Conformité," *Dictionnaire de L'Académie française* (1694), 365.
26. Jacques Roger, *The Life Sciences in Eighteenth-Century French Thought* (Stanford: Stanford University Press, 1997), 519. In 687n304, Roger cites Voltaire's criticism in the *Bibliothèque raisonnée*, July, August, and September 1752, in *Œuvres complètes de Voltaire*, 52 vols., edited by Louis Moland (Paris: Garnier, 1877–1885), 23:535–45.
27. François-Marie Arouet de Voltaire, *Eléments de la philosophie de Newton, Part 1, Chapter 1*, in *Œuvres complètes de Voltaire*, 52 vols., edited by Louis Moland (Paris: Garnier, 1877-1885), 22:403. Cited by Jacques Roger, *The Life Sciences in Eighteenth-Century French Thought* (Stanford: Stanford University Press, 1997), 517.
28. Ibid., 403-04. Ibid.
29. François-Marie Arouet de Voltaire, *Review of Charles Bonnet's Considérations sur les corps organisés* in *Mélanges*, April 4, 1764, in *Œuvres de Voltaire avec préfaces, avertissements, notes, etc.*, 70 vols., edited by M. Beuchot (Paris: Lefèvre, 1829–1840), 41:430–31. Ibid., 521.
30. "...avoir affirmé que les enfants se forment par attraction dans le ventre de la mère, que l'œil gauche attire la jambe droite..." François-Marie Arouet de Voltaire, *Histoire du docteur Akakia et du natif de Saint-Malo* in *Œuvres de Voltaire avec préfaces, avertissements, notes, etc.*, 70 vols., edited by M. Beuchot (Paris: Lefèvre, 1829–1840), 39:477.
31. Pierre-Louis Moreau de Maupertuis, *Essae de cosmologie* in *Œuvres*, 4 vols (Lyon: J.-M. Bruyset, 1756), 1:42–43.
32. François-Marie Arouet de Voltaire, *Bibliothèque raisonnée* in *Œuvres complètes de Voltaire*, 52 vols., edited by Louis Moland (Paris: Garnier, 1877–1885), 23:539. Cited by Jacques Roger, *The Life Sciences in Eighteenth-Century French Thought* (Stanford: Stanford University Press, 1997), 520.

33. Jacques Roger, *The Life Sciences in Eighteenth-Century French Thought* (Stanford: Stanford University Press, 1997), 520.
34. François-Marie Arouet de Voltaire, *Histoire du docteur Akakia et du natif de Saint-Malo* in *Œuvres de Voltaire avec préfaces, avertissements, notes, etc.*, 70 vols., edited by M. Beuchot (Paris: Lefèvre, 1829–1840), 39:473–91.
35. "Ce discours, rebattu d'après Lucrèce, est asez réfuté par la sensation donnée aux animaux, et par l'intelligence donnée à l'homme. Comment des combinaisons *que le hasard a produites* produiraient-elles cette sensation et cette intelligence..." François-Marie Arouet de Voltaire, "Athéisme," *Dictionnaire philosophique* in *Œuvres de Voltaire avec préfaces, avertissements, notes, etc.*, 70 vols., edited by M. Beuchot (Paris: Lefèvre, 1829–1840), 27:174–75.
36. "La disposition d'une aile de mouche, les organes d'un limaçon, suffisent pour vous atterrer." Ibid., 175.
37. "La tortue et le rhinocéros, et toutes les différentes espèces, prouvent également, dans leurs variétés infinies, la même cause, le même dessein, le même but, qui sont la conservation, la génération, et la mort. L'unité se trouve dans cette infinie variété; l'écaille et la peau rendent également témoignage." Ibid.
38. "Taisez-vous donc aussi, puisque vous ne concevez pas son utilité plus que moi..." Ibid., 176.
39. "Il y en a de venimeux, vous l'avez été vous-même." Ibid.
40. "Vous demandez pourquoi le serpent nuit. Et vous, pourquoi avez-vous nui tant de fois? pourquoi avez-vous été persécuteur, ce qui est le plus grand des crimes pour un philosophe?" Ibid.
41. "...mais les fripons! que sont-ils? des fripons." Ibid., 177.
42. "Fourbe, qui n'a ni honneur, ni foi, ni probité." "Fripon," *Dictionnaire de L'Académie française* (1762), 785.
43. "Trompeur, qui trompe avec finesse, avec adresse." "Fourbe," *Dictionnaire de L'Académie française* (1762), 773.
44. "*Jacob Ruhe*, chirugien à Berlin, est d'une de ces races. Né avec six doigts à chaque main et à chaque pied, il tient cette singularité de sa mère *Elisabeth Ruhe*, qui la tenait de sa mère *Elisabeth Horstmann*, de Rostock. Elisabeth Ruhe la transmit à quatre enfants des huit qu'elle eut de Jean Christian Ruhe, qui n'avait rien d'extraordinaire aux pieds ni aux mains. *Jacob Ruhe*, l'un de ces enfants sexdigitaires, épousa à Dantzic en 1733 Sophie-Louise de Thüngen, qui n'avait rien d'extraordinaire: il a eu six enfants; deux garçons ont été sexdigitaires. L'un d'eux, *Jacob Ernest*, a six doigts au pied gauche et cinq au droit: il avait à la main droite un sixième doigt, qu'on lui a coupé; à la gauche il n'a à la place du sixième doigt qu'une verrue." Pierre-Louis Moreau de Maupertuis, *Lettres* (1752), *Lettre 14*, cited in *Maupertuis: Le savant et le philosophe; Présentation et extraits*, edited by Emile Callot (Paris: Marcel Rivière et Cie, 1964), 158.
45. "Je veux bien croire que ces doigts surnuméraires dans leur première origine ne sont que des variétés accidentelles, dont j'ai essayé de donner la production dans la *Vénus Physique*: mais ces variétés une fois confirmées par un nombre suffisant de générations où les deux sexes les ont eues, fondent des espèces; et c'est peut-être ainsi que toutes les espèces se sont multipliées." Ibid., 158–59.
46. "Mais si l'on voulait regarder la continuation du sexdigitisme comme un effet du pur hasard, il faut voir quelle est la probabilité que cette variété accidentelle dans un premier parent ne se rétéra pas dans ses descendants.

Après une recherche que j'ai faite dans une ville qui a cent mille habitants, j'ai trouvé deux hommes qui avaient cette singularité. Supposons, ce qui est difficile, que trois autres me soient échappés; et que sur 20 000 hommes on puisse compter 1 sexdigitaire: la probabilité que son fils ou sa fille ne naîtra point avec le sexdigitisme est de 20 000 à 1: et celle que son fils et son petit-fils ne seront point sexdigitaires est de 20 000 fois 20 000, ou de 400 000 000 à 1: enfin, la probabilité que cette singularité ne se continuerait pas pendant trois générations consécutives serait de 8 000 000 000 000 à 1; nombres si grands que la certitude des choses les mieux démontrées en physique n'approche pas de ces probabilités." Ibid., 159.

47. "Null hypothesis," *Webster's Third New International Dictionary of the English Language Unabridged* (Springfield: Merriam-Webster Inc., 1993), 1548.
48. David Beeson, *Maupertuis: An Intellectual Biography* (Oxford: The Voltaire Foundation at the Taylor Institution, 1992), 235.
49. Bentley Glass, "Maupertuis, Pioneer of Genetics and Evolution" in *Forerunners of Darwin: 1745–1859*, edited by Bentley Glass, et. al. (Baltimore: The Johns Hopkins Press, 1959), 60.
50. Ibid., 64.
51. Ibid., 72. Glass cites from Maupertuis, *Letters* (1752), *Letter 14*.
52. *Dissertatio inauguralis metaphysica, de universali naturae systemate, pro gradu doctoris habita* (Erlangen, 1751). Very few copies of the work were published and the original Latin edition is virtually unattainable today.
53. "On ne fut plus en peine que pour savoir où placer ces magasins inépuisables d'individus..." Pierre-Louis Moreau de Maupertuis, *Essai sur la formation des corps organisés* (Berlin, 1754), 10.
54. "...& chacun pendant long-tems fut content de ses idées." Ibid.
55. Ibid., 11.
56. Pierre-Louis Moreau de Maupertuis, *Essai sur la formation des corps organisés* (Berlin, 1754), 14, 18–19.
57. "Je crois en voir la nécessité. Jamais on n'expliquera la formation d'aucun corps organisé, par les seules propriétés physiques de la matiere..." Ibid., 29.
58. "...il faudra bien en admettre encore de nouvelles, ou plutôt reconnoître les propriétés qui y sont." Ibid., 26.
59. Ibid., 40–41.
60. "Mais chaque élément, en déposant sa forme, & s'acumulant au corps qu'il va former, déposeroit-il aussi sa perception? Perdroit-il, afoibliroit-il le petit degré de sentiment qu'il avoit, ou l'augmenteroit-il par son union avec les autres, pour le profit du tout?" Ibid., 48.
61. "...mais chez nous il semble que de toutes les perceptions des élémens rassemblées, il en résulte une perception unique beaucoup plus forte, beaucoup plus parfaite qu'aucune des perceptions élémentaires, & qui est peut-être à chacune de ces perceptions dans le même raport que le corps organisé est à l'élément. Chaque élément, dans son union avec les autres, ayant confondu sa perception avec les leurs, & perdu le sentiment particulier du *soi*, le souvenir de l'état primitif des élémens nous manque, & notre origine doit être entièrement perdue pour nous." Ibid., 50–51.
62. "Les difficultés générales et communes aux deux systèmes ont été senties par un homme d'esprit, qui me paroît avoir mieux raisonné que tous ceux qui ont écrit avant lui sur cette matière, je veux parler de l'auteur de la *Vénus physique*, imprimé

Notes   285

en 1745; ce traité, quoique fort court, rassemble plus d'idées philosophiques qu'il n'y en a dans plusieurs gros volumes sur la génération...cet auteur est le premier qui ait commencé à se rapprocher de la vérité dont on étoit plus loin que jamais depuis qu'on avoit imaginé les œufs et découvert les animaux spermatiques." George-Louis Leclerc de Buffon, *Histoire naturelle, générale et particulière*, 15 vols. (Paris: Imprimerie royale, 1749–1767), 2:163–64. Cited by Pierre Brunet, *Maupertuis* (Paris: A. Blanchard, 1929, 2:329, and Bentley Glass, "Maupertuis, Pioneer of Genetics and Evolution," *Forerunners of Darwin: 1745–1859* (Baltimore: The Johns Hopkins Press, 1959), 77–78.

63. "...on ne niera pas qu'il ne faille l'embrasser comme une hypothèse essentielle au progrès de la physique expérimentale, à celui de la philosophie rationnelle, à la découverte et à l'explication des phénomènes qui dépendent de l'organisation." Denis Diderot, *De l'Interprétation de la nature, Pensée 12*, in *Œuvres complètes De Diderot*, edited by Jean Assézat and Maurice Tourneux (Paris: Garnier, 1875–1877), 2:16.

64. "...du docteur d'Erlangen, dont l'ouvrage, rempli d'idées singulières et neuves...Son objet est le plus grand que l'intelligence humaine puisse se proposer...L'hypothèse du docteur Baumann développera, si l'on veut, le mystère le plus incompréhensible de la nature...un sujet dont se sont occupés les premiers hommes dans tous les siècles...le fruit d'une méditation profonde, la tentative d'un grand philosophe." Ibid., 2:45, 48–49.

65. David Beeson, *Maupertuis: An Intellectual Biography* (Oxford: The Voltaire Foundation at the Taylor Institution, 1992), 249.

66. Denis Diderot, "Je lui demanderai donc si l'univers, ou la collection générale de toutes les molécules sensibles et pensantes, forme un tout...il monde, semblable à un grand animal, a une âme; que, le monde pouvant être infini, cette âme du monde, je ne dis pas est, mais peut être un système infini de perceptions, et que le monde peut être Dieu." Denis Diderot, *De l'Interprétation de la nature, Pensée 50*, in *Œuvres complètes De Diderot*, edited by Jean Assézat and Maurice Tourneux (Paris: Garnier, 1875–1877), 2:48. Cited in David Beeson, *Maupertuis: An Intellectual Biography* (Oxford: The Voltaire Foundation at the Taylor Institution, 1992), 250.

67. "Notre esprit, aussi borné qu'il est, trouvera-t-il jamais aucun système où toutes les consequences s'accordent...Tous nos systèmes, même les plus étendus, n'embrassent qu'une petite partie du plan qu'a suivie la suprême Intelligence; nous ne voyons ni le rapport des parties entr'elles, ni leur rapport avec le tout..." Pierre-Louis Moreau de Maupertuis, *Réponse aux objections de m. Diderot* in *Œuvres de Mr. de Maupertuis*, 4 vols. (Lyon: J.-M. Bruyset, 1756), 2:199. Ibid., 250.

68. Ibid., 251.

69. "Cette manière de raisonner, que M. Diderot appelle l'acte de la généralisation, & qu'il regarde comme la pierre de touche des systèmes, n'est qu'une espèce d'analogie, qu'on est en droit d'arrêter où l'on veut; incapable de prouver ni la fausseté ni la vérité d'un système." Maupertuis, *Œuvres de Mr de Maupertuis*, 4 vols. (Lyon: J.-M. Bruyset, 1756), 2:206. Ibid., 251.

70. Ibid.

CHAPTER SIX

Chapter 6 on Diderot copyright © 2007 from Diderot and the Metamorphosis of Species by Mary Efrosini Gregory. Reprinted by permission of Routledge, Inc., a division of Informa plc.

1. "Quand on voit les métamorphoses successives…approcher un règne d'un autre règne par des degrés insensibles, et peupler les confins des deux règnes…qui ne se sentirait porté à croire qu'il n'y a jamais eu qu'un premier être prototype de tous les êtres?" Diderot, *De l'Interprétation de la nature*, *Pensée 12*, in *Œuvres complètes de Diderot*, edited by Jean Assézat and Maurice Tourneux (Paris: Garnier, 1875–1877), 2:16.
2. Lester Crocker, "Diderot and Eighteenth-Century French Transformism" in *Forerunners of Darwin: 1745–1859*, edited by Bentley Glass, et al, (Baltimore: The Johns Hopkins Press, 1959), 114–43.
3. Diderot, *Letter on the Blind* in *Diderot's Early Philosophical Works*, translated and edited by Margaret Jourdain (Chicago and London: The Open Court Publishing Company, 1916), 112. "S'il n'y avait jamais eu d'êtres informes, vous ne manqueriez pas de prétendre qu'il n'y en aura jamais, et que je me jette dans des hypothèses chimériques; mais l'ordre n'est pas si parfait, continua Saunderson, qu'il ne paraisse encore de temps en temps des productions monstrueuses." Puis, se tournant en face du ministre, il ajouta: "Voyez-moi bien, monsieur Holmes, je n'ai point d'yeux. Qu'avions-nous fait à Dieu, vous et moi, l'un pour avoir cet organe, l'autre pour en être privé?" Diderot, *Lettre sur les aveugles* in *Œuvres complètes de Diderot*, edited by Jean Assézat and Maurice Tourneux (Paris: Garnier, 1875–1877), 1:310.
4. Denis Diderot, *Philosophic Thoughts* (1746), *Thought 18*, in *Diderot's Early Philosophical Works*, translated and edited by Margaret Jourdain (Chicago and London: The Open Court Publishing Company, 1916), 35. "Ce n'est pas de la main du métaphysicien que sont partis les grands coups que l'athéisme a reçus. Les méditations sublimes de Malebranche et de Descartes étaient moins propres à ébranler le matérialisme qu'une observation de Malpighi…Ce n'est que dans les ouvrages de Newton, de Muschenbroek, d'Hartzoeker et de Nieuwentit, qu'on a trouvé des preuves satisfaisantes de l'existence d'un être souverainement intelligent." Denis Diderot, *Pensées philosophiques* (1746), *Pensée 18*, in *Œuvres complètes de Diderot*, edited by Jean Assézat and Maurice Tourneux (Paris: Garnier, 1875–1877), 1:132–33.
5. Ibid. "Grâce aux travaux de ces grands hommes, le monde n'est plus un dieu, c'est une machine qui a ses roues, ses cordes, ses poulies, ses ressorts et ses poids." Ibid., 1:133.
6. Denis Diderot, *Philosophic Thoughts* (1746), *Thought 19*, in *Diderot's Early Philosophical Works*, translated and edited by Margaret Jourdain (Chicago and London: The Open Court Publishing Company, 1916), 35. "…c'est à la connaissance de la nature qu'il était réservé de faire de vrais déistes." Denis Diderot, *Pensées philosophiques* (1746), *Pensée 19*, in *Œuvres complètes de Diderot*, edited by Jean Assézat and Maurice Tourneux (Paris: Garnier, 1875–1877), 1:133.
7. Ibid. "…toutes les observations concourent à me démontrer que la putréfaction seule ne produit rien d'organisé…" Ibid.

Notes 287

8. "J'ouvre les cahiers d'un professeur célèbre…" Denis Diderot, *Lettres philosophiques*, Pensée 21, in *Œuvres complètes de Diderot*, edited by Jean Assézat and Maurice Tourneux (Paris: Garnier, 1875–1877), 1:135.
9. "Tous les commentateurs depuis Brière voient dans le 'professeur célèbre' D.F. Rivard, professeur de philosophie au collège de Beauvais." Diderot, *Pensées philosophiques* in *Œuvres philosophiques*, introduced and annotated by Paul Vernière (Paris: Garnier, 1998), 21, note 2.
10. "Diderot fut en effet son élève et vante plusieurs fois son sens pédagogique…Il se fit l'introducteur des mathématiques à l'université de Paris." Ibid., 21–22.
11. Denis Diderot, *Pensées philosophiques*, Pensée 21, in *Œuvres complètes de Diderot*, edited by Jean Assézat and Maurice Tourneux (Paris: Garnier, 1875–1877), 1:135–36.
12. Denis Diderot, *Philosophic Thoughts*, Thought 21, in *Diderot's Early Philosophical Works*, translated and edited by Margaret Jourdain (Chicago and London: The Open Court Publishing Company, 1916), 39–40. "Donc, l'esprit doit être plus étonné de la durée hypothétique du chaos que de la naissance réelle de l'univers." Denis Diderot, *Pensées philosophiques*, Pensée 21, in *Œuvres complètes de Diderot*, edited by Jean Assézat and Maurice Tourneux (Paris: Garnier, 1875–1877), 1:136.
13. Mary Efrosini Gregory, *Diderot and the Metamorphosis of Species* (New York and London: Routledge, 2007), 19–51.
14. "…si nous remontions à la naissance des choses et des temps, et que nous sentissions la matière se mouvoir et le chaos se débrouiller, nous rencontrerions une multitude d'êtres informes pour quelques êtres bien organisés." Denis Diderot, *Lettre sur les aveugles* in *Œuvres complètes de Diderot*, edited by Jean Assézat and Maurice Tourneux (Paris: Garnier, 1875–1877), 1:309.
15. Denis Diderot, *Letter on the Blind* in *Diderot's Early Philosophical Works*, translated and edited by Margaret Jourdain (Chicago and London: The Open Court Publishing Company, 1916), 111–12. "Je puis vous demander, par exemple, qui vous a dit à vous, à Leibnitz, à Clarke et à Newton, que dans les premiers instants de la formation des animaux, les uns n'étaient pas sans tête et les autres sans pieds? Je puis vous soutenir que ceux-ci n'avaient point d'estomac, et ceux-là point d'intestins; que tels à qui un estomac, un palais et des dents semblaient promettre de la durée, ont cessé par quelque vice du cœur ou des poumons; que les monstres se sont anéantis successivement; que toutes les combinaisons vicieuses de la matière ont disparu, et qu'il n'est resté que celles où le mécanisme n'impliquait aucune contradiction importante, et qui pouvaient subsister par elles-mêmes et se perpétuer." Denis Diderot, *Lettre sur les aveugles* in *Œuvres complètes de Diderot*, edited by Jean Assézat and Maurice Tourneux (Paris: Garnier, 1875–1877), 1:309.
16. sed quia multa modis multis primordia rerum
    ex infinito iam tempore percita plagis
    ponderibusque suis consuerunt concita ferri
    omnimodisque coire atque omnia pertemptare,
    quæcumque inter se possent

    but because many first-beginnings of things in many ways, struck with blows and carried along by their own weight from infinite time up to the present, have been accustomed to move and to meet in all manner of ways, and to try all combinations, whatsoever they could produce by coming

congressa creare, propterea fit uti magnum volgata per ævom, omne genus coctus et motus experiundo, tandem conveniant ea quæ convecta repente magnarum rerum fiunt exordia sæpe, terrai maris et cæli generique animantum. Titus Lucretius Carus, *De rerum natura*, translated by W.H.D. Rouse and revised by Martin Ferguson Smith (Cambridge, MA: Harvard University Press, 1924, 5.422–31, pp. 410, 412.	together, for this reason it comes to pass that being spread abroad through a vast time, by attempting every sort of combination and motion, at length those come together which, being suddenly brought together, often become the beginnings of great things, of earth and sea and sky and the generation of living creatures. Titus Lucretius Carus, *De rerum natura*, translated by W.H.D. Rouse and revised by Martin Ferguson Smith (Cambridge, MA: Harvard University Press, 1924), 5.422–31, pp. 411, 413.

17. Denis Diderot, *Lettre sur les aveugles* in *Œuvres complètes de Diderot*, edited by Jean Assézat and Maurice Tourneux (Paris: Garnier, 1875–1877), 1:309–11.
18. Denis Diderot, *Philosophic Thoughts, Thought 21*, in *Diderot's Early Philosophical Works*, translated and edited by Margaret Jourdain (Chicago and London: The Open Court Publishing Company, 1916), 112. "...que devenait le genre humain? il eût été enveloppé dans la dépuration générale de l'univers; et cet être orgueilleux qui s'appelle homme, dissous et dispersé entre les molécules de la matière, serait resté peut-être pour toujours, au nombre des possibles." Denis Diderot, *Lettre sur les aveugles* in *Œuvres complètes de Diderot*, edited by Jean Assézat and Maurice Tourneux (Paris: Garnier, 1875–1877), 1:310.
19. "Selon Jean Deprun, dans l'introduction à son grand ouvrage sur La Philosophie de l'inquiétude en France au XVIIIe siècle…les hommes…voyant le Dieu paternel et créateur s'éloigner d'eux de plus en plus, ils se sont sentis abandonnés; chassés du centre de l'univers dont ils n'étaient plus la fin, ils ont été immergés dans le flux changeant des phénomènes et dans les enchaînements inépuisables de causes et d'effets." Henri Coulet, "Diderot et le problème du changement," *Recherches sur Diderot et sur l'Encyclopédie* 2 (April 1987): 59.
20. Ibid., 63. Coulet cites Diderot, *Eléments de physiologie* in *Œuvres complètes*, 15 volumes, edited by Roger Lewinter (Paris: Le Club Français du Livre, 1969–1973), 13:764.
21. Lester Crocker, "The Idea of a 'Neutral' Universe" in *Diderot Studies* 21 (1983):67.
22. "Quand on voit les métamorphoses successives de l'enveloppe du prototype, quel qu'il ait été, approcher un règne d'un autre règne par des degrés insensibles, et peupler les confins des deux règnes (s'il est permis de se servir du terme de confins où il n'y a aucune division réelle), et peupler, dis-je, les confins des deux règnes, d'êtres incertains, ambigus, dépouillés en grande partie des formes, des qualités et des fonctions de l'un, et revêtus des formes, des qualités, des fonctions de l'autre, qui ne se sentirait porté à croire qu'il n'y a jamais eu qu'un premier être prototype de tous les êtres?" Denis Diderot, *De l'interprétation de la nature, Pensée 12*, in *Œuvres complètes de Diderot*, edited by Jean Assézat and Maurice Tourneux (Paris: Garnier, 1875–1877), 2:16.

23. "Omnes elementorum perceptiones conspirare, et in unam fortiorem et magis perfectam perceptionem coalescere videntur. Haec forte ad unamquamque ex aliis perceptionibus se habet in eadem ratione qua corpus organisatum ad elementum. Elementum quodvis, post suam cum aliis copulationem, cum suam perceptionem illarum perceptionibus confudit, et sui conscientiam perdidit primi elementorum status memoria nulla superest, et nostra nobis origo omnino abdita manet." Denis Diderot, *De l'interprétation de la nature*, Pensée 50, in *Œuvres complètes de Diderot*, edited by Jean Assézat and Maurice Tourneux (Paris: Garnier, 1875–1877), 2:47–48. The French translation of Maupertuis' *Inaugural Dissertation in Metaphysics*, entitled, *Essai sur la formation des corps organisés* (Berlin, 1754), rendered the passage thus: "Mais chez nous, il semble que de toutes les perceptions des éléments rassemblés, il en résulte une perception unique beaucoup plus forte, beaucoup plus parfaite qu'aucune des perceptions élémentaires, et qui est peut-être à chacune de ces perceptions dans le même rapport que le corps organisé à l'élément. Chaque élément, dans son union avec les autres, ayant confondu sa perception avec les leurs, et perdu le sentiment particulier du soi, le souvenir de l'état primitif des éléments nous manque, et notre origine doit être entièrement perdu pour nous."
24. Mary Efrosini Gregory, *Diderot and the Metamorphosis of Species* (New York and London: Routledge, 2007), 152.
25. "C'est qu'il a fallu que je fusse tel…" Diderot, *Rêve de d'Alembert* in *Œuvres complètes de Diderot*, edited by Jean Assézat and Maurice Tourneux (Paris: Garnier, 1875–1877), 2:137.
26. "Et si tout est un flux général, comme le spectacle de l'univers me le montre partout, que ne produiront point ici et ailleurs la durée et les vicissitudes de quelques millions de siècles?" Ibid.
27. "…garantissez-vous du sophisme de l'éphémère…" Ibid., 2:134.
28. "C'est celui d'un être passager qui croit à l'immortalité des choses." Ibid.
29. Diderot, *Letter on the Blind* in *Diderot's Early Philosophical Works*, translated and edited by Margaret Jourdain (Chicago and London: The Open Court Publishing Company, 1916), 113. "Je conjecture donc que, dans le commencement où la matière en fermentation faisait éclore l'univers, mes semblables étaient fort communs." Diderot, *Lettre sur les aveugles* in *Œuvres complètes de Diderot*, edited by Jean Assézat and Maurice Tourneux (Paris: Garnier, 1875–1877), 1:310.
30. "Suite indéfinie d'animalcules dans l'atome qui fermente, même suite indéfinie d'animalcules dans l'autre atome qu'on appelle la Terre. Qui sait les races d'animaux qui nous ont précédés? Qui sait les races d'animaux qui succéderont aux nôtres? Tout change, tout passe, il n'y a que le tout qui reste. Le monde commence et finit sans cesse; il est à chaque instant à son commencement et à sa fin; il n'en a jamais eu d'autre, et n'en aura jamais d'autre." Diderot, *Rêve de d'Alembert* in *Œuvres complètes de Diderot*, edited by Jean Assézat and Maurice Tourneux (Paris: Garnier, 1875–1877), 2:132.
31. "Corps qu'on regarde comme indivisible, à cause de sa petitesse. Démocrite & Epicure ont prétendu que le monde étoit composé d'atomes, que les corps se formoient par la rencontre fortuite des atomes." "Atome," *Dictionnaire de L'Académie française* (1762), 117.
32. "ATOME se dit aussi de cette petite poussière que l'on voit voler en l'air aux rayons du soleil." Ibid.

33. "Vous consentez donc que j'éteigne notre soleil?" Denis Diderot, *Entretien entre d'Alembert et Diderot* in *Œuvres complètes de Diderot*, edited by Jean Assézat and Maurice Touurneux (Paris: Garnier, 1875–1877), 2:111.
34. "D'autant plus volontiers que ce ne sera pas le premier qui se soit éteint." Ibid.
35. Denis Diderot, *Letter on the Blind* in *Diderot's Early Philosophical Works*, translated and edited by Margaret Jourdain (Chicago and London: The Open Court Publishing Company, 1916), 113. "Combien de mondes estropiés, manqués, se sont dissipés, se reforment et se dissipent peut-être à chaque instant dans des espaces éloignés, où je ne touche point..." Denis Diderot, *Lettre sur les aveugles* in *Œuvres complètes de Diderot*, edited by Jean Assézat and Maurice Tourneux (Paris: Garnier, 1875–1877), 1:310.
36. William Harvey, *Exercitatione de Generatione Animalium. Quibus accedunt Quædum de Partu: De Membranis ac Humoribus Uteri et de Conceptione* (London: Octavius Pulleyn, 1651), 148, and the English translation, *Anatomical Exercitations concerning the Generation of Living Creatures: To which are added Particular Discourses, of Births, and of Conceptions*, translated by Martin Llewellyn (London: Octavian Pullen, 1653), 272.
37. "...les molécules qui devaient former les premiers rudiments de mon geomètre étaient éparses dans les jeunes et frêles machines de l'une et l'autre..." Denis Diderot, *Entretien entre d'Alembert et Diderot* in *Œuvres complètes de Diderot*, edited by Jean Assézat and Maurice Tourneux (Paris: Garnier, 1875–1877), 2:109.
38. "...où le moindre brin ne peut être cassé, rompu., déplacé, manquant, sans conséquence facheuse pour le tout, devrait se nouer, s'embarrasser encore plus souvent dans le lieu de sa formation que mes soies sur ma tournette." Diderot, *Rêve de d'Alembert* in *Œuvres complètes de Diderot*, edited by Jean Assézat and Maurice Tourneux (Paris: Garnier, 1875–1877), 2:149.
39. The French translation of the original Latin is Pierre-Louis Moreau de Maupertuis, *Essai sur la formation des corps organisés* (Berlin, 1754), 37–38.
40. Ibid.
41. "Une chambre chaude, tapissée de petits cornets, et sur chacun de ces cornets une étiquette: guerriers, magistrats, philosophes, poëtes, cornet de courtisans, cornet de catins, cornet de rois." Diderot, *Rêve de d'Alembert* in *Œuvres complètes de Diderot*, edited by Jean Assézat and Maurice Tourneux (Paris: Garnier, 1875–1877), 2:131.
42. Lester Crocker, "Diderot and Eighteenth-Century French Transformism" in *Forerunners of Darwin: 1745–1859*, edited by Bentley Glass, et al, (Baltimore: The Johns Hopkins Press, 1959), 118.
43. Emita Hill, "Materialism and Monsters in Diderot's *Le Rêve de d'Alembert*," *Diderot Studies* 10 (1968): 67–93.
44. Emita Hill, "The Role of 'Le Monstre' in Diderot's Thought," *Studies on Voltaire and the Eighteenth Century* 97 (1972): 148–261.
45. Gerhardt Stenger, "L'ordre et les monstres dans la pensée philosophique, politique et morale de Diderot" in *Diderot et la question de la forme* (Paris: Presses Universitaires de France, 1999), 139–57.
46. Aurélie Suratteau, "Les hermaphrodites de Diderot" in *Diderot et la question de la forme* (Paris: Presses Universitaires de France, 1999), 105–37.

47. Johan Werner Schmidt, "Diderot and Lucretius: the *De Rerum Natura* and Lucretius' Legacy in Diderot's Scientific, Aesthetic and Ethical Thought," *Studies on Voltaire and the Eighteenth Century* 208 (1982): 183–294.
48. Christine M. Singh, "The Lettre sur les aveugles: Its Debt to Lucretius," *Studies in Eighteenth-Century French Literature Presented to Robert Niklaus*, edited by J.H. Fox, M.H. Waddicor and D.A. Watts (Exeter: University of Exeter, 1975), 233–42.
49. Henri Coulet, "Diderot et le problème du changement," *Recherches sur Diderot et sur l'Encyclopédie* 2 (April 1987): 59.
50. Ibid., 63. Coulet cites Diderot, *Eléments de physiologie* in *Œuvres complètes*, 15 volumes, edited by Roger Lewinter (Paris: Le Club Français du Livre, 1969–1973), 13:764.
51. "Mais si l'on voulait regarder la continuation du sexdigitisme comme un effet du pur hasard, il faut voir quelle est la probabilité que cette variété accidentelle dans un premier parent ne se rétéra pas dans ses descendants.

    Après une recherche que j'ai faite dans une ville qui a cent mille habitants, j'ai trouvé deux hommes qui avaient cette singularité. Supposons, ce qui est difficile, que trois autres me soient échappés; et que sur 20 000 hommes on puisse compter 1 sexdigitaire: la probabilité que son fils ou sa fille ne naîtra point avec le sexdigitisme est de 20 000 à 1: et celle que son fils et son petit-fils ne seront point sexdigitaires est de 20 000 fois 20 000, ou de 400 000 000 à 1: enfin, la probabilité que cette singularité ne se continuerait pas pendant trois générations consécutives serait de 8 000 000 000 000 à 1; nombres si grands que la certitude des choses les mieux démontrées en physique n'approche pas de ces probabilités." Pierre-Louis Moreau de Maupertuis, *Lettres* (1752), *Lettre 14*, cited in *Maupertuis: Le savant et le philosophe; Présentation et extraits*, edited by Emile Callot (Paris: Marcel Rivière et Cie, 1964), 159.
52. Bentley Glass, "Maupertuis, Pioneer of Genetics and Evolution" in *Forerunners of Darwin: 1745–1859*, edited by Bentley Glass, et al (Baltimore: The John Hopkins Press, 1959), 51–83.
53. Aram Vartanian, "Diderot and Maupertuis," *Revue internationale de philosophie* 148–49 (1984): 46–66.
54. Michèle Duchet, "L'anthropologie de Diderot," *Anthropologie et histoire au siècle des Lumières: Buffon, Voltaire, Rousseau, Helvétius, Diderot* (Paris: Maspéro, 1971), 407–75.
55. Jean Ehrard, "Diderot, l'Encyclopédie, et l'Histoire et théorie de la Terre," *Buffon 88* (Paris: Vrin, 1988), 135–42.
56. Arthur O. Lovejoy, "Buffon and the Problem of Species" in *Forerunners of Darwin, 1745–1859*, edited by Bentley Glass et al. (Baltimore: The Johns Hopkins Press, 1959), 84–113.
57. Jacques Roger. *Buffon: A Life in Natural History* (Ithaca: Cornell University Press, 1997).
58. Jacques Roger, "Diderot et Buffon en 1749," *Diderot Studies* 4 (1963): 221–36.
59. Jacques Roger, *The Life Sciences in Eighteenth-Century French Thought* (Stanford: Stanford University Press, 1997).
60. Aram Vartanian, "Buffon et Diderot," *Buffon 88* (Paris: Vrin, 1988), 119–33.
61. Jean E. Perkins, "Diderot and La Mettrie," *Studies on Voltaire and the Eighteenth Century* 10 (1959): 49–100.

62. Ann Thomson, "La Mettrie et Diderot," http://www.sigu7.jussieu.fr/diderot/travaux/revseance2.htm (Jan. 24, 2006).
63. Ann Thomson, "L'unité matérielle de l'homme chez La Mettrie et Diderot," *Colloque International Diderot* (1985): 61–68.
64. Aram Vartanian, "La Mettrie and Diderot Revisited: An Intertextual Encounter," *Diderot Studies* 21 (1983): 155–97.
65. Aram Vartanian, "Trembley's Polyp, La Mettrie, and Eighteenth-Century French Materialism," *Journal of the History of Ideas* 2 (1950): 259–86.
66. Marx W. Wartofsky, "Diderot and the Development of Materialist Monism," *Diderot Studies* 2 (1952): 279–329.

CHAPTER SEVEN

1. Jean-Jacques Rousseau, *Discourse on the Origin of Inequality* in *The Social Contract and Discourses*, translated and introduced by G.D.H. Cole (London: J.M. Dent and Sons, 1913), 177. "…je le supposerai conformé de tout temps, comme je le vois aujourd'hui, marchant à deux pieds, se servant de ses mains comme nous faisons des nôtres, portant ses regards sur toute la nature, et mesurant des yeux la vaste étendue du ciel." Jean-Jacques Rousseau, *Discours sur l'origine et les fondements de l'inégalité parmi les hommes* in *Œuvres complètes de Jean-Jacques Rousseau avec les notes de tous les commentateurs* (Paris: Dalibon, 1826), 1:244–45.
2. Jean-Jacques Rousseau, *Discourse on the Origin of Inequality* in *The Discourses and Other Early Political Writings*, edited and translated by Victor Gourevitch (Cambridge: Cambridge University Press, 2005), 192n3(2).
3. Ibid.
4. Jean-Jacques Rousseau, *Discourse on the Origin of Inequality* in *The Social Contract and Discourses*, translated and introduced by G.D.H. Cole (London: J.M. Dent and Sons, 1913), 168. "et comment l'homme viendra-t-il à bout de se voir tel que l'a formé la nature, à travers tous les changements que la succession des temps et des choses a dû produire dans sa constitution originelle, et de démêler ce qu'il tient de son propre fonds d'avec ce que les circonstances et ses progrès ont ajouté ou changé à son état primitif?" Jean-Jacques Rousseau, *Discours sur l'origine et les fondements de l'inégalité parmi les hommes* in *Œuvres complètes de Jean-Jacques Rousseau avec les notes de tous les commentateurs* (Paris: Dalibon, 1826), 1:229.
5. Ibid. "Semblable à la statue de Glaucus, que le temps, la mer et les orages avoient tellement défigurée qu'elle ressembloit moins à un dieu qu'à une bête féroce, l'âme humaine, altérée au sein de la société par mille causes sans cesse renaissantes, par l'acquisition d'une multitude de connoissances et d'erreurs, par les changements arrivés à la constitution des corps, et par le choc continuel des passions, a pour ainsi dire changé d'apparence au point d'être presque méconnoissable; et l'on n'y retrouve plus, au lieu d'un être agissant toujours par des principes certains et invariables, au lieu de cette céleste et majestueuse simplicité dont son auteur l'avoit empreinte, que le difforme contraste de la passion qui croit raisonner, et de l'entendement en délire." Ibid., 1:229–30.

6. Ibid. "Ce qu'il y a de plus cruel encore c'est que tous les progrès de l'espèce humaine l'éloignant sans cesse de son état primitif…" Ibid., 1:230.
7. Ibid., 177. "Je n'examinerai pas si, comme le pense Aristote, ses ongles allongés ne furent point d'abord des griffes crochues; s'il n'étoit point velu comme un ours; et si, marchant à quatre pieds, ses regards dirigés vers la terre, et bornés à un horizon de quelques pas, ne marquoient point à la fois le caractère et les limites de ses idées. Je ne pourrois former sur ce sujet que des conjectures vagues et presque imaginaires. L'anatomie comparée a fait encore trop peu de progrès, les observations des naturalistes sont encore trop incertaines, pour qu'on puisse établir sur de pareils fondements la base d'un raisonnement solide…" Ibid., 1:244.
8. Jean-Jacques Rousseau, *Discourse on the Origin of Inequality* in *The Discourses and Other Early Political Writings*, edited and translated by Victor Gourevitch (Cambridge: Cambridge University Press, 2005), 191–92n3(2). Ibid., 1:348–49.
9. Jean-Jacques Rousseau, *Discourse on the Origin of Inequality* in *The Social Contract and Discourses*, translated and introduced by G.D.H. Cole (London: J.M. Dent and Sons, 1913), 177. "…je vois un animal moins fort que les uns, moins agile que les autres, mais, à tout prendre, organisé le plus avantageusement de tous…" Ibid., 1:245.
10. "Parcourant ensuite successivement & par ordre les différens objets qui composent l'Univers, & se mettant à la tête de tous les êtres créez, il verra avec étonnement qu'on peut descendre par des degrés presqu'insensibles, de la créature la plus parfaite jusqu'à la matière la plus informe, de l'animal le mieux organisé jusqu'au minéral le plus brut…" Georges-Louis Leclerc de Buffon, *Premier discours* (1749) in *Histoire naturelle, générale et particulière*, 15 volumes (Paris: Imprimerie royale, 1749–1767), 1:12.
11. Jean-Jacques Rousseau, *Discourse on the Origin of Inequality* in *The Social Contract and Discourses*, translated and introduced by G.D.H. Cole (London: J.M. Dent and Sons, 1913), 178. "Accoutumés dès l'enfance aux intempéries de l'air et à la rigueur des saisons, exercés à la fatigue…" Jean-Jacques Rousseau, *Discours sur l'origine et les fondements de l'inégalité parmi les hommes* in *Œuvres complètes de Jean-Jacques Rousseau avec les notes de tous les commentateurs* (Paris: Dalibon, 1826), 1:245.
12. Ibid. "La nature en use précisément avec eux comme la loi de Sparte avec les enfants des citoyens; elle rend forts et robustes ceux qui sont bien constitués, et fait périr tous les autres…" Ibid., 1:246.
13. Ibid. "…différente en cela de nos sociétés, où l'état, en rendant les enfants onéreux aux pères, les tue indistinctement avant leur naissance." Ibid.
14. Ibid., 182. "Le cheval, le chat, le taureau, l'âne même, ont la plupart une taille plus haute, tous une constitution plus robuste, plus de vigueur, de force et de courage dans les forêts que dans nos maisons…" Ibid., 1:253.
15. Ibid. "Il en est ainsi de l'homme même: en devenant sociable et esclave il devient foible, craintif, rampant…" Ibid.
16. Ibid., 184. "Je ne vois dans tout animal qu'une machine ingénieuse, à qui la nature a donné des sens pour se remonter elle-même, et pour se garantir, jusqu'à un certain point, de tout ce qui tend à la détruire ou à la déranger. J'apperçois précisément les mêmes choses dans la machine humaine, avec cette différence que la nature seule fait tout dans les opérations de la bête, au lieu que l'homme concourt aux siennes en

qualité d'agent libre. L'un choisit ou rejette par instinct, et l'autre par un acte de liberté..." Ibid., 1:255–56.

17. "...il doit se ranger lui-même dans la classe des animaux, auxquels il ressemble par tout ce qu'il a de matériel." Georges-Louis Leclerc de Buffon, *Premier discours* (1749) in *Histoire naturelle, générale et particulière*, 15 volumes (Paris: Imprimerie royale, 1749–1767), 1:12.

18. Aristotle, *History of the Animals*, translated by A.L. Peck (Cambridge, MA: Harvard University Press, 2001), 1.1.488a5–10. p. 15.

19. "Le premier pas et le plus difficile que nous ayions à faire pour parvenir à la connoissance de nous-mêmes, est de reconnoître nettement la nature des deux substances qui nous composent; dire simplement que l'une est inétendue, immatérielle, immortelle, et que l'autre est étendue, matérielle et morelle..." Georges-Louis Leclerc de Buffon, *Histoire naturelle de l'homme, Chapter 1*, in *Histoire naturelle, générale et particulière*, 15 volumes (Paris: Imprimerie royale, 1749–1767), 2:430.

20. "L'existence de notre âme nous est démontrée, ou plutôt nous ne faisons qu'un, cette existence et nous: être et penser, sont pour nous la même chose, cette vérité est intime et plus qu'intuitive, elle est indépendante de nos sens, de notre imagination, de notre mémoire, et de toutes nos autres facultés relatives." Ibid., 2:432.

21. Jacques Roger, *Buffon: A Life in Natural History*, translated by Sarah Lucille Bonnefoi and edited by L. Pearce Williams (Ithaca: Cornell University Press, 1997), 256–57.

22. Jean-Jacques Rousseau, *Discourse on the Origin of Inequality* in *The Social Contract and Discourses*, translated and introduced by G.D.H. Cole (London: J.M. Dent and Sons, 1913), 169. "...j'ai hasardé quelques conjectures...car ce n'est pas une légère entreprise...de bien connoître un état qui n'existe plus, qui n'a peut-être point existé, qui probablement n'existera jamais..." Jean-Jacques Rousseau, *Discours sur l'origine et les fondements de l'inégalité parmi les hommes* in *Œuvres complètes de Jean-Jacques Rousseau avec les notes de tous les commentateurs* (Paris: Dalibon, 1826), 1:230–31.

23. Jacques Roger, *Buffon: A Life in Natural History*, translated by Sarah Lucille Bonnefoi and edited by L. Pearce Williams (Ithaca: Cornell University Press, 1997), 257.

24. Ibid.

25. Ibid., 258–59.

26. Otis Fellows, "Buffon and Rousseau: Aspects of a Relationship," *PMLA* 75, no. 3 (June 1960): 184.

27. Ibid., 190.

28. Ibid. "...peut-être verroit-il clairement que la vertu appartient à l'homme sauvage plus qu'à l'homme civilisé, et que le vice n'a pris naissance que dans la société." Georges-Louis Leclerc de Buffon, *Variétés dans l'espèce humaine* in *Histoire naturelle, générale et particulière*, 15 volumes (Paris: Imprimerie royale, 1749–1767), 3:492–93.

29. Ibid.

30. Ibid.

31. Jean-Jacques Rousseau, *Discourse on the Origin of Inequality* in *The Discourses and Other Early Political Writings*, edited and translated by Victor Gourevitch (Cambridge: Cambridge University Press, 2005), 189–190n2. Jean-Jacques

Rousseau, *Discours sur l'origine et les fondements de l'inégalité parmi les hommes* in *Œuvres complètes de Jean-Jacques Rousseau avec les notes de tous les commentateurs* (Paris: Dalibon, 1826), 1:346n. Rousseau cites Buffon, *De la nature de l'homme* in *Histoire naturelle*, 1749 edition, 2:429.

32. Jean Morel, "Recherches sur les sources du *Discours de l'inégalité*" in *Annales de la Société Jean-Jacques Rousseau*, 1909, pp. 179ff.
33. Jean-Jacques Rousseau, *Discourse on the Origin of Inequality, Note 10*, in *The Discourses and Other Early Political Writings*, edited and translated by Victor Gourevitch (Cambridge: Cambridge University Press, 2005), 205. "On trouve, dit le traducteur de l'*Histoire des Voyages* dans le royaume de Congo, quantité de ces grands animaux qu'on nomme *orangs-outangs* aux Indes orientales, qui tiennent comme le milieu entre l'espèce humaine et les babouins." Jean-Jacques Rousseau, *Discours sur l'origine et les fondements de l'inégalité parmi les hommes* in *Œuvres complètes de Jean-Jacques Rousseau avec les notes de tous les commentateurs* (Paris: Dalibon, 1826), 1:368n.
34. "...quelque ressemblance qu'il y ait entre l'Hottentot et le singe, l'intervalle qui les sépare est immense, puisqu'à l'intérieur il est rempli par la pensée et au dehors par la parole." Georges-Louis Leclerc de Buffon, *Nomenclature of Monkeys* (1776) in *Histoire naturelle, générale et particulière*, 15 volumes (Paris: Imprimerie royale, 1749–1767), 14:32.
35. Jean-Jacques Rousseau, *Discourse on the Origin of Inequality, Note 10*, in *The Discourses and Other Early Political Writings*, edited and translated by Victor Gourevitch (Cambridge: Cambridge University Press, 2005), 211. "...je dis que quand de pareils observateurs affirmeront d'un tel animal que c'est un homme, et d'un autre que c'est une bête, il faudra les en croire..." Jean-Jacques Rousseau, *Discours sur l'origine et les fondements de l'inégalité parmi les hommes* in *Œuvres complètes de Jean-Jacques Rousseau avec les notes de tous les commentateurs* (Paris: Dalibon, 1826), 1:376n.
36. "Je l'avoue, si l'on ne devoit juger que par la forme, l'espèce du singe pourroit être prise pour une variété dans l'espèce humaine...quelque ressemblance qu'il y ait entre l'Hottentot et le singe, l'intervalle qui les sépare est immense, puisqu'à l'intérieur il est rempli par la pensée et au dehors par la parole." Georges-Louis Leclerc de Buffon, *Nomenclature of Monkeys* (1776) in *Histoire naturelle, générale et particulière*, 15 volumes (Paris: Imprimerie royale, 1749–1767), 14:32.
37. Lord Monboddo, *Of the Origin and Progress of Language*, 2nd ed. (Edinburgh, 1774), 1:290.
38. Otis Fellows, "Buffon and Rousseau: Aspects of a Relationship," *PMLA* 75, no. 3 (June 1960): 192. "Mais non: il est certain, par la révélation, que tous les animaux ont également participé à la grâce de la création..." Georges-Louis Leclerc de Buffon, *L'asne* (1753) in *Histoire naturelle, générale et particulière*, 15 volumes (Paris: Imprimerie royale, 1749–1767), 4:383.
39. Jean-Jacques Rousseau, *Discourse on the Origin of Inequality, Note 10*, in *The Discourses and Other Early Political Writings*, edited and translated by Victor Gourevitch (Cambridge: Cambridge University Press, 2005), 206–07. "Dapper confirme que...Cette bête...est si semblable à l'homme, qu'il est tombé dans l'esprit à quelques voyageurs qu'elle pouvoit être sortie d'une femme et d'un singe...on trouve dans la description de ces prétendus monstres des conformités frappantes avec l'espèce humaine, et des différences moindres que celles qu'on pourroit

assigner d'homme à homme." Jean-Jacques Rousseau, *Discours sur l'origine et les fondements de l'inégalité parmi les hommes* in *Œuvres complètes de Jean-Jacques Rousseau avec les notes de tous les commentateurs* (Paris: Dalibon, 1826), 1:369n–70n.

40. Otis Fellows, "Buffon and Rousseau: Aspects of a Relationship," *PMLA* 75, no. 3 (June 1960): 191n21. Fellows cites Arthor O. Lovejoy, "Monboddo and Rousseau" in *Essays in the History of Ideas* (Baltimore: Johns Hopkins Press, 1948), 40 and 51.
41. Francis Moran III, "Of Pongos and Men: *Orangs-Outang* in Rousseau's *Discourse on Inequality*," *The Review of Politics* 57, no. 4 (Autumn 1995): 641.
42. Ibid., 643. In 643n4 Moran cites Victor Gourevitch, "Rousseau's Pure State of Nature," *Interpretation* 16 (1988): 23–60.
43. Ibid., 652–53.
44. Benoît de Maillet, *Telliamed*, translated by Albert O. Carozzi (Urbana, IL: University of Illinois Press, 1968), 201.
45. Francis Moran III, "Of Pongos and Men: *Orangs-Outang* in Rousseau's *Discourse on Inequality*," *The Review of Politics* 57, no. 4 (Autumn 1995): 657.
46. Francis Moran III, "Between Primates and Primitives: Natual man as the Missing Link in Rousseau's *Decond Discourse*," *Journal of the History of Ideas* 54, no. 1 (January 1993): 38. Moran cites Rousseau, *Discourse on the Origin and Foundation of Inequality among Men* in *Jean-Jacques Rousseau: The First and Second Discourses*, ed. Roger D. Masters (New York, 1964), 150.
47. Ibid. Ibid., 204.
48. Ibid., 38–39.
49. Jean Starobinski, "Rousseau and Buffon" in *Jean-Jacques Rousseau: Transparency and Obstruction*, translated by Arthur Goldhammer and introduced by Robert Mossisey (Chicago: University of Chicago Press, 1988), 323–32.
50. Ibid., 325.
51. Ibid., 326.
52. Ibid.
53. Ibid., 326–27.
54. Ibid., 327.
55. Ibid., 329.
56. Ibid.
57. Ibid., 330.
58. Leonard Sorenson, "Natural Inequality and Rousseau's Political Philosophy in his *Discourse on Inequality*," *The Western Political Quarterly* 43, no. 4 (December 1990): 763.
59. Ibid., 764.
60. Ibid., 784.
61. David Gauthier, *Rousseau: The Sentiment of Existence* (Cambridge: Cambridge University Press, 2006), 2–19.
62. Ibid., 10.
63. Ibid.
64. Ibid., 17.
65. Ibid.
66. Ibid.
67. Ibid.

CHAPTER EIGHT

1. François-Marie Arouet de Voltaire, *A Defence of My Uncle, Translated from the French of M. de Voltaire*, Chapter 19 ("Of Mountains and Shells") (London: S. Bladon, 1768), 105. "On a beau me dire que le porphyre est fait de pointes d'oursin, je le croirai quand je verrai que le marbre blanc est fait de plumes d'autruche." François-Marie Arouet de Voltaire, *La Défense de mon oncle* (1767), Chapître 19 ("Des montagnes et des coquilles"), in *Œuvres de Voltaire avec préfaces, avertissements, notes, etc.*, 70 vols., edited by M. Beuchot (Paris: Lefèvre, 1829–1840), 43:374.
2. "Système de ceux qui n'ayant aucun culte particulier, & rejetant toute sorte de révélation, croient seulement un souverain Etre. *Etre soupconné de déisme.*" "Déisme," *Dictionnaire de L'Académie française* (1762), 488.
3. "Celui ou celle qui reconnoît un Dieu, mais qui ne reconnoît aucune Religion révélée. *C'est un Déiste.*" "Déiste," *Dictionnaire de L'Académie française* (1762), 488.
4. "Deism," *Oxford English Dictionary Online*, 1, http://dictionary.oed.com (Aug. 21, 2006).
5. "Nous sommes des êtres intelligents; or, des êtres intelligents ne peuvent avoir été formés par un être brut, aveugle, insensible: il y a certainement quelque différence entre les idées de Newton et des crottes de mulet. L'intelligence de Newton venait donc d'une autre intelligence." François-Marie Arouet de Voltaire, "Athéisme," *Section 2, Dictionnaire philosophique* in *Œuvres de Voltaire avec préfaces, avertissements, notes, etc.*, 70 vols., edited by M. Beuchot (Paris: Lefèvre, 1829–1840), 27:170–71.
6. "J'ai cependant connu des mutins qui disent qu'il n'y a point d'intelligence formatrice, et que le mouvement seul a formé par lui-même tout ce que nous voyons et tout ce que nous sommes. Ils vous disent hardiment: La combinaison de cet univers était possible, puisqu'elle existe: donc il était possible que le mouvement seul l'arrangeât." Ibid., 27:172.
7. "La disposition d'une aile de mouche, les organes d'un limaçon, suffisent pour vous atterrer." Ibid., 27:175.
8. "Mais quand on la regarde attentivement, ce grand fantôme s'évanouit…" François-Marie Arouet de Voltaire, "Chaîne des êtres créés," *Dictionnaire philosophique* in *Œuvres de Voltaire avec préfaces, avertissements, notes, etc.*, 70 vols., edited by M. Beuchot (Paris: Lefèvre, 1829–1840), 27:560.
9. "Cette chaîne, cette gradation prétendue n'existe pas plus dans les végétaux et dans les animaux…" Ibid., 27:561.
10. "…la preuve en est qu'il y a des espèces de plantes et d'animaux qui sont détruites." Ibid.
11. "Nous n'avons plus de murex." Ibid.
12. "Il était défendu aux Juifs de manger du griffon et de l'ixion; ces deux espèces ont probablement disparu de ce monde…" Ibid.
13. "…où donc est la chaîne?" Ibid.
14. "…il est visible qu'on en peut détruire. Les lions, les rhinocéros commencent à devenir fort rares." Ibid.

15. "Il est probable qu'il y a eu des races d'hommes qu'on ne retrouve plus." Ibid., 27:562.
16. "N'y a-t-il pas donc visiblement un vide entre le singe et l'homme?" Ibid.
17. Arthur O. Lovejoy, *The Great Chain of Being* (Cambridge, MA, and London: Harvard University Press, 2001), 252.
18. Ibid. "N'y a-t-il pas visiblement un vide entre le singe et l'homme? n'est-il pas aisé d'imaginer un animal à deux pieds sans plumes…" François-Marie Arouet de Voltaire, "Chaîne des êtres créés," *Dictionnaire philosophique* in *Œuvres de Voltaire avec préfaces, avertissements, notes, etc.*, 70 vols., edited by M. Beuchot (Paris: Lefèvre, 1829–1840), 27:562.
19. Ibid. "Par-delà l'homme, vous logez dans le ciel, divin Platon, une file de substances célestes…" Ibid.
20. Ibid. "Mais vous, quelle raison avez-vous d'y croire?" Ibid.
21. Ibid. "Quelle gradation, je vous prie, entre vos planètes! la Lune est quarante fois plus petite que notre globe. Quand vous avez voyagé de la Lune dans le vide, vous trouvez Vénus; elle est environ aussi grosse que la terre…Où est la gradation prétendue?" Ibid., 27:562–63.
22. François-Marie Arouet de Voltaire, "Great Chain of Being," *Philosophical Dictionary*, edited and translated by Theodore Besterman (London: Penguin Books, 2004), 109. "Et puis, comment voulez-vous que dans de grands espaces vides il y ait une chaîne qui lie tout?" Ibid., 27:563.
23. Arthur O. Lovejoy, *The Great Chain of Being* (Cambridge, MA, and London: Harvard University Press, 2001), 253. "…mais pourquoi, et comment une existence infinie? Newton a démontré le vide, qu'on n'avait fait que supposer jusqu'à lui. S'il y a du vide dans la nature, le vide peut donc être hors de la nature. Quelle nécessité que les êtres s'étendent à l'infini? que serait-ce que l'infini en étendue? Il ne peut exister non plus qu'en nombre." François-Marie Arouet de Voltaire, *Il faut prendre un parti, ou le Principe d'action* (1772) in *Œuvres de Voltaire avec préfaces, avertissements, notes, etc.*, 70 vols., edited by M. Beuchot (Paris: Lefèvre, 1829–1840), 47:76–77.
24. Ibid., 365n15. "…il est démontré que les corps célestes font leurs révolutions dans l'espace non résistant. Tout l'espace n'est pas rempli. Il n'y a donc pas une suite de corps depuis un atome jusqu'à la plus reculée des étoiles; il peut donc y avoir des intervalles immenses entre les êtres sensibles, comme entre les insensibles. On ne peut donc assurer que l'homme soit nécessairement placé dans un des chaînons attachés l'un à l'autre par une suite non interrompue." François-Marie Arouet de Voltaire, *Poème sur le désastre de Lisbonne* (1756) in *Œuvres de Voltaire avec préfaces, avertissements, notes, etc.*, 70 vols., edited by M. Beuchot (Paris: Lefèvre, 1829–1840), 12:193–95note a.
25. "La chaîne universelle n'est point, comme on l'a dit, une gradation suivie qui lie tous les êtres. Il y a probablement une distance immense entre l'homme et la brute, entre l'homme et les substances supérieures; il y a l'infini entre Dieu et toutes les substances. Les globes qui roulent autour de notre soleil n'ont rien de ces gradations insensibles, ni dans leur grosseur, ni dans leurs distances, ni dans leurs satellites." Ibid., 12:193–94.
26. "La seule difficulté qui restait était de savoir comment il y avait eu de la farine avant qu'il y eût des hommes." François-Marie Arouet de Voltaire, *Questions sur les*

miracles (1766), *Lettre 5*, in *Œuvres de Voltaire avec préfaces, avertissements, notes, etc.*, 70 vols., edited by M. Beuchot (Paris: Lefèvre, 1829–1840), 42:202.

27. François-Marie Arouet de Voltaire, *A Defence of My Uncle, Translated from the French of M. de Voltaire, Chapter 19* ("Of Mountains and Shells") (London: S. Bladon, 1768), 105–06. I took the liberty of changing the translation so that *subtile* is translated as "penetrating" and *cannelée,* "channeled." "Les germes sont inutiles: tout naîtra de soi-même. On bâtit sur cette expérience prétendue un nouvel univers, comme nous faisions un monde, il y a cent ans, avec la matière subtile, la globuleuse et la cannelée." François-Marie Arouet de Voltaire, *La Défense de mon oncle* (1767), *Chapître 19* ("Des montagnes et des coquilles"), in *Œuvres de Voltaire avec préfaces, avertissements, notes, etc.*, 70 vols., edited by M. Beuchot (Paris: Lefèvre, 1829–1840), 43:374–75.

28. "Le Voltaire en plaisantera tant qu'il voudra, mais l'Anguillard a raison; j'en crois mes yeux; je les vois; combien il y en a! comme ils vont! comme ils viennent! comme ils frétillent!" Diderot, *Rêve de d'Alembert* in *Œuvres complètes de Diderot*, edited by Jean Assézat and Maurice Tourneux (Paris: Garnier, 1875–1877), 2:131.

29. Jacques Roger, *The Life Sciences in Eighteenth-Century French Thought*, edited by Keith R. Benson and translated by Robert Ellrich (Stanford: Stanford University Press, 1997), 522–29 and 687–89, notes 320–71.

30. Ibid.

31. Leonardo da Vinci, *Studies of Fossils and Notes and Drawings on the Motion of Water* in *Codex Hammer* (ex Leicester), Bill Gates Collection, Seattle, WA.

32. Nicolaus Steno, *De Solido intra Solidum Naturalitur Contento Dissertationis Prodromus* (Florence: Typographia sub signo Stellæ, 1669). [*The Prodromus to a Dissertation Concerning Solids Naturally Contained within Solids* (London: J. Winter, 1671). English translation.]

33. Robert Hooke, *Discourse of Earthquakes* in *The Posthumous Works of Robert Hooke, M.D.*, ed. Richard Waller (London: S. Smith and B. Walford, 1705), 327.

34. Antonio Vallisneri, *De' corpi marini, che su' monti si trovano* (Domenico Lovisa: Venice, 1721). [marine bodies found on the mountains]

35. "En lisant une lettre italienne sur les changemens arrivez au globe terrestre, imprimée à Paris cette année (1746)…les poissons pétrifiez ne sont, son avis, que des poissons rares, rejetez de la table des Romains, parce qu'ils n'étoient pas frais; & à l'égard des coquilles ce sont, dit-il, les pélerins de Syrie qui ont rapporté dans le temps des croisades celles des mers du levant qu'on trouve actuellement pétrifées en France, en Italie & dans les autres états de la chrétienté; pourquoi n'a-t-il pas ajouté que ce sont les singes qui ont transporté les coquilles au sommet des hautes montagnes & dans tous les lieux où les hommes ne peuvent habiter, cela n'eût rien gâté & eût rendu son explication encore plus vraisemblable. Comment se peut-il que des personnes éclairées & qui se piquent même de philosophie, aient encore des idées aussi fausses sur ce sujet! Nous ne nous contenterons donc pas d'avoir dit qu'on trouve des coquilles pétrifiées dans presque tous les endroits de la terre où l'on a fouillé, & d'avoir rapporté les témoignages des auteurs d'Histoire Naturelle; comme on pourroit les soupçonner d'apercevoir, en vûe de quelques systèmes, des coquilles où il n'y en a point, nous croyons devoir encore citer les voyageurs qui en ont remarqué par hasard, & dont les yeux moins exercez n'ont pû reconnoître que les coquilles entières & bien conservées; leur témoignage sera peut-être d'une plus

grande autorité auprès des gens qui ne sont pas à portée de s'assurer par eux-mêmes de la vérité des faits, & de ceux qui ne connoissent ni les coquilles, ni les pétrifications, & qui n'étant pas en état d'en faire la comparaison, pourroient douter que les pétrifications fussent en effet de vraies coquilles, & que ces coquilles se trouvassent entassées par millions dans tous les climats de la terre.

Tout le monde peut voir par ses yeux les bancs de coquilles qui sont dans les collines des environs de Paris..." Georges-Louis Leclerc de Buffon. *Histoire naturelle, générale et particulière*, 15 vols. (Paris: Imprimerie royale, 1749–1767), 1:281–82.

36. François-Marie Arouet de Voltaire, *A Defence of My Uncle, Translated from the French of M. de Voltaire, Chapter 19* ("Of Mountains and Shells") (London: S. Bladon, 1768), 97. François-Marie Arouet de Voltaire, *La Défense de mon oncle* (1767), Chapître 19 ("Des montagnes et des coquilles"), in *Œuvres de Voltaire avec préfaces, avertissements, notes, etc.*, 70 vols., edited by M. Beuchot (Paris: Lefèvre, 1829–1840), 43:369.

37. "FLUX se dit aussi De l'écoulement des excrémens devenus trop fluides, & signifie, Dévoiement." "Flux," *Dictionnaire de L'Académie française* (1762), 755.

38. François-Marie Arouet de Voltaire, *A Defence of My Uncle, Translated from the French of M. de Voltaire, Chapter 19* ("Of Mountains and Shells") (London: S. Bladon, 1768), 97–98. François-Marie Arouet de Voltaire, *La Défense de mon oncle* (1767), Chapître 19 ("Des montagnes et des coquilles"), in *Œuvres de Voltaire avec préfaces, avertissements, notes, etc.*, 70 vols., edited by M. Beuchot (Paris: Lefèvre, 1829–1840), 43:369.

39. Ibid., 98. Ibid.
40. Ibid. Ibid., 43:369–70.
41. Ibid. Ibid., 43:370.
42. Ibid., 98-99. Ibid.
43. Ibid., 99. Ibid.
44. Ibid., 99-100. Ibid., 43:370–71.
45. Ibid., 100. Ibid., 43:371.
46. Ibid., 100-101. Ibid., 43:371–72.
47. Ibid., 104. "Mais j'ai vu aussi sous vingt pieds de terre des monnaies romaines, des anneaux de chevaliers, à plus de neuf cent milles de Rome, et je n'ai point dit: Ces anneaux, ces espèces d'or et d'argent, ont été fabriqués ici. Je n'ai point dit non plus: Ces huîtres sont nées ici. J'ai dit: Des voyageurs ont appporté ici des anneaux, de l'argent, et des huîtres." Ibid., 43:373.
48. Ibid. "Quand je lus, il y a quarante ans, qu'on avait trouvé dans les Alpes des coquilles de Syrie, je dis, je l'avoue, d'un ton un peu goguenard, que ces coquilles avaient été apparemment apportées par des pèlerins qui revenaient de Jérusalem." Ibid., 43:373–74.
49. Ibid., 105. "On a beau me dire que le porphyre est fait de pointes d'oursin, je le croirai quand je verrai que le marbre blanc est fait de plumes d'autruche." Ibid., 43:374.
50. "On trouve dans quelques endroits de ce globe des amas de coquillages; on voit dans quelques autres des huîtres pétrifiées: de là on a conclu que, malgré les lois de la gravitation et celles des fluides, et malgré la profondeur du lit de l'Océan, la mer avait couvert toute la terre il y a quelques millions d'années." François-Marie Arouet de Voltaire, *Des singularités de la nature* (1768), Chapître 12 ("Des

coquilles, et des systèmes bâtis sur des coquilles"), in *Œuvres de Voltaire avec préfaces, avertissements, notes, etc.*, 70 vols., edited by M. Beuchot (Paris: Lefèvre, 1829–1840), 44:246.

51. "Mille endroits sont remplis de mille débris de testacées, de crustacées, de pétrifications. Mais remarquons, encore une fois, que ce n'est presque jamais ni sur la croupe ni dans les flancs de cette continuité de montagnes dont la surface du globe est traversée..." Ibid., 44:249

52. "...c'est à quelques lieues de ces grands corps, c'est au milieu des terres, c'est dans des cavernes, dans des lieux où il est très vraisemblable qu'il y avait de petits lacs qui ont disparu, de petites rivières dont le cours est changé...mais de véritables corps marins, c'est ce que vous ne voyez jamais. S'il y en avait, pourquoi n'aurait on jamais vu d'os de chiens marins, de requins, de baleines?" Ibid., 44:249–50.

53. "Je ne nie pas, encore une fois, qu'on ne rencontre à cent milles de la mer quelques huîtres pétrifiées, des conques, des univalves, des productions qui ressemblent parfaitement aux productions marines; mais est-on bien sûr que le sol de la terre ne peut enfanter ces fossiles? La formation des agates arborisées ou herborisées ne doit-elle pas nous faire suspendre notre jugement? Un arbre n'a point produit l'agate qui représente parfaitement un arbre; la mer peut aussi n'avoir point produit ces coquilles fossiles qui ressemblent à des habitations de petits animaux marins." Ibid., 44:251.

54. "N'y a-t-il pas là de quoi étonner du moins ceux qui affirment que tous les coquillages qu'on rencontre dans quelques endroits de la terre y ont été déposés par la mer?" Ibid., 44:253.

55. "Sur ce que j'ai écrit, *page 281,* au sujet de la Lettre italienne, dans laquelle il est dit *que ce sont les Pélerins & autres, qui dans le temps des croisades ont rapporté de Syrie des coquilles que nous trouvons dans le sein de la terre en France,* &c. on a pu trouver, comme je le trouve moi même, que je n'ai pas traité M. de Voltaire assez sérieusement; j'avoue que j'aurois mieux fait de laisser tomber cette opinion que de la relever par une plaisanterie, d'autant que ce n'est pas mon ton, & c'est peut-être la seule qui soit dans mes écrits. M. de Voltaire est un homme qui par la supériorité de ses talens, mérite les plus grands égards. On m'apporta cette Lettre italienne dans le temps même que je corrigeois la feuille de mon Livre où il en est question; je ne lûs cette Lettre qu'en partie, imaginant que c'étoit l'ouvrage de quelque Erudit d'Italie, qui d'après ses connoissances historiques, n'avoit suivi que son préjugé, sans consulter la Nature; & ce ne fut qu'après l'impression de mon volume sur la Théorie de la Terre, qu'on m'assura que la Lettre étoit de M. de Voltaire; j'eus regret alors à mes expressions. Voilà la vérité, je la déclare autant pour M. de Voltaire, que pour moi-même & pour la postérité à laquelle je ne voudrois pas laisser douter de la haute estime que j'ai toujours eue pour un homme aussi rare & qui fait tant d'honneur à son siècle." Georges-Louis Leclerc de Buffon, *Histoire naturelle, générale et particulière; Supplément*, 17 vols. (Paris: Imprimerie royale, 1774–1789), 5:285–86.

56. "L'autorité de M. de Voltaire ayant fait impression sur quelques personnes, il s'en est trouvé qui ont voulu vérifier par eux-mêmes si les objections contre les coquilles, avoient quelque fondement, & je crois devoir donner ici l'extrait d'un Mémoire qui m'a été envoyé & qui me paroît été fait que dans cette vue." Ibid., 5:286.

57. "En parcourant différentes provinces du Royaume & même de l'Italie, 'j'ai vu, dit le P. Chabenat, des pierres figurées de toutes parts, & dans certains endroits en si grande quantité, & arrangées de façon qu'on ne peut s'empêcher de croire que ces parties de la Terre n'aient autrefois été le lit de la mer. J'ai vu des coquillages de toute espèce, & qui sont parfaitement semblables à leurs analogues vivans. J'en ai vu de la même figure & de la même grandeur: cette observation m'a paru suffisante pour me persuader que tous ces individus étoient de différens âges, mais qu'ils étoient de la même espèce. J'ai vu des cornes d'ammon depuis un demi-pouce jusqu'à près de trois pieds de diamètre. J'ai vu des pétoncles de toutes grandeurs, d'autres bivalves & des univalves également. J'ai vu outre cela des bélemnites, des champignons de mer, &c.

La forme & la quantité de toutes ces pierres figurées, nous prouvent presque invinciblement qu'elles étoient autrefois des animaux qui vivoient dans la mer. La coquille sur-tout dont elles sont couvertes, semble ne laisser aucun doute, parce que dans certaines, elle se trouve aussi luisante, aussi fraîche & aussi naturelle que dans les vivans; si elle étoit séparé du noyau, on ne croiroit pas qu'elle fût pétrifiée...Tout ceci sembloit me dire fort intelligiblement que ce pays-ci avoit été anciennement le lit de la mer, qui par quelque révolution soudaine, s'en est retirée & y a laissé ses productions comme dans beaucoup d'autres endroits. Cependant je suspendois mon jugement à cause des objections de M. de Voltaire. Pour y répondre, j'ai voulu joindre l'expérience à l'observation.'" Ibid., 5:286–88.

58. "Le P. Chabenat rapporte ensuite plusieurs expériences pour prouver que les coquilles qui se trouvent dans le sein de la terre sont de la même nature que celles de la mer; je ne les rapporte pas ici, parce qu'elles n'apprennent rien de nouveau, & que personne ne doute de cette identité de nature entre les coquilles fossiles & les coquilles marines. Enfin le P. Chabenat conclut & termine son Mémoire en disant: 'on ne peut donc pas douter que toutes ces coquilles qui se trouvent dans le sein de la terre, ne soient de vraies coquilles & des dépouilles des animaux de la mer qui couvroit autrefois toutes ces contrées, & par conséquent les objections de M. de Voltaire ne soient mal fondées.'" Ibid., 5:288.

59. Martin J.S. Rudwick, *The Meaning of Fossils: Episodes in the History of Palaeontology*, 2nd ed. (Chicago and London: The University of Chicago Press, 1972), 88. In 99n51 Rudwick cites Francis C. Haber, *The Age of the World: Moses to Darwin* (Baltimore: Johns Hopkins Press, 1959), 107–08.

60. Shirley A. Roe, "Voltaire vs. Needham: Atheism, Materialism, and the Generation of Life," *Journal of the History of Ideas* 46 (1985):65, reprinted in *Philosophy, Religion and Science in the Seventeenth and Eighteenth Centuries*, edited by John W. Yolton (Rochester: University of Rochester Press, 1990), 417.

61. "La philosophie ayant compris l'impossibilité où elle étoit d'expliquer mécaniquement la formation des êtres organisés, a imaginé heureusement qu'ils existoient déjà en petit, sous la forme de *germes*, ou de *corpuscules organiques*. Et cette idée a produit deux hypothèses qui plaisent beaucoup à la raison." Charles Bonnet, *Considérations sur les corps organisés*, 2 vols. (Amsterdam: M.-M. Rey, 1762), 1:1.

62. "Je me contenterai de rappeler à l'esprit de mes lecteurs l'étonnant appareil de *fibres*, de *membranes*, de *vaisseaux*, de *ligamens*, de *tendons*, de *muscles*, de *nerfs*, de *veines*, d'*artères* etc. qui entrent dans la composition du corps d'un animal. Je les prierai de considérer attentivement la structure, les rapports et le jeu de toutes ces

parties. Je leur demanderai ensuite, s'ils conçoivent qu'un tout aussi composé, aussi lié, aussi harmonique, puisse être formé par le simple concours de molécules mues, ou dirigées, suivant certaines loix à nous inconnues. Je les prierai de me dire s'ils ne sentent point la nécessité où nous sommes d'admettre que cette admirable machine a été d'abord dessinée en petit par la même main qui a tracé le plan de l'univers." Ibid., 1:85.

63. "...nous ne croyons pas que ses *Considérations* puissent répandre beaucoup de jour sur cette grande et ténébreuse question, le désespoir des philosophes anciens et modernes; mais elles décèlent du moins un esprit très sage et très éclairé." François-Marie Arouet de Voltaire, *Review of Charles Bonnet's Considérations sur les corps organisés* in *Gazette littéraire de l'Europe, April 14, 1764*, in *Œuvres complètes de Voltaire avec préfaces, avertissements, notes, etc.*, 70 vols., edited by M. Beuchot (Paris: Lefèvre, 1829–1840), 41:427.

64. "Les anciens avaient voulu deviner comme nous les secrets de la nature, mais ils n'avaient point de fil pour se guider dans les détours de ce labyrinthe immense." Ibid., 41:427–28.

65. "On a fini par rester dans le doute; ce qui arrive toujours quand on veut remonter aux premières causes." Ibid., 41:430.

66. "L'auteur de la *Vénus physique* a eu recours à l'attraction; il a prétendu que, dans les principes féconds de l'homme et de la femme mêlés ensemble, la jambe gauche du fœtus attire la jambe droite sans se méprendre; qu'un œil attire un œil en laissant le nez entre deux, qu'un lobe du poumon est attiré par l'autre lobe, etc." Ibid., 41:430–31.

67. "Il semble qu'il en faille revenir à l'ancienne opinion que tous les germes furent formés à-la-fois par la main qui arrangea l'univers; que chaque germe contient en lui tous ceux qui doivent naître de lui, que toute génération n'est qu'un développement; et, soit que les germes des animaux soient contenus dans les mâles ou dans les femelles, il est vraisemblable qu'ils existent dès le commencement des choses, ainsi que la terre, les mers, les éléments, les astres." Ibid., 41:431–32.

68. François-Marie Arouet de Voltaire, "Atheist, Atheism," *Philosophical Dictionary*, edited and translated by Theodore Besterman (London: Penguin Books, 2004), 57–58. "Ajoutons surtout qu'il y a moins d'athées aujourd'hui que jamais, depuis que les philosophes ont reconnu qu'il n'y a aucun être végétant sans germe, aucun germe sans dessein, etc., et que le blé ne vient point de pourriture. Des géomètres non philosophes ont rejeté les causes finales, mais les vrais philosophes les admettent; et, comme l'a dit un auteur connu, un catéchiste annonce Dieu aux enfants, et Newton le démontre aux sages." François-Marie Arouet de Voltaire, "Athéisme," Section 4, *Dictionnaire philosophique* in *Œuvres de Voltaire avec préfaces, avertissements, notes, etc.*, 70 vols., edited by M. Beuchot (Paris: Lefèvre, 1829–1840), 27:189.

69. "Et il est démontré aujourd'hui aux yeux et à la raison qu'il n'est ni de végétal ni d'animal qui n'ait son germe. On le trouve dans l'œuf d'une poule comme dans le gland d'un chêne. Une puissance formatrice préside à tous ces développements d'un bout de l'univers à l'autre." François-Marie Arouet de Voltaire, *Des singularités de la nature* in *Œuvres complètes de Voltaire avec préfaces, avertissements, notes, etc.*, 70 vols., edited by M. Beuchot (Paris: Lefèvre, 1829–1840), 44:271.

70. "Il y a deux manières de parvenir à la notion d'un être qui préside à l'univers. La plus naturelle et la plus parfaite pour les capacités communes, est de considérer non seulement l'ordre qui est dans l'univers, mais la fin à laquelle chaque chose paraît se rapporter." Voltaire, *Traité de métaphysique, Chapître 2*, in *Œuvres complètes de Voltaire avec préfaces, avertissements, notes, etc.*, 70 vols., edited by M. Beuchot (Paris: Lefèvre, 1829–1840), 37:284.
71. "On m'a ajouté que jamais homme un peu instruit n'a avancé que les espèces non mélangées dégénérassent..." Ibid., 37:282.
72. "Il me semble alors que je suis assez bien fondé à croire qu'il en est des hommes comme des arbres; que les poiriers, les sapins, les chênes, et les abricotiers ne viennent point d'un même arbre, et que les blancs barbus, les nègres portant laine, les jaunes portant crins, et les hommes sans barbe, ne viennent pas du même homme." Ibid., 37:282–83.
73. "Toutes ces différentes races d'hommes produisent ensemble des individus capables de perpétuer, ce qu'on ne peut pas dire des arbres d'espèces différentes; mais y a-t-il eu un temps où il n'existait qu'on ou deux individus de chaque espèce? c'est ce que nous ignorons complètement." Ibid., 37:283.
74. "Il faut ici surtout raisonner de bonne foi, et ne point chercher à se tromper soi-même; quand on voit une chose qui a toujours le même effet, qui n'a uniquement que cet effet, qui est composée d'une infinité d'organes, dans lesquels il y a une infinité de mouvements qui tous concourent à la même production, il me semble qu'on ne peut, sans une secrète répugnance, nier une cause finale. Le germe de tous les végétaux, de tous les animaux, est dans ce cas: ne faut-il pas être un peu hardi pour dire que tout cela ne se rapporte à aucune fin?" Ibid., 37:295.
75. "Voici enfin une nouvelle richesse de la nature, une espèce qui ne ressemble pas tant à la nôtre que les barbets aux lévriers." François-Marie Arouet de Voltaire, *Relation touchant un Maure blanc amené d'Afrique à Paris en 1744* (1744) in *Œuvres complètes de Voltaire avec préfaces, avertissements, notes, etc.*, 70 vols., edited by M. Beuchot (Paris: Lefèvre, 1829–1840), 38:523.
76. Ibid., 38:522–24.
77. "...si nous pensons valoir beaucoup mieux qu'eux, nous nous trompons assez lourdement." Ibid., 38:524.
78. "Qu'est-ce qu'un vrai théiste? c'est celui qui dit à Dieu: *Je vous adore, et je vous sers*; c'est celui qui dit au Turc, au Chinois, à l'Indien, et au Russe: *Je vous aime*." François-Marie Arouet de Voltaire, "Théisme," *Dictionnaire philosophique* (1764), in *Œuvres de Voltaire avec préfaces, avertissements, notes, etc.*, 70 vols., edited by M. Beuchot (Paris: Lefèvre, 1829–1840), 32:351.
79. François-Marie Arouet de Voltaire, "Theist," *Philosophical Dictionary*, edited and translated by Theodore Besterman (London: Penguin Books, 2004), 386. "Il a des frères depuis Pékin jusqu'à la Cayenne, et il compte tous les sages pour ses frères." François-Marie Arouet de Voltaire, "Théiste," *Dictionnaire philosophique* in *Œuvres de Voltaire avec préfaces, avertissements, notes, etc.*, 70 vols., edited by M. Beuchot (Paris: Lefèvre, 1829–1840), 32:352.
80. François-Marie Arouet de Voltaire, *A Defence of My Uncle, Translated from the French of M. de Voltaire*, Chapter 18 ("Of Men of Different Colours") (London: S. Bladon, 1768), 91. "Il ne pensait pas que les huîtres d'Angleterre fussent engendrées des crocodiles du Nil, ni que les girofliers des îles Moluques tirassent leur origine des sapins des Pyrénées." François-Marie Arouet de Voltaire, *La*

Défense de mon oncle (1767), *Chapître 18* ("Des hommes de différentes couleurs"), in *Œuvres de Voltaire avec préfaces, avertissements, notes, etc.*, 70 vols., edited by M. Beuchot (Paris: Lefèvre, 1829–1840), 43:365–66.

81. François-Marie Arouet de Voltaire, *A Defence of My Uncle, Translated from the French of M. de Voltaire*, Chapter 19 ("Of Mountains and Shells") (London: S. Bladon, 1768), 100–01. "Ne perdez point de vue cette grande vérité que la nature ne se dément jamais. Toutes les espèces restent toujours les mêmes. Animaux, végétaux, minéraux, métaux, tout est invariable dans cette prodigieuse variété. Tout conserve son essence." François-Marie Arouet de Voltaire, *La Défense de mon oncle* (1767), *Chapître 19* ("Des montagnes et des coquilles"), in *Œuvres de Voltaire avec préfaces, avertissements, notes, etc.*, 70 vols., edited by M. Beuchot (Paris: Lefèvre, 1829–1840), 43:371–72.

82. Jacques Roger, *Buffon: A Life in Natural History* (Ithaca and London: Cornell University Press, 1997), 346.

83. Benoît de Maillet, *Telliamed: Or, Discourses between an Indian Philosopher, and a French Missionary, on the Diminution of the Sea, the Formation of the Earth, the Origin of Men and Animals, and other Curious Subjects, relating to Natural History and Philosophy* (London: Jacob Loyseau, 1750), 224. "Les petits ailerons qu'ils avoient sous le ventre, & qui, comme leurs nageoires, leur avoient aidé à se promener dans la mer, devinrent des pieds, & leur servirent à marcher sur la terre." Benoît de Maillet, *Telliamed ou Entretiens d'un philosophe indien avec un missionnaire françois sur la diminution de la mer, la formation de la terre, l'origine de l'homme, &c.*, Chapter 6, "Sixième journée. De l'origine de l'homme & des animaux, & de la propagation des espèces par les semences," 2 vols. (Amsterdam: L'Honoré et fils, 1748), 2:140.

84. Ibid., 246. "Pour revenir aux diverses espèces d'hommes, ceux qui ont des queues peuvent ils être les fils de ceux qui n'en ont point? Comme les singes à queue ne descendent certainement point de ceux qui sont sans queue, ne seroit-il pas naturel de penser de même, que les hommes qui naissent avec des queues sont d'une espèce diverse de ceux qui n'en ont jamais eu?" Ibid., 2:174.

85. "…qui m'apprit que les montagnes et les hommes sont produits par les eaux de la mer. Il y eut d'abord de beaux hommes marins qui ensuite devinrent amphibies. Leur belle queue fourchue se changea en cuisses et en jambes. J'étais encore tout plein des Métamorphoses d'Ovide, et d'un livre où il était démontré que la race des hommes était bâtarde d'une race de babouns: j'aimais autant descendre d'un poisson que d'un singe." François-Marie Arouet de Voltaire, *L'Homme aux 40 écus* in *Œuvres complètes de Voltaire avec préfaces, avertissements, notes, etc.*, 70 vols., edited by M. Beuchot (Paris: Lefèvre, 1829–1840), 34:43.

86. "La grande chaîne…paraît une pièce essentielle à la machine du monde, comme les os le sont aux quadrupèdes et aux bipèdes. C'est autour de leurs faîtes que s'assemblent les nuages et les neiges, qui de là, se répandant sans cesse, forment tous les fleuves et toutes les fontaines, dont on a si longtemps et si faussement attribué la source à la mer…Les chaînes de ces montagnes qui couvrent l'un et l'autre hémisphère ont une utilité plus sensible. Elles affermissent la terre; elles servent à l'arroser; elles renferment à leurs bases tous les métaux, tous les minéraux. Qu'il soit permis de remarquer à cette occasion que toutes les pièces de la machine de ce monde semblent faites l'une pour l'autre." François-Marie Arouet de Voltaire, *Des singularités de la nature* in *Œuvres complètes de Voltaire avec*

préfaces, avertissements, notes, etc., 70 vols., edited by M. Beuchot (Paris: Lefèvre, 1829–1840), 44:235–36.

87. "Le grand Etre qui a formé l'or et le fer, les arbres, l'herbe, l'homme et la fourmi, a fait l'océan et les montagnes. Les hommes n'ont pas été des poissons, comme le dit Maillet; tout a été probablement ce qu'il est par des lois immuables." Ibid., 44:241.

88. "Il est arrivé aux coquilles la même chose qu'aux anguilles: elles ont fait éclore des systèmes nouveaux…Si la mer a été partout, il y a eu un temps où le monde n'était peuplé que de poissons. Peu à peu les nageoires sont devenues des bras; la queue fourchue, s'étant allongée, a formé des cuisses et des jambes; enfin les poissons sont devenus des hommes, et tout cela s'est fait en conséquence des coquilles qu'on a déterrées. Ces systèmes valent bien l'horreur du vide, les formes substantielles, la matière globuleuse, subtile, cannelée, striée, la négation de l'existence des corps, la baguette divinatoire de Jacques Aimard, l'harmonie préétablie, et le mouvement perpétuel." Ibid., 44:246–47.

89. Blaise Pascal, *Thoughts*, translated by W.F. Trotter, Brunschvicg numbering system (New York: P.F. Collier & Son, 1910), fragment 693. "En voyant l'aveuglement et la misère de l'homme, en regardant tout l'univers muet, et l'homme sans lumière, abandonné à lui-même et comme égaré dans ce recoin de l'univers, sans savoir qui l'y a mis, ce qu'il y est venu faire, ce qu'il deviendra en mourant, incapable de toute connaissance, j'entre en effroi, comme un homme qu'on aurait porté endormi dans une île déserte et effroyable et qui s'éveillerait sans connaître où il est, et sans moyen d'en sortir." Blaise Pascal, *Pensées*, in *Œuvres de Blaise Pascal*, edited by Léon Brunschvicg, Pierre Boutroux, and Félix Gazier (Paris: Librairie Hachette & Cie, 1904–1914), fragment 693 (Lafuma 198; Sellier 229).

90. Paul Johnson, *A History of Christianity* (London: Weidenfeld and Nicolson, 1976), 353.

91. "Substantial," 3, *Oxford English Dictionary Online*, http://www.oed.com (September 1, 2006).

92. "Form," 4a, *Oxford English Dictionary Online*, http://www.oed.com (September 1, 2006).

93. Chambers' definition is cited in "Substantial," 3, *Oxford English Dictionary Online*, http://www.oed.com (September 1, 2006).

94. "Globuleux," *Encyclopédie, ou Dictionnaire raisonné des sciences, des arts et des métiers*, edited by Denis Diderot and Jean Le Rond d'Alembert (Paris: Briasson, David, Le Breton, Durant; Neuchâtel: S. Faulche, 1751–1765), 7:715.

95. "Matière subtile, est le nom que les Cartésiens donnent à une *matière* qu'ils supposent traverser & pénétrer les pores de tous les corps, & remplir ces pores de façon à ne laisser aucun vuide ou interstices entr'eux. *Voyez* Cartésianisme. Mais en vain ils ont recours à cette machine pour étayer leur sentiment d'un plein absolu…il faudroit pour qu'elle dût remplir les vuides de tous eles autres corps, qu'elle fût elle-même entièrement destituée de vuide; c'est-à-dire parfaitement solide, beaucoup plus solide, par exemple que l'or, & par conséquent, qu'elle fût beaucoup plus pesante que ce métal, & qu'elle résistât davantage (*voyez* Résistance); ce qui ne sauroit s'accorder avec les phénomènes." Jean Le Rond d'Alembert, "Matière subtile," *Encyclopédie, ou Dictionnaire raisonné des sciences, des arts et des métiers*, edited by Denis Diderot and Jean Le Rond d'Alembert (Paris: Briasson, David, Le Breton, Durant; Neuchâtel: S. Faulche, 1751–1765), 10:191.

96. Ibid.
97. Ibid.
98. René Descartes, *Principles of Philosophy*, translated and annotated by Valentine Rodger Miller and Reese P. Miller (Dordrecht, The Netherlands: Kluwer Academic Publisher, 1991), 86–87 and 131–33.
99. François-Marie Arouet de Voltaire, *A Defence of My Uncle, Translated from the French of M. de Voltaire, Chapter 19* ("Of Mountains and Shells") (London: S. Bladon, 1768), 105–06. I took the liberty of changing the translation so that *subtile* is translated as "penetrating" and *cannelée*, "channeled." "Les germes sont inutiles: tout naîtra de soi-même. On bâtit sur cette expérience prétendue un nouvel univers, comme nous faisions un monde, il y a cent ans, avec la matière subtile, la globuleuse et la cannelée." François-Marie Arouet de Voltaire, *La Défense de mon oncle* (1767), *Chapître 19* ("Des montagnes et des coquilles"), in *Œuvres de Voltaire avec préfaces, avertissements, notes, etc.*, 70 vols., edited by M. Beuchot (Paris: Lefèvre, 1829–1840), 43:374–75.
100. Jacques Roger, *The Life Sciences in Eighteenth-Century French Thought*, edited by Keith R. Benson and translated by Robert Ellrich (Stanford: Stanford University Press, 1997), 522.

CHAPTER NINE

1. Julien Offray de La Mettrie, *Man a Machine* in *Man a Machine, including Frederick the Great's "Eulogy" on La Mettrie and Extracts from La Mettrie's "The Natural History of the Soul,"* notes by Gertrude Carman Bussey (Chicago: The Open Court Publishing Company, 1912), 100. "En un mot seroit-il absolument impossible d'apprendre une Langue à cet Animal? Je ne le crois pas." Julien Offray de La Mettrie, *L'Homme-machine* in *Œuvres philosophiques* (London: Jean Nourse, 1751), 26.
2. "Avez-vous vu au Jardin du Roi, sous une cage de verre, un orang-outang qui a l'air d'un saint Jean qui prêche au désert?…Le cardinal de Polignac lui disait un jour: 'Parle, et je te baptise.'" Denis Diderot, *Suite de l'Entretien* in *Œuvres complètes de Diderot*, edited by Jean Assézat and Maurice Tourneux (Paris: Garnier, 1875–1877), 2:190.
3. Philippo Pigafetta, *A Reporte of the Kingdome of Congo, a Region of Africa*, translated from the Italian by Abraham Hartwell (London: John Wolfe, 1597), 89.
4. Samuel Purchas, *Purchas his Pilgrimage, or Relatons of the World and the Religions Observed in all Ages and Places Discovered, from the Creation unto this Present* (London: William Stansby, 1613), 466.
5. Samuel Purchas, *Purchas his Pilgrimes* (London: William Stansby, 1625), 2:982.
6. Francis Moran, "Between Primates and Primitives: Natural Man as the Missing Link in Rousseau's Second Discourse," *Journal of the History of Ideas* 54, no. 1 (January 1993), 37–58.
7. François Leguat, *The Voyage of François Leguat* (1708), edited by Pasfield Oliver, 2 vols (London, 1891), 2:234. Cited by Francis Moran, "Between Primates and

Primitives: Natural Man as the Missing Link in Rousseau's Second Discourse," *Journal of the History of Ideas* 54, no. 1 (January 1993), 41.
8. Daniel Beeckman, *A Voyage to and from the Island of Borneo* (London, 1718), 37. Ibid.
9. Moran cites sources using the following editions: Purchas, *Hakluytus Posthumus*, 6:398; Tyson, *Orang-Outang or the Anatomy of a Pygmie compared with that of a Monkey, an Ape, and a Man* (1699), cited in Ashley Montagu, "Edward Tyson, M.D., F.D.S. 1650-1708," *Memoirs of the American Philosophical Society*, 20 (1943), 261; François Leguat, *The Voyage of François Leguat* (1708), edited by Pasfield Oliver, 2 vols (London, 1891), 2:234; Daniel Beeckman, *A Voyage to and from the Island of Borneo* (London, 1718), 37; William Smith, *A New Voyage to Guinea* (1744), in abbé Prévost, *Histoire générale des voyages*, 4:240; Benoît de Maillet, *Telliamed* (1748), translated by Albert O. Carozzi (Chicago, 1968), 201; and Buffon, "Natural History of the Orang-Outangs, or the Pongo and Jocko" (1766) in *Natural History: General and Particular*, edited by William Smellie, 8 vols (London, 1781), 8:86.
10. Ibid., 40. Moran cites Edward Tyson, *Orang-Outang or the Anatomy of a Pygmie Compared with that of a Monkey, an Ape, and a Man* (1699), cited in Ashley Montagu, "Edward Tyson, M.D., F.D.S., 1650–1708," *Memoirs of an American Philosophical Society*, 20 (1943): 244.
11. Ibid. Moran cites John Green, *A New General Collection of Voyages and Travels*, 4 vols. (London, 1745–1747), 2:718.
12. Ibid., 43.
13. Ibid., 44. Moran cites Edward Tyson, *Orang-Outang or the Anatomy of a Pygmie compared with that of a Monkey, an Ape, and a Man* (1699), cited in Ashley Montagu, "Edward Tyson, M.D., F.D.S., 1650–1708," *Memoirs of an American Philosophical Society*, 20 (1943): 257.
14. Ibid. Ibid.
15. "Singe," *Encyclopédie, ou Dictionnaire raisonné des sciences, des arts et des métiers*, edited by Denis Diderot and Jean Le Rond d'Alembert (Paris: Briasson, David, Le Breton, Durant; Neuchâtel: S. Faulche, 1751–1765), 15:208–10.
16. "La plûpart de ces animaux ont plus de rapport avec l'homme que les autres quadrupèdes, surtout pour les dents, les oreilles, les narines &c. Ils ont des cils dans les deux paupières, & deux mamelles sur la poitrine. Les femelles ont pour la plûpart des menstrues comme les femmes. Les piés de devant ont beaucoup de rapport à la main de l'homme; les piés de derrière ont aussi la forme d'une main, car les quatre doigts sont plus longs que ceux du pié de devant, & le pouce est long, gros & fort écarté du premier doigt; aussi se servent-ils des piés de derrière comme de ceux de devant pour saisir & empoigner." Ibid., 15:208.
17. "Il y a plusieurs espèces de *singes*, qui ne diffèrent entr'elles que par la grandeur; elles ont beaucoup de rapport à l'homme par la face, les oreilles & les ongles." Ibid.
18. "L'homme des bois, *ourang outand bout*; cet animal est des Indes orientales; il ressemble plus à l'homme qu'aucune autre espèce de *singe*; son poil est court & assez doux." Ibid.
19. "Le *pongo* (dit en substance André Batell, dans les voyages de Purchass, *l. VII. c. iij. p. 974.*) a plus de cinq piés: il est de la hauteur d'un homme ordinaire, mais deux fois plus gros. Il a le visage sans poil, & ressemblant à celui d'un homme, les yeux assez grands quoiqu'enfoncés, & des cheveux qui lui couvrent la tête & les

épaules…Ces animaux grimpent sur les arbres pour y passer la nuit: ils s'y bâtissent même des espèces d'abris contre les pluies dont ce pays est inondé pendant l'été. Ils ne vivent que de fruits & de plantes: ils couvrent leurs morts de feuilles & de branches; ce que les Nègres regardent comme une sorte de sépulture. Lorsque les *pongos* trouvent le matin les feux que les Nègres allument la nuit, en voyageant à travers de ces forêts, on les voit s'en approcher avec une apparence de plaisir. Néanmoins, ils n'ont jamais imaginé de les entretenir en y jettant du bois. Aussi les Nègres assurent-ils que les *pongos* n'ont une classe supérieure à celle des animaux." Jaucourt, "Pongo," *Encyclopédie, ou Dictionnaire raisonné des sciences, des arts et des métiers*, edited by Denis Diderot and Jean Le Rond d'Alembert (Paris: Briasson, David, Le Breton, Durant; Neuchâtel: S. Faulche, 1751–1765), 13:25.
20. Robert Wokler, "The Ape Debates in Enlightenment Anthropology," *Studies on Voltaire and the Eighteenth Century* 192 (1980): 1164–75.
21. Ibid., 1168.
22. Ibid., 1169. Wokler cites Claude Perrault, *Suite des mémoires pour servir à l'histoire des animaux* (Paris, 1676), 126.
23. Ibid., 1168. Wokler cites Edward Tyson, *Orang-outang, sive, Homo Sylvestris, or, The anatomy of a Pygmie compared with that of a Monkey, an Ape, and a Man* (London, 1699), iii and 55, and also refers the reader to his own article, Wokler, "Tyson and Buffon on the orang-utan," *Studies on Voltaire and the Eighteenth Century* 155 (1976): 2301–19.
24. Julien Offray de La Mettrie, *Man a Machine* in *Man a Machine, including Frederick the Great's "Eulogy" on La Mettrie and Extracts from La Mettrie's "The Natural History of the Soul,"* notes by Gertrude Carman Bussey (Chicago: The Open Court Publishing Company, 1912), 100. "Parmi les Animaux, les uns apprennent à parler & à chanter; ils retiennent des airs, & prennent tous les sons, aussi exactement qu'un Musicien." Julien Offray de La Mettrie, *L'Homme-machine* in *Œuvres philosophiques* (London: Jean Nourse, 1751), 25.
25. Ibid. "Les autres, qui montrent cependant plus d'esprit, tels que le Singe, n'en peuvent venir à bout. Pourquoi cela, si ce n'est par un vice des organes de la parole?" Ibid.
26. Ibid. "Mais ce vice est-il tellement de conformation, qu'on n'y puisse apporter aucun remède? En un mot seroit-il absolument impossible d'apprendre une Langue à cet Animal? Je ne le crois pas." Ibid.
27. Ibid. "Cet Animal nous ressemble si fort, que les Naturalistes l'ont appelé *Homme Sauvage*, ou Homme *des bois*. Je le prendrois aux mêmes conditions des Ecoliers d'Amman; c'est-à-dire, que je voudrois qu'il ne fût ni trop jeune, ni trop vieux; car ceux qu'on nous apporte en Europe, sont communément trop âgés. Je choisirois celui qui auroit la physionomie la plus spirituelle, & qui tiendroit le mieux dans mille petites opérations, ce qu'elle m'auroit promis." Ibid., 26.
28. Ibid., 100–01. "Enfin, ne me trouvant pas digne d'être son Gouverneur, je le mettrois à l'Ecole de l'excellent Maître que je viens de nommer, ou d'un autre aussi habile, s'il en est." Ibid.
29. Ibid., 101. I took the liberty of changing the translation of *singe* to "ape" everywhere for the purpose of consistency, as La Mettrie did not distinguish between monkeys and apes. "Pourquoi donc l'éducation des Singes seroit-elle impossible? Pourquoi ne pourroit-il enfin, à force de soins, imiter, à l'exemple des sourds, les mouvemens nécessaires pour prononcer? Je n'ose décider si les organes

de la parole du Singe ne peuvent, quoi qu'on fasse, rien articuler; mais cette impossibilité absolue me surprendroit, à cause de la grande Analogie du Singe et de l'Homme, & qu'il n'est point d'Animal connu jusqu'à présent, dont le dedans & le dehors lui ressemblent d'une manière si frappante." Ibid., 26–27.

30. Ibid., 103. "Non seulement je défie qu'on me cite aucune expérience vraiment concluante, qui décide mon projet impossible & ridicule; mais la similitude de la structure & des opérations du Singe est telle, que je ne doute presque point, si on exerçoit parfaitement cet Animal, qu'on ne vînt enfin à bout de lui apprendre à prononcer, & par conséquent à savoir une langue. Alors ce ne seroit plus ni un Homme Sauvage, ni un Homme manqué: ce seroit un Homme parfait, un petit Homme de Ville, avec autant d'étoffe ou de muscles que nous mêmes, pour penser & profiter de son éducation." Ibid., 28.

31. Ibid. "Des Animaux à l'Homme, la transition n'est pas violente; les vrais Philosophes en conviendront. Qu'étoit l'Homme, avant l'invention des Mots & la connoissance des Langues? Un Animal de son espèce, qui avec beaucoup moins d'instinct naturel, que les autres, dont alors il ne se croioit pas Roi, n'étoit distingué du Singe & des autres Animaux, que comme le Singe l'est lui-même; je veux dire, par un physionomie qui annonçoit plus de discernement...Les Mots, les Langues, les Loix, les Sciences, les Beaux Arts sont venus...On a dressé un Homme, comme un Animal..." Ibid., 28–29.

32. Leonora Cohen Rosenfield, *From Beast-Machine to Man-Machine; Animal Soul in French Letters from Descartes to La Mettrie* (New York: Octagon Books, 1968), 3. "Ex animalium quibusdam actionibus valde perfectis, suspicamur ea liberum arbitrium non habere." René Descartes, *Cogitationes privatæ, January 1619*, in *Œuvres de Descartes*, edited by Charles Adam and Paul Tannery, 12 vols. (Paris: L. Cerf, 1897–1913), 10:219.

33. Ibid., 6-7. Rosenfield cites René Descartes, *Chapter 5* of the *Discourse on Method* in *The Method, Meditations and Philosophy of Descartes*, translated by John Veitch (New York: Tudor Publishing Company, 1937), 189–90.

34. Ibid., 7. Ibid., 190.

35. Georges-Louis Leclerc de Buffon, *A Dissertation on the Nature of Animals* (1753) in *Natural History, General and Particular*, translated by William Smellie, 9 vols., 2nd edition (London: W. Strahan and T. Cadell, 1785), 3:210. "L'animal a deux manières d'être, l'état de mouvement et l'état de repos, la veille et le sommeil, qui se succèdent alternativement pendant toute la vie; dans le premier état, tous les ressorts de la machine animale sont en action; dans le second, il n'y en a qu'une partie, et cette partie qui est en action pendant le sommeil, est aussi en action pendant la veille..." Georges-Louis Leclerc de Buffon, *Discours sur la nature des animaux* (1753) in *Histoire naturelle, générale et particulière*, 15 volumes (Paris: Imprimerie royale, 1749–1767), 4:6.

36. Pierre-Louis Moreau de Maupertuis, *Essaie de cosmologie* in *Œuvres*, 4 vols (Lyon: J.-M. Bruyset, 1756), 1:42–43.

37. Georges-Louis Leclerc de Buffon, *A Dissertation on the Nature of Animals* (1753) in *Natural History, General and Particular*, translated by William Smellie, 9 vols., 2nd edition (London: W. Strahan and T. Cadell, 1785), 3:224. "Cette question est d'autant plus difficile à résoudre, qu'étant par notre nature différens des animaux, l'âme a part à presque tous nos mouvemens, et peut-être à tous, et qu'il nous est très-difficile de distinguer les effets de l'action de cette substance spirituelle, de

Notes 311

ceux qui sont produits par les seules forces de notre être matériel: nous ne pouvons en juger que par analogie et en comparant à nos actions les opérations naturelles des animaux; mais comme cette substance spirituelle n'a été accordée qu'à l'homme, et que ce n'est que par elle qu'il pense et qu'il réfléchit; que l'animal est au contraire un être purement matériel, qui ne pense ni ne réfléchit, et qui cependant agit et semble se déterminer, nous ne pouvons pas douter que le principe de la détermination du mouvement ne soit dans l'animal un effet purement méchanique, et absolument dépendant de son organisation." Georges-Louis Leclerc de Buffon, *Discours sur la nature des animaux* (1753) in *Histoire naturelle, générale et particulière*, 15 volumes (Paris: Imprimerie royale, 1749–1767), 4:22–23.

38. Ibid., 2:236. "…ils n'inventent, ils ne perfectionnent rien, ils ne réfléchissent par conséquent sur rien, ils ne font jamais que les mêmes choses, de la même façon…" Ibid., 4:39.
39. Ibid., 2:237–38. "On peut expliquer…toutes les actions des animaux, quelque compliquées qu'elles puissent paroître, sans qu'il soit besoin de leur accorder, ni la pensée, ni la réflexion, leur sens intérieur suffit pour produire tous leurs movemens." Ibid., 4:40.
40. Ibid., 2:225. "Le sens intérieur de l'animal est, aussi-bien que ses sens extérieurs, un organe, un résultat de méchanique, un sens purement matériel." Ibid., 4:24.
41. Ibid. "Nous avons, comme l'animal, ce sens intérieur matériel, et nous possédons de plus un sens d'une nature supérieure et bien différente, qui réside dans la substance spirituelle qui nous anime et nous conduit." Ibid.
42. Ibid., 2:249. "Cette puissance de réfléchir ayant été refusée aux animaux, il est donc certain qu'ils ne peuvent former d'idées, et que par conséquent leur conscience d'existence est moins sûre et moins étendue que la nôtre; car ils ne peuvent avoir aucune idée du temps, aucune connaissance du passé, aucune notion de l'avenir: leur conscience d'existence est simple, elle dépend uniquement des sensations qui les affectent actuellement, et consiste dans le sentiment intérieur que ces sensations produisent." Ibid., 4:53–54.
43. Ibid., 2:251. "…les animaux n'ont aucune connaissance du passé, aucune idée du temps, et que par conséquent ils n'ont pas la mémoire." Ibid., 4:55.
44. Jacques Roger, *Buffon: A Life in Natural History* (Ithaca: Cornell University Press, 1997), 254. Roger cites from Georges-Louis Leclerc de Buffon, *Discours sur la nature des animaux* (1753) in *Histoire naturelle, générale et particulière*, 15 volumes (Paris: Imprimerie royale, 1749–1767), 4:41, 44, 47, 85.
45. Ibid. Ibid., 4:22.
46. Ibid., 258–59. Ibid., 4:110, and *Nomenclature des singes* (1766), 14:32
47. Georges-Louis Leclerc de Buffon, *The Nomenclature of Apes* (1766) in *Natural History, General and Particular*, translated by William Smellie, 9 vols., 2nd edition (London: W. Strahan and T. Cadell, 1785), 8:64–66. "On verra dans l'histoire de l'orang-outang, que si l'on ne faisoit attention qu'à la figure on pourroit également regarder cet animal comme le premier des singes ou le dernier des hommes parce qu'à l'exception de l'âme, il ne lui manque rien de tout ce que nous avons, et parce qu'il diffère moins de l'homme pour le corps, qu'il ne diffère des autres animaux auxquels on a donné le même nom de singe. L'âme, la pensée, la parole ne dépendent donc pas de la forme ou de l'organisation du corps; rien ne prouve mieux que c'est un don particulier, et fait à l'homme seul, puisque l'orang-outang qui ne parle ni ne pense, néanmoins le corps, les membres, les sens, le cerveau et la langue

entièrement semblables à l'homme...le Créateur n'a pas voulu faire pour le corps de l'homme un modèle absolument différent de celui de l'animal; il a compris sa forme, comme celle de tous les animaux dans un plan général; mais en même temps qu'il lui a départi cette forme matérielle semblable à celle du singe, il a pénétré ce corps animal de son souffle divin; s'il eût fait la même faveur, je ne dis pas au singe, mais à l'espèce la plus vile, à l'animal qui nous paroît le plus mal organisé, cette espèce seroit bientôt devenue la rivale de l'homme; vivifiée par l'esprit, elle eût primé sur les autres; elle eût pensé, elle eût parlé: quelque ressemblance qu'il y ait donc entre l'Hottentot et le singe, l'intervalle qui les sépare est immense, puisqu'à l'intérieur il est rempli par la pensée et au dehors par la parole." Georges-Louis Leclerc de Buffon, *Nomenclature des singes* (1766) in *Histoire naturelle, générale et particulière*, 15 volumes (Paris: Imprimerie royale, 1749–1767), 14:30, 32.

48. Jacques Roger, *Buffon: A Life in Natural History* (Ithaca: Cornell University Press, 1997), 259. Roger cites *Nomenclature des singes* (1766), 14:31–32.

49. Georges-Louis Leclerc de Buffon, *The Nomenclature of Apes* (1766) in *Natural History, General and Particular*, translated by William Smellie, 9 vols., 2nd edition (London: W. Strahan and T. Cadell, 1785), 8:66. "Qui pourra jamais dire en quoi l'organisation d'un imbécile diffère de celle d'un autre homme?" Georges-Louis Leclerc de Buffon, *Nomenclature des singes* (1766) in *Histoire naturelle, générale et particulière*, 15 volumes (Paris: Imprimerie royale, 1749–1767), 14:32.

50. Jean-Jacques Rousseau, *Essay on the Origin of Languages, Chapter 1*, in *the Discourses and other Early Political Writings*, edited and translated by Victor Gourevitch (Cambridge: Cambridge University Press, 2005), 252.

51. Jean-Jacques Rousseau, *Discourse on the Origin of Inequality* in *The Social Contract and Discourses*, translated and introduced by G.D.H. Cole (London: J.M. Dent and Sons, 1913), 191. "Le premier langage de l'homme, le langage le plus universel, le plus énergique, et le seul dont il eut besoin avant qu'il fallût persuader des hommes assemblés, est le cri de la nature: Comme ce cri n'étoit arraché que par une sorte d'instinct dans les occasions pressantes, pour implorer du secours dans les grands dangers ou du soulagement dans les maux violents, il n'étoit pas d'un grand usage dans le cours ordinaire de la vie; où règnent des sentiments plus modérés. Quand les idées des hommes commencèrent à s'étendre et à se multiplier, et qu'il s'établit entre eux une communication plus étroite, ils cherchèrent des signes plus nombreux et un langage plus étendu; ils multiplièrent les inflexions de la voix, et y joignirent les gestes qui, par leur nature, sont plus expressifs, et dont le sens dépend moins d'une détermination antérieure." Jean-Jacques Rousseau, *Discours sur l'origine et les fondements de l'inégalité parmi les hommes* in *Œuvres complètes de Jean-Jacques Rousseau avec les notes de tous les commentateurs* (Paris: Dalibon, 1826), 1:267.

52. Ibid., 192. "D'ailleurs les idées générales ne peuvent s'introduire dans l'esprit qu'à l'aide des mots, et l'entendement ne les saisit que par des propositions. C'est une des raisons pourquoi les animaux ne sauroient se former de telles idées ni jamais acquérir la perfectibilité qui en dépend. Quand un singe va hésiter d'une noix à l'autre, pense-t-on qu'il ait l'idée générale de cette sorte de fruit, et qu'il compare son archétype à ces deux individus? Non, sans doute; mais la vue de l'une de ces noix rappelle à sa mémoire les sensations qu'il a reçues de l'autre; et ses yeux, modifiés d'une certaine manière, annoncent à son goût la modification qu'il va

recevoir. Toute idée générale est purement intellectuelle; pour peu que l'imagination s'en mêle, l'idée devient aussitôt particulière." Ibid., 1:269–70.
53. "L'intermédiaire entre l'homme et les autres animaux, c'est le singe." *Eléments de physiologie* in *Œuvres: Philosophie*, edited by Laurent Versini (Paris: Robert Laffont, 1994), 1:1278.
54. Denis Diderot, *Conversation between D'Alembert and Diderot* in *Rameau's Nephew and D'Alembert's Dream*, translated and introduced by Leonard Tancock (London: Penguin Books, 1966), 158–59. "Cet animal se meut, s'agite, crie; j'entends ses cris à travers la coque; il se couvre de duvet; il voit. La pesanteur de sa tête, qui oscille, porte sans cesse son bec contre la paroi intérieure de sa prison; la voilà brisée; il en sort, il marche, il vole, il s'irrite, il fuit, il approche, il se plaint, il souffre, il aime, il désire, il jouit; il a toutes vos affections; toutes vos actions, il les fait. Prétendrez-vous, avec Descartes, que c'est une pure machine imitative? Mais les petits enfants se moqueront de vous, et les philosophes vous répliqueront que si c'est là une machine, vous en êtes une autre." Denis Diderot, *Entretiens entre d'Alembert et Diderot* in *Œuvres complètes de Diderot*, edited by Jean Assézat and Maurice Tourneux (Paris: Garnier, 1875–1877), 2:115.
55. Denis Diderot, *Sequel to the Conversation* in *Rameau's Nephew and D'Alembert's Dream*, translated and introduced by Leonard Tancock (London: Penguin Books, 1966), 231–32. "…c'est qu'ils ont vu dans la basse-cour de l'archduc un infâme lapin qui servait de coq à une vingtaine de poules infâmes qui s'en accommodaient; ils ajouteront qu'on leur a montré des poulets couverts de poils et provenus de cette bestialité. Croyez qu'on s'est moqué d'eux." Denis Diderot, *Suite de l'Entretien* in *Œuvres complètes de Diderot*, edited by Jean Assézat and Maurice Tourneux (Paris: Garnier, 1875–1877), 2:189.
56. Ibid., 232. "Et la question de leur baptême?…Ferait un beau charivari en Sorbonne." Ibid., 190.
57. Ibid., 233. "Avez-vous vu au Jardin du Roi, sous une cage de verre, un orang-outang qui a l'air d'un saint Jean qui prêche au désert?…Oui, je l'ai vu…Le cardinal de Polignac lui disait un jour: 'Parle, et je te baptise.'" Ibid.
58. "L'intermédiaire entre l'homme et les autres animaux, c'est le singe." *Eléments de physiologie* in *Œuvres: Philosophie*, edited by Laurent Versini (Paris: Robert Laffont, 1994), 1:1278.
59. Denis Diderot, *Supplement to the Voyage of Bougainville* in *Political Writings*, translated and edited by John Hope Mason and Robert Wokler (Cambridge: Cambridge University Press, 1992), 40. "Le Taïtien touche à l'origine du monde, et l'Européen touche à sa vieillesse. L'intervalle qui le sépare de nous est plus grand que la distance de l'enfant qui naît à l'homme décrépit." Denis Diderot, *Supplément au Voyage de Bougainville* in *Œuvres complètes de Diderot*, edited by Jean Assézat and Maurice Tourneux (Paris: Garnier, 1875–1877), 2:212.

CHAPTER TEN

1. "Que si tu te crois autorisé à m'opprimer, parce que tu es plus fort et plus adroit que moi, ne te plains donc pas quand mon bras vigoureux ouvrira ton sein pour y

chercher ton cœur..." Denis Diderot, *Histoire des deux Indes* in *Œuvres: Politique*, edited by Laurent Versini (Paris: Robert Laffont, 1995), 3:739 [Guillaume-Thomas Raynal, *Histoire philosophique et politique des établissemens et du commerce des européens dans les deux Indes, Book 11, Chapter 24* (Geneva: Jean-Léonard Pellet, 1780), 3:195].

2. François Bernier, *A New Division of the Earth, according to the Different Species or Races of Men who Inhabit It*, translated by T. Bendyshe, in *Memoirs Read before the Anthropological Society of London* 1 (1863–1864): 360–64 [François Bernier, *Nouvelle division de la Terre, pour les différentes espèces ou races d'hommes qui l'habitent* in *Journal des sçavans*, April 24, 1684, 133–40].

3. Georges-Louis Leclerc de Buffon, *The Ass* (1753) in *Natural History, General and Particular*, translated by William Smellie, 9 vols., 2nd edition (London: W. Strahan and T. Cadell, 1785), 3:406. "...on peut aussi réunir en une seule espèce deux successions d'individus qui se reproduisent en se mêlant..." Georges-Louis Leclerc de Buffon, *L'Asne* (1753) in *Histoire naturelle, générale et particulière*, 15 volumes (Paris: Imprimerie royale, 1749–1767), 4:385.

4. Ibid., 3:407. "L'espèce n'étant donc autre chose qu'une succession constante d'individus semblables et qui se reproduisent..." Ibid., 4:386.

5. Ibid., 3:406. "...on peut toujours tirer une ligne de séparation entre deux espèces, c'est-à-dire, entre deux successions d'individus qui se reproduisent et ne peuvent se mêler..." Ibid., 4:385.

6. Ibid. "...chaque espèce, chaque succession d'individus qui se reproduisent et ne peuvent se mêler, sera considérée à part et traitée séparément, et nous ne nous servirons des familles, des genres, des ordres et des classes, pas plus que ne s'en sert la Nature." Ibid., 4:386.

7. Ibid., 3:408. "...mais ces différences de couleur et de dimension dans la taille n'empêchent pas que le Nègre et le Blanc, le Lappon et le Patagon, le géant et le nain, ne produisent ensemble des individus qui peuvent eux-mêmes se reproduire, et que par conséquent ces hommes, si différens en apparence, ne soient tous d'ue seule et même espèce, puisque cette reproduction constante est ce qui constitue l'espèce." Ibid., 4:387.

8. "...c'est la succession constante & le renouvellement noninterrompu de ces individus qui la constituent...on peut toujours tirer une ligne de séparation entre deux *espèces*, c'est-à-dire entre deux successions d'individus qui se reproduisent & ne peuvent se mêler, comme l'on peut aussi réunir en une seule *espèce* deux successions d'individus qui se reproduisent en se mêlant...L'*espèce* n'étant donc autre chose qu'une succession constante d'individus semblables & qui se reproduisent...M. de Buffon, *hist. nat. gen. & part. &c. tom. IV. p. 784 & suiv.*" Denis Diderot, "Espèce" (1755), *Encyclopédie, ou Dictionnaire raisonné des sciences, des arts et des métiers*, edited by Denis Diderot and Jean Le Rond d'Alembert (Paris: Briasson, David, Le Breton, Durant; Neuchâtel: S. Faulche, 1751–1765), 5:956–57.

9. Robert Bernasconi, ed., *Race* (Malden, MA: Blackwell Publishers, Inc., 2001, 15. Bernasconi advises that the Latin text of 1758 is quoted by T. Bendyshe, "The History of Anthropology," Memoirs Read before the Anthropological Society of London, vol. 1, 1863–1864, pp. 424–26. An English translation is in Winthrop D. Jordan *White over Black* (Baltimore: Penguin, 1969), 220–21.

Notes 315

10. Carl Linnaeus, *Systema naturæ per regna tria naturæ, secundum classes, ordines, genera, species, cum characteribus, differentiis synonymis, locis* (Stockholm: Laurentius Salvius, 1758), 1:20.
11. Ibid., 1:22–24.
12. "De ce qui précede il suit que dans tout le nouveau continent que nous venons de parcourir, il n'y a qu'une seule & même face d'hommes, plus ou moins basanés. Les Américains sortent d'une même souche. Les Européens sortent d'une même souche. Du nord au midi on apperçoit les mêmes variétés dans l'un & l'autre hémisphere. Tout concourt donc à prouver que le genre *humain* n'est pas composé d'espèces essentiellement différentes. La différence des blancs aux bruns vient de la nourriture, des moeurs, des usages, des climats; celle des bruns aux noir a la même cause. Il n'y a donc eu originairement qu'une seule race d'hommes, qui s'étant multipliée & répandue sur la surface de la terre, a donné à la longue toutes les variétés dont nous venons de faire mention…" Diderot, "Humaine espèce (*Hist. nat.*)" (1765), *Encyclopédie, ou Dictionnaire raisonné des sciences, des arts et des métiers*, edited by Denis Diderot and Jean Le Rond d'Alembert (Paris: Briasson, David, Le Breton, Durant; Neuchâtel: S. Faulche, 1751–1765), 8:348.
13. Georges-Louis Leclerc de Buffon, *Of the Degeneration of animals* (1766) in *Natural History, General and Particular*, translated by William Smellie, 9 vols., 2nd edition (London: W. Strahan and T. Cadell, 1785), 7:394–95. "…s'il arrivoit, dis-je, que l'homme fût contraint d'abandonner les climats qu'il a autrefois envahis pour se réduire à son pays natal, il reprendroit avec le temps ses traits originaux, sa taille primitive et sa couleur naturelle: le rappel de l'homme à son climat amèneroit cet effet, le mélange des races l'amèneroit aussi et bien plus promptement; le blanc avec la Noire, ou le Noir avec la Blanche produisent également un Mulâtre dont la couleur est brune, c'est-à-dire, mêlée de blanc et de noir; ce Mulâtre avec un Blanc produit un second Mulâtre moins brun que le premier; et si ce second Mulâtre s'unit de même à un individu de race blanche, le troisème Mulâtre n'aura plus qu'une nuance légère de brun qui disparoîtra tout-à-fait dans les générations suivantes: il ne faut donc que cent cinquante ou deux cents ans pour laver la peau d'un Nègre par cette voie du mélange avec le sang du Blanc, mais il faudroit peut-être un assez grand nombre de siècles pour produire ce même effet par la seule influence du climat." Georges-Louis Leclerc de Buffon, *De la dégénération des animaux* (1766) in *Histoire naturelle, générale et particulière*, 15 volumes (Paris: Imprimerie royale, 1749–1767), 14:313.
14. Ibid., 7:392. "…et lorsqu'après des siècles écoulés, des continens traversés et des générations déjà dégénérées par l'influence des différentes terres, il a voulu s'habituer dans les climats extrêmes, et peupler les sables du Midi et les glaces du Nord; les changemens sont devenus si grands et si sensibles, qu'il y auroit lieu de croire que le Nègre, le Lappon et le Blanc forment des espèces différentes, si d'un côté l'on n'étoit assuré qu'il n'y a eu qu'un seul Homme de créé, et de l'autre que ce Blanc, ce Lappon et ce Nègre, si dissemblans entr'eux, peuvent cependant s'unir ensemble et propager en commun la grande et unique famille de notre genre humain…" Ibid., 14:311.
15. "Il me semble alors que je suis assez bien fondé à croire qu'il en est des hommes comme des arbres; que les poiriers, les sapins, les chênes, et les abricotiers ne viennent point d'un même arbre, et que les blancs barbus, les nègres portant laine, les jaunes portant crins, et les hommes sans barbe, ne viennent pas du même

homme." François-Marie Arouet de Voltaire, *Traité de métaphysique, Chapître 2,* in *Œuvres complètes de Voltaire avec préfaces, avertissements, notes, etc.,* 70 vols., edited by M. Beuchot (Paris: Lefèvre, 1829–1840), 37:282–83.

16. "Je me demande donc pardon aux Physiciens modernes, si je ne puis admettre les systemes qu'ils ont si ingéniusement imaginés. Car je ne suis pas de ceux qui croient qu'on avance la Physique en s'attachant à un systeme malgré quelque phénomene qui lui est évidemment incompatible; & qui, ayant remarqué quelqu'endroit d'où suit nécessairement la ruine de l'édifice, achevent cependant de le bâtir, & l'habitent avec autant de sécurité, que s'il étoit le plus solide." Pierre-Louis de Maupertuis, *Vénus physique* (The Hague, 1745), 96–97.

17. "La chymie en a depuis reconnu la nécessité; & les chymistes les plus fameux aujourd'hui, admettent l'Attraction, & l'étendent plus loin que n'ont fait les astronomes. Pourquoi, si cette force existe dans la Nature, n'auroît-elle pas lieu dans la formation du corps des animaux?" Ibid., 104.

18. "Si les premiers hommes blancs qui en virent des noirs, les avoient trouvés dans les forêts, peut-être ne leur auroient-ils pas accordé le nom d'hommes. Mais ceux qu'on trouva dans de grandes villes, qui étoient gouvernés par de sages Reines, qui faisoient fleurir les Arts & les Sciences, dans des temps où presque tous les autres peuples étoient des barbares, ces Noirs-là, auroient bien pu ne pas vouloir regarder les Blancs comme leurs frères." Ibid., 119–20.

19. "Tous ces peuples que nous venons de parcourir, tant d'hommes divers, sont-ils sortis d'une même mère? Il ne nous est pas permis d'en douter. Ce qui nous reste à examiner, c'est comment d'un seul individu, il a pu naître tant d'especes si différentes. Je vais hasarder sur cela quelques conjectures." Ibid., 134.

20. "Ces variétés, si on les pouvoit suivre, auroient peut-être leur origine dans quelqu'ancêtre inconnu. Elles se perpetuent par des générations répétées d'individus qui les ont; & s'effacent par des générations d'individus qui ne les ont pas. Mais, ce qui est peut-être encore plus étonnant, c'est après une interruption de ces variétés, de les voir reparoître; de voir l'enfant qui ne ressemble ni à son pere ni à sa mère, naître avec les traits de son ayeul." Ibid., 138–39.

21. "La Nature contient le fonds de toutes ces variétés: mais le hasard ou l'art les mettent en œuvre. C'est ainsi que ceux dont l'industrie s'applique à satisfaire le goût des curieux, sont, pour ainsi dire, créateurs d'espèces nouvelles. Nous voyons paroître des races de chiens, de pigeons, de serins qui n'étoient point auparavant dans la nature. Ce n'ont été d'abord que es individus fortuits; l'art & les générations répétées en ont fait des espèces." Ibid., 140.

22. "Il me paroît donc démontré que s'il naît des noirs de parens blancs, ces naissances sont incomparablement plus rares que les naissances d'enfans blancs de parens noirs. Cela suffiroit peut-être pour faire penser que le blanc est la couleur des premiers hommes; & que ce n'est que par quelque accident que le noir est devenu une couleur héréditaire aux grandes familles qui peuplent la Zone torride; parmi lesquelles cependant la couleur primitive n'est pas si parfaitement effacée qu'elle ne reparoisse quelquefois." Ibid., 165.

23. "Voici enfin une nouvelle richesse de la nature, une espèce qui ne ressemble pas tant à la nôtre que les barbets aux lévriers." François-Marie Arouet de Voltaire, *Relation touchant un Maure blanc amené d'Afrique à Paris en 1744* (1744) in *Œuvres complètes de Voltaire avec préfaces, avertissements, notes, etc.,* 70 vols., edited by M. Beuchot (Paris: Lefèvre, 1829–1840), 38:523.

24. "...si nous pensons valoir beaucoup mieux qu'eux, nous nous trompons assez lourdement." Ibid., 38:524.
25. "Ne pouroit-on pas expliquer par-là comment de deux seuls individus, la multiplication des espèces les plus dissemblables auroit pu s'ensuivie? Elles n'auroient dû leur première origine qu'à quelques productions fortuites dans lesquelles les parties élémentaires n'auroient pas retenu l'ordre qu'elles tenoient dans les animaux pères & mères: chaque degré d'erreur auroit fait une nouvelle espèce; & à force d'écarts répétés seroit venue la diversité infinie des animaux que nous voyons aujourd'hui, qui s'accroîtra peut-être encore avec le temps, mais à laquelle peut-être la suite des siècles n'apporte que des accroissements imperceptibles." Maupertuis, *Essai sur la formation des corps organisés* (Berlin, 1754), 40–41.
26. Robert Bernasconi, editor, *Race* (Malden, MA: Blackwell Publishers, Inc., 2001), 11.
27. *Le Code Noir, ou recueil d'édits, déclarations et arrêts, concernant la discipline & le commerce des esclaves nègres des Isles françaises de l'Amérique* in *Recueils de réglemens, édits, déclarations et arrêts, concernant le commerce, l'administration de la justice, & la police des colonies françaises de l'Amérique, & les engagés, avec la Code Noir et l'addition audit code* (Paris: Libraires Associés, 1745), 2:81.
28. Ibid., 2:82 - 83.
29. Ibid., 2:85–86.
30. Ibid., 2:86.
31. Ibid., 2:92.
32. Ibid., 2:92–93.
33. Ibid., 2:95.
34. Charles-Louis de Secondat de Montesquieu, *Persian Letters, Letter 161* (London: Penguin Books, 2004), 280.
35. "On comptait, en 1757, dans la Saint-Domingue française, environ trente mille personnes, et cent mille esclaves nègres ou mulâtres, qui travaillaient aux sucreries, aux plantations d'indigo, de cacao, et qui abrègent leur vie pour flatter nos appétits nouveaux, en remplissaant nos nouveaux besoins, que nos pères ne connaissaient pas. Nous allons acheter ces nègres à la côte de Guinée, à la côte d'Or, à celle d'Ivoire. Il y a trente ans qu'on avait un beau nègre pour cinquante livres; c'est à peu près cinq fois moins qu'un bœuf gras...Nous leur disons qu'ils sont hommes comme nous...et ensuite on les fait travailler comme des bêtes de somme; on les nourrit plus mal: s'ils veulent s'enfuir, on leur coupe une jambe, et on leur fait tourner à bras l'arbre des moulins à sucre, lorsqu'on leur a donné une jambe de bois. Après cela nous osons parler du droit des gens!" François-Marie Arouet de Voltaire, *Essai sur les mœurs, Chapter 152*, in *Œuvres complètes de Voltaire avec préfaces, avertissements, notes, etc.*, 70 vols., edited by M. Beuchot (Paris: Lefèvre, 1829–1840), 17:450–51.
36. "C'est donc aller directement contre le droit des gens & contre la nature, que de croire que la religion chrétienne donne à ceux qui la professent, un droit de réduire en servitude ceux qui ne la professent pas, pour travailler plus aisément à sa propagation. Ce fut pourtant cette manière de penser qui encouragea les destructeurs de l'Amérique dans leurs crimes; & ce n'est pas la seule fois que l'on se soit servi de la religion contre ses propres maximes, qui nous apprennent que la qualité de prochain s'étend sur tout l'univers." Louis de Jaucourt, "Esclavage"

(1755), *Encyclopédie, ou Dictionnaire raisonné des sciences, des arts et des métiers*, edited by Denis Diderot and Jean Le Rond d'Alembert (Paris: Briasson, David, Le Breton, Durant; Neuchâtel: S. Faulche, 1751–1765), 5:938.
37. "Traité des nègres. (*Commerce d'Afrique.*) c'est l'achat des nègres que font les Européens sur les côtes d'Afrique, pour employer ces malheureux dans leurs colonies en qualité d'esclaves. Cet achat de nègres, pour les réduire en esclavage, est un négoce qui viole la religion, la morale, les lois naturelles, & tous les droits de la nature humaine." Louis de Jaucourt, "Traité des nègres" (1765), *Encyclopédie, ou Dictionnaire raisonné des sciences, des arts et des métiers*, edited by Denis Diderot and Jean Le Rond d'Alembert (Paris: Briasson, David, Le Breton, Durant; Neuchâtel: S. Faulche, 1751–1765), 16:532.
38. "Les rois, les princes, les magistrats ne sont point les propriétaires de leurs sujets, ils ne sont donc pas en droit de disposer de leur liberté, & de les vendre pour esclaves." Ibid.
39. "D'un autre côté, aucun homme n'a droit de les acheter ou de s'en rendre le maître; les hommes & leur liberté ne sont point un objet de commerce; ils ne peuvent être ni vendus, ni achetés, ni payés à aucun prix." Ibid.
40. "J'ai vu ces vastes et malheureuses contrées qui ne semblent destinées qu'à couvrir la terre de troupeaux d'esclaves. A leur vil aspect j'ai détourné les yeux de dédain, d'horreur et de pitié; et, voyant la quatrième partie de mes semblables changée en bêtes pour le service des autres, j'ai gémi d'être homme." Jean-Jacques Rousseau, *Julie, ou La Nouvelle Héloïse, Lettre 3, A Madame Orbe*, in *Œuvres complètes de Jean-Jacques Rousseau avec les notes de tous les commentateurs* (Paris: Dalibon, 1826), 9:180–81.
41. Jean-Jacques Rousseau, *The Social Contract, Book 1, Chapter 4*, in *The Social Contract and Discourses*, translated and introduced by G.D.H. Cole (London: J.M. Dent and Sons, 1913), 9. "Puisqu'aucun homme n'a une autorité naturelle sur son semblable, & puisque la force ne produit aucun droit, restent donc les conventions pour base de toute autorité légitime parmi les hommes." Jean-Jacques Rousseau, *Du Contrat social* (Amsterdam: Marc-Michel Rey, 1762), 13.
42. Ibid., 10. "Dire qu'un homme se donne gratuitement, c'est dire une chose absurde & inconcevable; un tel acte est illégitime & nul, par cela seul que celui qui le fait n'est pas dans son bon sens. Dire la même chose de tout un peuple, c'est supposer un peuple de fous: la folie ne fait pas droit…Renoncer à sa liberté c'est renoncer à sa qualité dhomme, aux droits de l'humanité, même à ses devoirs…une telle renonciation est incompatible avec la nature de l'homme, & c'est ôter toute moralité à ses actions que d'ôter toute liberté à sa volonté." Ibid., 15–16.
43. Ibid., 13. "Ainsi, de quelque sens qu'on envisage les choses, le droit d'esclavage est nul, non seulement parce qu'il est illégitime, mais parce qu'il est absurde et ne signifie rien. Ces mots, esclavage, et droit, sont contradictoires; ils s'excluent mutuellement. Soit d'un homme à un homme, soit d'un homme à un peuple, ce discours sera toujours également insensé: *Je fais avec toi une convention toute à ta charge et toute à mon profit, que j'observerai tant qu'il me plaira, et que tu observeras tant qu'il me plaira.*" Ibid., 22.
44. "La liberté, est la liberté de soi." Denis Diderot, *Histoire des deux Indes* in *Œuvres: Politique*, edited by Laurent Versini (Paris: Robert Laffont, 1995), 3:738 ["La liberté, est la propriété de soi." Guillaume-Thomas Raynal, *Histoire philosophique*

et politique des établissemens et du commerce des européens dans les deux Indes, Book 11, Chapter 24 (Geneva: Jean-Léonard Pellet, 1780), 3:194].

45. "La liberté naturelle est le droit que la nature a donné à tout homme de disposer de soi, à sa volonté." Ibid. ["La liberté naturelle, est le droit que la nature a donné à tout homme de disposer de soi, à sa volonté." Ibid.].

46. "Hommes ou démons, qui que vous soyez, oserez-vous justifier les attentats contre mon indépendance par le droit du plus fort? Quoi! celui qui veut me rendre esclave n'est point coupable; il use de ses droits. Où sont-ils ces droits? Qui leur a donné un caractère assez sacré pour faire taire les miens? Je tiens de la nature le droit de me défendre; elle ne t'a pas donc donné celui de m'attaquer. Que si tu te crois autorisé à m'opprimer, parce que tu es plus fort et plus adroit que moi, ne te plains donc pas quand mon bras vigoureux ouvrira ton sein pour y chercher ton cœur..." Ibid., 3:739 [Ibid., 3:195].

47. "Mais les nègres sont une espèce d'hommes nés pour l'esclavage. Ils sont bornés, fourbes, méchants; ils conviennent eux-mêmes de la supériorité de notre intelligence, et reconnaissent presque la justice de notre empire." Ibid., 3:740 [Ibid., 3:197].

48. "Les nègres sont bornés, parce que l'esclavage brise tous les ressorts de l'âme. Ils sont méchants, pas assez avec vous. Ils sont fourbes, parce qu'on ne doit pas la vérité à ses tyrans. Ils reconnaissent la supériorité de notre esprit, parce que nous avons perpétué leur ignorance; la justice de notre empire, parce que nous avons abusé de leur faiblesse. Dans l'impossibilité de maintenir notre supériorité par la force, une criminelle politique s'est rejetée sur la ruse. Vous êtes presque parvenus à leur persuader qu'ils étaient une espèce singulière, née pour l'abjection et la dépendance, pour le travail et le châtiment. Vous n'avez rien négligé pour dégrader ces malheureux, et vous leur reprochez ensuite d'être vils." Ibid. [Ibid.].

49. "D'ailleurs, il est aussi absurde qu'atroce d'oser avancer que la plupart des malheureux achetés en Afrique sont des criminels. A-t-on peur qu'on n'ait pas assez de mépris pour eux, qu'on ne les traite pas avec assez de dureté? & comment suppose-t-on qu'il existe un pays où il se commette tant de crimes, et où cependant il se fasse si exacte justice" Marie-Jean-Antoine-Nicolas de Caritat, marquis de Condorcet, *Réflexions sur l'esclavage des nègres, Chapter 2* (Neufchatel: Société typographique, 1781), 8.

50. Louis-Antoine de Bougainville, *Voyage autour du monde* (Paris: Saillant et Nyon, 1771).

51. Denis Diderot, *Supplément to the Voyage of Bougainville* in *Political Writings*, translated and edited by John Hope Mason and Robert Wokler (Cambridge: Cambridge University Press, 1992), 37. "Comment explique-t-il le séjour de certains animaux dans des îles séparées de tout continent par des intervalles de mer effrayants? Qui est-ce qui a porté là le loup, le renard, le chien, le cerf, le serpent?" Denis Diderot, *Supplément au Voyage de Bougainville* in *Œuvres complètes de Diderot*, edited by Jean Assézat and Maurice Tourneux (Paris: Garnier, 1875–1877), 2:209.

52. Ibid. "Qui sait l'histoire primitive de notre globe? Combien d'espaces de terre, maintenant isolés, étaient autrefois continus? Le seul phénomène sur lequel on pourrait former quelque conjecture, c'est la direction de la masse des eaux qui les a séparés." Ibid.

53. Ibid., 38. "A l'inspection du lieu qu'elle occupe sur le globe, il n'est personne qui ne se demande qui est-ce qui a placé là des hommes? Quelle communication les liait autrefois avec le reste de leur espèce?" Ibid., 2:210.
54. Ibid. "...et de là tant d'usages d'une cruauté nécessaire et bizarre, dont la cause s'est perdue dans la nuit des temps, et met les philosophes à la torture. Une observation assez constante, c'est que les institutions surnaturelles et divines se fortifient et s'éternisent, en se transformant, à la longue, en lois civiles et nationales; et que les institutions civiles et nationales se consacrent, et dégénèrent en préceptes surnaturels et divins." Ibid.
55. Ibid. "N'était-il pas au Paraguay au moment même de l'expulsion des jésuites?" Ibid.
56. Ibid. "...ces cruels Spartiates en jaquette noire en usaient avec leurs esclaves Indiens, comme les Lacédémoniens avec les Ilotes; les avaient condamnés à un travail assidu; s'abreuvaient de leur sueur, ne leur avaient laissé aucun droit de propriété; les tenaient sous l'abrutissement de la superstition; en exigeaient une vénération profonde; marchaient au milieu d'eux, un fouet à la main, et en frappaient indistinctement tout âge et tout sexe." Ibid.
57. Charles-Louis de Secondat de Montesquieu, *The Spirit of the Laws, Book 15, Chapter 17*, translated and edited by Anne M. Cohler, Basia Carolyn Miller, and Harold Samuel Stone (Cambridge: Cambridge University Press, 1989), 260. Charles-Louis de Secondat de Montesquieu, *De l'esprit des loix, Book 15, Chapter 17*, in 2 volumes (Edinburgh: G. Hamilton and J. Balfour, 1750), 2:343.
58. Plutarch, *Lycurgus* in *Lives*, translated by Bernadotte Perrin, 11 vols. (Cambridge, MA: Harvard University Press, 2005), 28.1–4, pp. 1:289, 1:291.
59. Denis Diderot, *Supplément to the Voyage of Bougainville* in *Political Writings*, translated and edited by John Hope Mason and Robert Wokler (Cambridge: Cambridge University Press, 1992), 39. "Ce sont de bonnes gens qui viennent à vous, et qui vous embrassent en criant *Chaoua*." Denis Diderot, *Supplément au Voyage de Bougainville* in *Œuvres complètes de Diderot*, edited by Jean Assézat and Maurice Tourneux (Paris: Garnier, 1875–1877), 2:211.
60. Ibid., 40. "Le Taïtien touche à l'origine du monde, et l'Européen touche à sa vieillesse. L'intervalle qui le sépare de nous est plus grand que la distance de l'enfant qui naît à l'homme décrépit." Ibid., 2:212.

CONCLUSION

1. "L'intermédiaire entre l'homme et les autres animaux, c'est le singe." *Eléments de physiologie* in *Œuvres: Philosophie*, edited by Laurent Versini (Paris: Robert Laffont, 1994), 1:1278.

Bibliography

PRIMARY SOURCES

Aristotle. *History of Animals: Books I-III*. Translated by A.L. Peck. Cambridge, MA: Harvard University Press, 1965.
———. *History of Animals: Books VII-X*. Edited by D. M. Balme. Cambridge, MA: Harvard University Press, 1991.
———. *Metaphysics: Books I-IX*. Translated by Hugh Tredennick. Cambridge, MA: Harvard University Press, 1933.
———. *Metaphysics: Books X-XIV*. Translated by Hugh Tredennick. Cambridge, MA: Harvard University Press, 1935.
———. *Parts of Animals*. Translated by A. L. Peck. Cambridge, MA: Harvard University Press, 1937.
Buffon, Georges-Louis Leclerc de. *Histoire naturelle, general et particulière*. 15 vols. Paris: Imprimerie royale, 1749–1767.
———. *Histoire naturelle, générale et particulière; Supplément* 17 vols. Paris: Imprimerie royale, 1774–1789.
———. *Natural History, General and Particular*. Translated by William Smellie. 9 vols. 2nd Edition. London: W. Strahan and T. Cadell, 1785. *Le Code noir, ou recueil d'édits, déclarations et arrêts, concernant la discipline & le commerce des esclaves nègres des Isles françaises de l'Amérique* in *Recueils de réglemens, édits,déclarations et arrêts, concernant le commerce, l'administration de la justice, & la police des colonies françaises de l'Amérique, & les engagés, avec la code Noir et l'addition audit code*. Vol. 2. Paris: Librairies Associez, 1745.
Descartes, René. *Œuvres et Lettres*. Edited by André Bridoux. Paris: Gallimard, 1953.
———. *The Philosophical Writings of Descartes*. Translated by John Cottingham, *et al*. Cambridge: Cambridge University Press, 1991.
———. *Principles of Philosophy*. Translated and annotated by Valentine Rodger Miller and Reese P. Miller. Dordrecht: Kluwer Academic Publisher, 1991.
Dictionnaire de L'Académie française. 1st ed. Paris: Baptiste Coignard, 1694.
Dictionnaire de L'Académie française. 4th ed. Paris: Brunet, 1762.
Dictionnaire de L'Académie française. 5th ed. Paris: J. J. Smits, 1798.
Diderot, Denis. *Correspondance inédite*. 2 vols. Edited by André Babelon. Paris: Gallimard, 1931.
———. *Diderot's Early Philosophical Works*. Translated and edited by Margaret Jourdain. Chicago and London: The Open Court Publishing Company, 1916.

———. *Œuvres complètes de Diderot*. 20 vols. Edited by Jean Assézat and Maurice Tourneux. Paris: Garnier, 1875–1877.
———. *Œuvres: Philosophie*. Vol. 1. Edited by Laurent Versini. Paris: Robert Laffont, 1994.
———. *Œuvres philosophiques*. Edited by Paul Vernière. Paris: Garnier, 1998.
———. *Œuvres: Politique*. Vol. 3. Edited by Laurent Versini. Paris: Robert Laffont, 1995.
———. *Thoughts on the Interpretation of Nature and Other Philosophical Works*. Introduced and annotated by David Adams. Manchester: Clinamen Press, 1999.
Diodorus Siculus. *Library of History, Books 2.35–4.58*. Translated by C.H. Oldfather. Cambridge, MA: Harvard University Press, 2000.
Encyclopédie, ou Dictionnaire raisonné des sciences, des arts et des métiers. 17 vols. Edited by Denis Diderot and Jean Le Rond d'Alembert. Paris: Briasson, David, Le Breton, Durand; Neuchâtel: S. Faulche, 1751–1765.
Fontenelle, Bernard Le Bouyer de. *A Conversation on the Plurality of Worlds*. London: J. Dursley, A. Millard, E. Jobson, D. Evans, and R. Newton, 1783.
———. *Entretiens sur la pluralité des mondes*. Paris: Michel Brunet, 1724.
Harvey, William. *Anatomical Exercitations concerning the Generation of Living Creatures: To which are added Particular Discourses, of Births, and of Conceptions*. Translated by Martin Llewellyn. London: Octavian Pullen, 1653.
———. *Exercitatione de Generatione Animalium. Quibus accedunt Quædum de Partu: De Membranis ac Humoribus Uteri et de Conceptione*. London: Octavius Pulleyn, 1651.
Herodotus. *The Persian Wars: Books III-IV*. Translated by A.D. Godley. Cambridge, MA: Harvard University Press, 1921.
Hooke, Robert. *Discourse of Earthquakes* in *The Posthumous Works of Robert Hooke, M.D.* Edited by Richard Waller. London: S. Smith and B. Walford, 1705.
La Mettrie, Julien Offray de. *Machine Man and Other Writings*. Translated and edited by Ann Thomson. Cambridge: Cambridge University Press, 1996.
———. *Man a Machine and Man a Plant*. Translated by Richard A. Watson and Maya Rybalka, and introduced and annotated by Justin Leiber. Indianapolis: Hackett Publishing Company, Inc., 1994.
———. *Man a Machine, including Frederick the Great's "Eulogy" on La Mettrie and Extracts from La Mettrie's "The Natural History of the Soul."* Notes by Getrude Carman Bussey. Chicago: The Open Court Publishing Company, 1912.
———. *Œuvres philosophiques*. London: Jean Nourse, 1751.
Leibniz, Gottfried Wilhelm. *Opera omnia*. 6 vols. Edited by Louis Dutens. Geneva: De Tournes, 1768.
Lucretius Carus, Titus. *De rerum natura*. Translated by W.H.D. Rouse. Revised by Martin F. Smith. Cambridge, MA: Harvard University Press, 1924.
———. *On the Nature of Things*. Translated by R.E. Latham. Revised and introduced by John Godwin. London: Penguin Books, 1994.
Maillet, Benoît de. *Telliamed: Or, Discourses between an Indian Philosopher, and a French Missionary, on the Diminution of the Sea, the Formation of the Earth, the Origin of Men and Animals, and other Curious Subjects, relating to Natural History and Philosophy*. London: Jacob Loyseau, 1750.
———. *Telliamed, ou Entretiens d'un philosophe indien avec un missionnaire français sur la diminution de la mer, la formation de la terre, l'origine de l'homme, etc.* 2 vols. Amsterdam: L'Honoré et fils, 1748.

Maupertuis, Pierre-Louis Moreau de. *Essai sur la formation des corps organisés*. Berlin, 1754.
———. *Maupertuis: Le savant et le philosophe; Présentation et extraits*. Edited by Emile Callot. Paris: Marcel Rivière et Cie, 1964.
———. *Œuvres de Mr de Maupertuis*. 4 vols. Lyon: J.-M. Bruyset, 1756.
———. *Vénus physique*. 1745.
Mela, Pomponius. *De chorographia*. Introduced and annotated by Piergiorgio Parroni. Rome: Edizioni di Storia e Letteratura, 1984.
———. *Geography/De situ orbis A.D. 43*. Translated by Paul Berry. Lewiston, NY: The Edwin Mellen Press, 1997.
Montesquieu, Charles-Louis de Secondat de. *De l'esprit des lois*. Introduced and annotated by J. Ehrard. Paris: Editions Sociales, 1969.
———. *De l'esprit des loix*. 2 vols. Edinburgh: G. Hamilton and J. Balfour, 1750.
———. *Lettres persanes*. Cologne: Pierre Marteau, 1754.
———. *Lettres persanes*. Annotated by Antoine Adam. Geneva: Librairie Droz, 1954.
———. *Persian Letters*. London: Penguin Books Ltd, 2004.
———. *The Spirit of the Laws*. Translated and edited by Anne M. Cohler, Basia Carolyn Miller, and Harold Samuel Stone. Cambridge: Cambridge University Press, 1989.
Needham, John Turberville. *An Account of Some New Microscopical Discoveries founded on an Examination of the Calamary and its Wonderful Milt-vessels*. London: F. Needham, 1745.
———. *Observations upon the Generation, Composition, and Decomposition of Animal and Vegetable Substances. Communicated in a Letter to Martin Folkes, Esq; President of the Royal Society* (London, 1749).
Oxford English Dictionary Online. http://dictionary.oed.com (Jan. 24, 2006).
Pascal, Blaise. *Œuvres de Blaise Pascal*. Edited by Léon Brunschvicg, Pierre Boutroux, and Félix Gazier. Paris: Librairie Hachette & Cie, 1904–1914.
———. *Thoughts*. Translated by W.F. Trotter. Brunschvicg numbering system. New York: P.F. Collier and Son Company, 1910.
Pigafetta, Philippo. *A Reporte of the Kingdome of Congo, a Region of Africa*. Translated from the Italian by Abraham Hartwell. London: John Wolfe, 1597.
Plutarch. *Lives*. Translated by Bernadotte Perrin. 11 vols. Cambridge, MA: Harvard University Press, 2005.
Purchas, Samuel. *Purchas his Pilgrimage, or Relations of the World and the Religions Observed in all Ages and Places Discovered, from the Creation unto this Present*. London: William Stansby, 1613.
———. *Purchas his Pilgrimes*. London: William Stansby, 1625.
Raynal, Guillaume-Thomas. *Histoire philosophique et politique des établissemens et du commerce des européens dans les deux Indes*. 5 vols. Geneva: Jean-Léonard Pellet, 1780.
Rousseau, Jean-Jacques. *The Discourses and Other Early Political Writings*. Edited and translated by Victor Gourevitch. Cambridge: Cambridge University Press, 2005.
———. *Discours sur l'origine et les fondements de l'inégalité parmi les hommes* in *Œuvres complètes de Jean-Jacques Rousseau avec les notes de tous les commentateurs*. Paris: Dalibon, 1826, 1:211–392.
———. *Du Contrat social*. Amsterdam: Marc-Michel Rey, 1762.
———. *Julie, ou La Nouvelle Héloïse* in *Œuvres complètes de Jean-Jacques Rousseau avec les notes de tous les commentateurs*. Vols. 8–10. Paris: Dalibon, 1826.

———. *The Social Contract and Discourses*. Translated and introduced by G.D.H. Cole. London: J.M. Dent and Sons, 1913.
Voltaire, François-Marie Arouet de. *A Defence of My Uncle, Translated from the French of M. de Voltaire*. London: S. Bladon, 1768.
———. *Œuvres complètes de Voltaire*. 52 vols. Edited by Louis Moland. Paris: Garnier, 1877–1885.
———. *Œuvres de Voltaire avec préfaces, avertissements, notes, etc.* 72 vols. Edited by M. Beuchot. Paris: Lefèvre, 1829–1840.
Wafer, Lionel. *A New Voyage and Description of the Isthmus of America*. London: James Knapton, 1699.

SECONDARY SOURCES

Audidière, Sophie. "Helvétius et Diderot: Matérialisme et physiologie." http://www.sigu7.jussieu.fr/diderot/travaux/revseance5.htm (Jan. 24, 2006).
———. "La *Lettre sur les aveugles* et l'éducation des sens." *Recherches sur Diderot et sur l'Encyclopédie* 28 (April 2000): 67–82.
Badash, Lawrence. "The Age-of-the-Earth Debate." *Scientific American* 261, no. 2 (August 1989): 78–83.
Barsanti, Guilio. "Buffon et l'image de la nature: De l'échelle des êtres à la carte géographique et à l'arbre généalogique." *Buffon 88*. Edited by Jean Gayon. Paris: Vrin, 1992, 255–96.
———. "Linné et Buffon: Deux visions différentes de la nature et de l'histoire naturelle." *Revue de synthèse* 105, no. 113–14 (1984): 83–111.
Beeson, David. *Maupertuis: An Intellectual Biography* (Oxford: The Voltaire Foundation at the Taylor Institution, 1992).
Belaval, Yvon. "L'horizon matérialiste du *Rêve de d'Alembert*." *Diderot und die Aufklärung*. Edited by Herbert Dieckmann. Munich: Krauss International Publications, 1980.
———. "Sur le matérialisme de Diderot." *Europäische, Aufklärung; Herbert Dieckmann zum 60; Geburtstag*. Munich: Fink Verlag, 1967, 9–21.
———. "Trois lectures du *Rêve de d'Alembert*." *Diderot Studies* 18 (1975): 15–32.
Benitez, Miguel. "Benoît de Maillet et l'origine de la vie dans la mer: Conjecture amusante ou hypothèse scientifique?" *Revue de synthèse* 113–14 (1984): 37–54.
Bernasconi, Robert, editor. *Race*. Malden, MA: Blackwell Publishers, Inc., 2001.
Bernasconi, Robert and Sybil Cook, editors. *Race and Racism in Continental Philosophy*. Bloomington: Indiana University Press, 2003.
Betts, C. J. "The Function of Analogy in Diderot's *Rêve de d'Alembert*." *Studies on Voltaire and the Eighteenth Century*. 185 (1980): 267–81.
Biographie universelle ancienne et moderne. Edited by Michaud. 45 vols. Paris: Ch. Delagrave et Cie, 1870–1873.
Bourdier, Frank. "Trois siècles d'hypothèses sur l'origine et la transformation des êtres vivants (1550–1859). *Revue d'histoire des sciences* 13 (1960): 1–44.
Bourdin, Jean-Claude. "Diderot et la langue du matérialisme." *Les materialists philosophiques*. Paris: Kimé, 1997, 89–109.
———. *Diderot: Le matérialisme*. Paris: Presses Universitaires de France, 1998.

———. "Matérialisme et scepticisme chez Diderot." *Recherches sur Diderot et sur l'Encyclopédie* 26 (1999): 85–97.
———. "Le matérialisme dans la *Lettre sur les aveugles*." *Recherches sur Diderot et sur l'Encyclopédie* 28 (April 2000): 83–96.
———. *Les matérialistes au XVIIIe siècle*. Paris: Payot & Rivages, 1996, 193–266.
Bowler, Peter J. "Bonnet and Buffon: Theories of Generation and the Problem of Species." *Journal of the History of Biology* 6, no. 2 (Fall 1973): 259–81.
———. *Evolution: The History of an Idea*. Berkeley: University of California Press, 2003.
Brunet, Pierre. *Maupertuis*. 2 vols. Paris: A. Blanchard, 1929.
———. "La notion d'infini mathématique chez Buffon." *Archeion* 13 (1931): 24–39.
Callot, Emile. "Diderot." *La philosophie de la vie au XVIIIe siècle*. Paris: Marcel Rivière, 1986, 245–316.
Caro, Edme-Marie. "De l'idée transformiste dans Diderot." *Revue des Deux Mondes*, October 15, 1879.
Carozzi, Albert V. "De Maillet's *Telliamed* (1748): An Ultra-Neptunian Theory of the Earth" in *Toward a History of Geology*. Edited by Cecil J. Schneer. Cambridge, MA: MIT Press, 1969, 80–91.
Cassini, Paolo. "Diderot et les philosophes de l'antiquité." *Colloque International de Diderot* (1985): 33–43.
Cassirer, Ernst. *The Philosophy of the Enlightenment*. Oxford: Oxford University Press, 1951.
Charrak, André. "Géométrie et métaphysique dans la *Lettre sur les aveugles* de Diderot." *Recherches sur Diderot et sur l'Encyclopédie* 28 (April 2000): 43–53.
Cherni, Amor. *Buffon: La nature et son histoire*. Paris: Presses Universitaires de France, 1998.
Chouillet, Jacques. "Matière et mémoire dans l'œuvre de Diderot." *Revue de métaphysique et de morale*. April–June 1984, 214–25.
———. "Le personnage du sceptique dans les premières œuvres de Diderot (1745–1747)." *Dix-huitième siècle* 1 (1963): 195–211.
———. "Unité et diversité dans l'œuvre philosophique de Diderot." *Beiträge zur Romanischer philologie*. 1985, 187–94.
Cohen, Claudine. "Benoît de Maillet et la diffusion de l'histoire naturelle à l'aube des Lumières." *Revue d'histoire des sciences* 44, no. 3–4 (1991): 323–42.
Corbey, Raymond, and Bert Theunissen, eds. *Ape, Man, Apeman: Changing Views since 1600*. Leiden: Dept of Prehistory, Leiden University, 1993.
Coulet, Henri. "Diderot et le problème du changement." *Recherches sur Diderot et sur l'Encyclopédie* 2 (1987): 59–67.
Courtois, Jean-Patrice, "Le Physique et le moral dans la théorie du climat chez Montesquieu" in *Le Travail des Lumières pour Georges Benrekassa*. Edited by Caroline Jacot Grapa, et al. Paris: Honoré Champion, 2002, 139–56.
Crampe-Casnabet, Michèle. "Les articles AME dans l'*Encyclopédie*." *Recherches sur Diderot et sur l'Encyclopédie* 25 (October 1998): 91–99.
———. "Qu'appelle-t-on sentir?" *Recherches sur Diderot et sur l'Encyclopédie* 28 (April 2000): 55–66.
Cresson, André. *Diderot, sa vie, son œuvre, avec un exposé de sa philosophie*. Paris: Presses Universitaires de France, 1949.

Crocker, Lester. "Diderot and Eighteenth-Century French Transformism" in *Forerunners of Darwin, 1745–1859*, edited by Bentley Glass *et al.* Baltimore: The Johns Hopkins Press, 1959, 114–43.

———. "The Idea of a 'Neutral' Universe." *Diderot Studies* 21 (1983): 45–76.

———. "Pensée XIX of Diderot." *Modern Language Notes* 67 (1950): 433–39.

———. "Toland et le matérialisme de Diderot." *Revue d'Histoire de littérature de la France.* July-September 1953, 289–95.

Crombie, Alistair C. "P.-L. Moreau de Maupertuis, Précurseur du transformisme." *Revue de synthèse* 78 (1957): 35–56.

Daudin, Henri. *De Linné à Lamarck: Méthodes de la classification et idée de série en botanique et en zoologie (1740–1790).* Paris: Alcan, 1926.

Dédéyan, Charles. *Diderot et la pensée anglaise.* Florence: Leo S. Olschki, 1987.

Desné, Roland. *Les Matérialistes français de 1750 à 1800.* Paris: Buchet-Chastel, 1965.

Dieckmann, Herbert. "The Influence of Francis Bacon on Diderot's *Interprétation de la nature.*" *The Romanic Review* 34 (1943): 303–30.

———. "The Metaphoric Structure of the *Rêve de d'Alembert.*" *Diderot Studies* 17 (1973): 15–24.

———. "Natural History from Bacon to Diderot: A Few Guideposts." *Essays on the Age of Enlightenment in Honor of Ira O. Wade.* Geneva: Droz, 1977, 93–112.

Dougherty, Frank W.P. "Missing Link, Chain of Being, Ape and Man in the Enlightenment: The Argument of the Naturalists" in *Ape, Man, Apeman: Changing Views since 1600.* Edited by Raymond Corbey and Bert Theunissen. Leiden: Dept. of Prehistory, Leiden University, 1995, 63–70.

Duchesneau, François. "Buffon et la physiologie." *Buffon 88.* Edited by Jean Gayon. Paris: Vrin, 1992, 451–62.

———. "Diderot et la physiologie de la sensibilité." *Dix-huitième siècle* 31 (1999): 195–216.

Duchet, Michèle. "L'anthropologie de Diderot." *Anthropologie et histoire au siècle des Lumières: Buffon, Voltaire, Rousseau, Helvétius, Diderot.* Paris: Maspéro, 1971, 407–*Diderot et l'Histoire des deux Indes: ou, L'écriture fragmentaire.* Paris: A.-G. Nizet, 1978.

Duflo, Colas. "Diderot et la conséquence de l'athéisme." *Les athéismes philosphiques.* Paris: Kimé, 2000, 63–78.

———. "La fin du finalisme. Les deux natures: Holmes et Saunderson." *Recherches sur Diderot et sur l'Encyclopédie* 28 (April 2000): 107–31.

Dulieu, Louis. "Le mouvement scientifique montpelliérien au XVIII siècle." *Revue d'histoire des sciences* 11 (1958): 227–49.

Duris, Pascal, and Gabriel Gohau. *Histoire des sciences de la vie.* Paris: Nathan Université, Coll., 1997.

Eddy, John H. Jr. "Buffon, Organic Alterations, and Man." *Studies in the History of Biology* 7 (1984): 1–45.

———. "Buffon, Organic Change, and the Races of Man." Ph.D diss., University of Oklahoma, 1977.

———. "Buffon's *Histoire naturelle*: History? A Critique of Recent Interpretations." *Isis* 85, no. 4 (December 1994): 644–61.

Ehrard, Jean. "Diderot, l'*Encyclopédie*, et l'*Histoire et théorie de la Terre.*" *Buffon 88.* Edited by Jean Gayon. Paris: Vrin, 1992, 135–42.

———. *L'idée de nature en France à l'aube des Lumières.* Paris: Albin Michel, 1994.

———. "Matérialisme et naturalisme: Les sources occultistes de la pensée de Diderot." *Cahiers de l'Association Internationale des Etudes Françaises* 13 (1961): 189–201.
Encyclopædia Britannica. 15th edition. Chicago: Encyclopædia Britannica, Inc., 1976.
Fagot, Anne. "Le transformisme de Maupertuis" in *Actes de la journée Maupertuis*. Paris: Vrin, 1975, 163–78.
Farber, Paul Lawrence. "Buffon and the Concept of Species." *Journal of the History of Biology* 5, no. 2 (Fall 1972): 259–84.
———. "Buffon and Daubenton: Divergent Traditions within the *Histoire naturelle*." *Isis* 66, no. 1 (March 1975): 63–74.
———. "Buffon's Concept of Species." Ph.D diss., Indiana University, 1970.
———. "Research Traditions in Eighteenth-Century Natural History in *Lazzaro Spallanzani e la biologia del Settecento: Teorie, esperimenti, istituzioni scientifiche*. Edited by Giuseppe Montalenti and Paolo Rossi. Florence: Leo S. Olschki, 1982, 397–403.
Farley, John. *The Spontaneous Generation Controversy from Descartes to Oparin*. Baltimore: The Johns Hopkins University Press, 1977.
———. "The Spontaneous Generation Controversy (1700–1860): The Origin of Parasitic Worms." *Journal of the History of Biology* 5, no. 1 (March 1972): 95–125.
Fellows, Otis. "Buffon and Rousseau: Aspects of a Relationship." *PMLA* 75, no. 3 (June 1960): 184–96.
———. "Buffon's Place in the Enlightenment." *Studies on Voltaire and the Eighteenth Century* 25 (1963): 603–29.
———. *Diderot*. Boston: Twayne Publishers, 1977.
———. *From Voltaire to La Nouvelle Critique: Problems and Personalities*. Introduced by Norman L. Torrey. Geneva: Droz, 1970.
———. "Voltaire and Buffon: Clash and Conciliation." *Symposium* 9, no. 2 (Fall 1955): 222–35.
Fellows, Otis E. and Milliken, Stephen F. *Buffon*. New York: Twayne Publishers, Inc., 1972.
Fellows, Otis E. and Norman L. Torrey. *The Age of Enlightenment*. New York: Appleton-Century-Crofts, 1971.
Fischer, Jean-Louis. "L'hybridologie et la zootaxie du siècle des Lumières à l'Origine des espèces." *Revue de synthèse* 101–02 (1981): 47–72.
Fontenay, Elisabeth de. *Diderot ou le matérialisme enchanté*. Paris: Grasset, 1981.
Frautschi, Richard L. "Les articles anonymes de l'*Encyclopédie* et le 'style' de Diderot." *Revue internationale de philosophie*, 103 (1973): 66–72.
———. "The Authorship of Certain Unsigned Articles in the *Encyclopédie*: A First Report," *Computer Studies in the Humanities and Verbal Behavior*, 1970, 66–76. [Computers attribute "Hylopathianisme," "Métempsychose," "Origénistes," "Probabilité," and "Resurrection" to Diderot.]
Gauthier, David. *Rousseau: The Sentiment of Existence*. Cambridge: Cambridge University Press, 2006.
Glass, Bentley O. "The Germination of the Idea of Biological Species" in *Forerunners of Darwin, 1745–1859*, edited by Bentley Glass *et al*. Baltimore: The Johns Hopkins Press, 1959, 30–48.
———. "Heredity and Variation in the Eighteenth-Century Concept of the Species" in *Forerunners of Darwin, 1745–1859*, edited by Bentley Glass *et al*. Baltimore: The Johns Hopkins Press, 1959, 144–72.
———. "Maupertuis, Pioneer of Genetics and Evolution" in *Forerunners of Darwin, 1745–1859*, edited by Bentley Glass *et al*. Baltimore: The Johns Hopkins Press, 1959, 51–83.

Gohau, Gabriel. "La 'Théorie de la Terre' de 1749." *Buffon 88*. Edited by Jean Gayon. Paris: Vrin, 1992, 343–52.

Goyard-Fabre, Simone. *La philosophie des lumières en France*. Paris: Klincksieck, 1972.

Gregory, Mary Efrosini. *Diderot and the Metamorphosis of Species*. New York and London: Routledge, 2007.

Groethuysen, Bernard. "La Pensée de Diderot." *La Grande Revue* 22 (November 25, 1913): 322–41.

Groult, Martine. "De l'œuvre de d'Alembert au rêve de Diderot; Brouillon succint de travail destiné à la discussion." http://www.sigu7.jussieu.fr/diderot/travaux/revseance1.htm (Jan. 24, 2006).

Guedon, Jean-Claude. "Chimie et matérialisme: La stratégie antinewtonienne de Diderot." *Dix-huitième siècle* 11 (1980): 185–200.

Haber, Francis C. "Fossils and Early Cosmologies" in *Forerunners of Darwin, 1745–1859*, edited by Bentley Glass *et al*. Baltimore: The Johns Hopkins Press, 1959, 3–29.

———. "Fossils and the Idea of a Process of Time in Natural History" in *Forerunners of Darwin, 1745–1859*, edited by Bentley Glass *et al*. Baltimore: The Johns Hopkins Press, 1959, 222–61.

Hankins, Thomas L. *Science and the Enlightenment*. Cambridge: Cambridge University Press, 1985.

Hastings, Hester. *Man and Beast in French Thought of the Eighteenth Century*. Baltimore: The Johns Hopkins Press, 1936.

Hazard, Paul. "Diderot." *La pensée européenne au XVIIIe siècle*. Paris: Hachette, 1963, 370–81.

Hill, Emita. "Materialism and Monsters in Diderot's *Le Rêve de d'Alembert*." *Diderot Studies* 10 (1968): 67–93.

———. "The Role of 'Le Monstre' in Diderot's Thought." *Studies on Voltaire and the Eighteenth Century* 97 (1972): 148–261.

Hodge, Jonathan. "Two Cosmogonies ('Theory of the Earth' and 'Theory of Generation') and the Unity of Buffon's Thought." *Buffon 88*. Edited by Jean Gayon. Paris: Vrin, 1992, 241–54.

Horowitz, Asher. "Laws and Customs Thrust Us Back into Infancy: Rousseau's Historical Anthropology." *Review of Politics* 52, no. 2 (Spring 1990): 215–41.

Ibrahim, Annie. "Le matérialisme de Diderot: Formes et forces dans l'ordre des vivants" in *Diderot et la question de la forme*. Paris: Presses Universitaires de France, 1999, 87–103.

———. "Matière des métaphores, métaphores de la matière." *Recherches sur Diderot et sur l'Encyclopédie* 26 (April 1999): 125–33.

———. "Une philosophie d'aveugle: la matière fait de l'esprit." *Recherches sur Diderot et sur l'Encyclopédie* 28 (April 2000): 97–106.

Janet, Paul. "La philosophie de Diderot: Le dernier mot d'un matérialiste." *Nineteenth Century and After* 9 (April 1881): 695–708.

Johnson, Paul. *A History of Christianity*. London: Weidenfeld and Nicolson, 1976.

Jones, Christopher S. "Politcizing Travel and Climatizing Philosophy: Watsuji, Montesquieu and the European Tour. *Japan Forum* 14, no. 1 (March 1, 2002): 41–62.

Jouary, Jean-Paul. "Denis Diderot ou le matérialisme en chantier." *La pensée* 239 (1984): 46–59.

———. *Diderot et la matière vivante*. Paris: Messidor, 1992.

Kaitaro, Timo. *Diderot's Holism: Philosophical Anti-Reductionism and its Medical Background*. Frankfurt am Main: Peter Lang, 1997.
Lefebvre, Henri. *Diderot*. Paris: Hier et Aujourd'hui, 1949.
———. *Diderot ou les affirmations fondamentales du matérialisme*. Paris: Arche, 1983.
Lenoble, Robert. "L'évolution de l'idée de 'nature' du XVIe au XVIIIe siècle." *Revue de métaphysique et de morale* 58, no. 1–2 (January-June 1953): 108–29.
Lepenies, Wolf. "De l'histoire naturelle à l'histoire de la nature." *Dix-huitième siècle* 11 (1979): 175–84.
Le Ru, Véronique. "La *Lettre sur les aveugles* et le bâton de la raison." *Recherches sur Diderot et sur l'Encyclopédie* 28 (April 2000): 25–41.
———. "Le mécanisme cartésien 'traduit' par Diderot ou le problème du serin et de la serinette." http://www.sigu7.jussieu.fr/diderot/travaux/revseance4.htm (Jan. 24, 2006).
Lojkine, Stéphane. "Le matérialisme biologique du *Rêve de d'Alembert*." *Littératures* 30 (1994): 27–49.
Lovejoy, Arthur O. "Buffon and the Problem of Species" in *Forerunners of Darwin, 1745–1859*, edited by Bentley Glass *et al.* Baltimore: The Johns Hopkins Press, 1959, 84–113.
———. *The Great Chain of Being: A Study of the History of an Idea*. Cambridge, MA: Harvard University Press, 1936.
———. "Monboddo and Rousseau" in *Essays in the History of Ideas*. Baltimore: The Johns Hopkins Press, 1948, 38–61.
Loy, J. Robert. *Montesquieu*. New York: Twayne Publishers, Inc., 1968.
Lyon, John, and Phillip R. Sloan, eds. and trans. *From Natural History to the History of Nature; Readings from Buffon and His Critics*. Notre Dame: University of Notre Dame Press, 1981.
Markovits, Francine. "Mérian, Diderot et l'aveugle" in *J.-B. Mérian: Sur le problème de Molyneux*. Paris: Flammarion, 1984, 193–289.
Martin-Haag, Eliane. "Droit naturel et histoire dans la philosophie de Diderot." *Recherches sur Diderot et sur l'Encyclopédie* 26 (1999): 37–47.
———. "De la revalorisation de la chimie de l'*Encyclopédie* au *Rêve* de d'Alembert, ou de G.-F. Venel à Diderot. http://www.sigu7.jussieu.fr/diderot/travaux/revseance3.htm (Jan. 24, 2006).
May, Georges. *Quatre visages de Denis Diderot*. Paris: Boivin, 1951.
Mayer, Jean. *Diderot: Homme de science*. Rennes: Imprimerie bretonne, 1959.
Montesquieu's Science of Politics: Essays on The Spirit of Laws. Edited by David W. Carrithers, Michael A. Mosher, and Paul A. Rahe. Lanham, MD: Rowman and Littlefield Publishers, Inc., 2001.
Moran, Francis. "Between Primates and Primitives: Natural Man as the Missing Link in Rousseau's *Second Discourse*." *Journal of the History of Ideas* 54, no. 1 (January 1993): 37–58.
———. "Of Pongos and Men: Orangs-Outang in Rousseau's *Discourse on Inequality*." *Review of Politics* 57, no. 4 (Autumn 1995): 641–64.
Moravia, Sergio. "The Enlightenment and the Sciences of Man." *History of Science* 18 (1980): 247–68.
Mornet, Daniel. *Diderot, l'homme et l'œuvre*. Paris: Boivin, 1941.
———. *Les Sciences de la nature en France au XVIIIe siècle*. Paris: Armand Colin, 1911.
Mortier, Roland. "Holbach et Diderot: Affinités et divergences." *Revue de l'université de Bruxelles* (1972–1973): 223–37.

———. "Note sur un passage du *Rêve de d'Alembert*: Réaumur et le problem de l'hybridation." *Revue d'histoire des sciences* (1960): 309–16.
———. "A propos du sentiment de l'existence chez Diderot et Rousseau." *Diderot Studies* 6 (1964): 183–95.
———. "Rhétorique et discours scientifique dans *Le Rêve de d'Alembert*." *Le Cœur et la Raison*. Oxford: The Voltaire Foundation, 1990, 236–49.
Oxford English Dictionary Online. http://oed.com (April 3, 2006).
Paitre, Fernand. *Diderot biologiste*. Geneva: Slatkine Reprints, 1971.
Parmentier, Marc. "Le problème de Molyneux de Locke à Diderot." *Recherches sur Diderot et sur l'Encyclopédie* 28 (April 2000): 13–23.
Pennebaker, James W., et al. "Stereotypes of Emotional Expressiveness of Northerners and Southerners: A Cross-Cultural Test of Montesquieu's Hypotheses." *Journal of Personality and Social Psychology* 70, no. 2 (February 1996): 372–80.
Perkins, Jean E. "Diderot and La Mettrie." *Studies on Voltaire and the Eighteenth Century* 10 (1959): 49–100.
Perol, Lucette. "Quelques racines encyclopédiques du *Rêve de d'Alembert*" in *L'Encyclopédie et Diderot*. Edited by Degar Mass and Peter-Eckhard Knabe. Cologne: Verlag Köln, 1985, 229–45.
Pichot, André. *Histoire de la notion de la vie*. Paris: Gallimard, 1993.
Pomeau, René. *La Religion de Voltaire*. Paris: Nizet, 1956.
Powell, James Lawrence. *Mysteries of Terra Firma: The Age and Evolution of the Earth*. http://www.powells.com/biblio?show=HARDCOVER:SALE:068487282X:10.98&page=excerpt#page (September 11, 2006).
Proust, Jacques. "Diderot et la philosophie du polype." *Revue des sciences humaines* 182 (April-June 1981): 21–30.
———. "Diderot et le système des connaissances humaines." *Studies on Voltaire and the Eighteenth Century* 256 (1988): 117–27.
———. "Source et portée de la théorie de la sensibilité généralisée dans *Le Rêve de d'Alembert*" in *La Quête de bonheur et l'expression de la douleur*. Edited by Carminella Biondi, Carmenila Imbroscio, et al. Geneva: Droz, 1995.
———. "Variations sur un thème de l'*Entretien avec d'Alembert*." *Revue des sciences humaines* 112 (October-December 1963): 435–54.
Pucci, Suzanne. "Metaphor and Metamorphosis in Diderot's *le Rêve de d'Alembert*." *Symposium* 25 (Winter 1981–1982): 325–39.
Pulskamp, Richard J. "Jean Le Rond d'Alembert on Probability and Statistics." http://www.cs.xu.edu/math/Source/Dalembert/index.html (April 21, 2006).
Rappaport, Rhoda. "Borrowed Words: Problems of Vocabulary in Eighteenth-Century Geology." *British Journal for the History of Science* 15 (1982): 27–44.
———. "Geology and Orthodoxy: The Case of Noah's Flood in Eighteenth-Century Thought." *British Journal for the History of Science* 11 (1978): 1–18.
———. *When Geologists Were Historians, 1665–1750*. Ithaca: Cornell University Press, 1997.
Rey, Roselyne. "Buffon et le vitalisme." *Buffon 88*. Edited by Jean Gayon. Paris: Vrin, 1992, 399–413.
———. "Naissance et développement du vitalisme en France, de la deuxième moitié du XVIIIe siècle à la fin du Ier empire." Ph. D diss., Université de Paris-I, 1987.
Ritterbush, Philip C. *Overtures to Biology: The Speculation of the Eighteenth-Century Naturalists*. New Haven: Yale University Press, 1964.

Roe, Shirley A. "Buffon and Needham: Diverging Views on Life and Matter." *Buffon 88*. Edited by Jean Gayon. Paris: Vrin, 1992, 439–50.

———. "Voltaire vs. Needham: Atheism, Materialism, and the Generation of Life," *Journal of the History of Ideas* 46 (1985): 65–87, reprinted in *Philosophy, Religion and Science in the Seventeenth and Eighteenth Centuries*, edited by John W. Yolton. Rochester: University of Rochester Press, 1990, 417–39.

Roger, Jacques. *Buffon: A Life in Natural History*. Translated by Sarah Lucille Bonnefoi. Edited by L. Pearce Williams. Ithaca: Cornell University Press, 1997.

———. "Buffon et le transformsime." *La recherche* 138 (November 1982): 1246–54.

———. *Buffon: Un Philosophe au Jardin du Roi*. Paris: Fayard, 1989.

———. "Buffon et la théorie de l'anthropologie." *Enlightenment Studies in Honour of Lester G. Crocker*. Edited by A.J. Bingham and V. Topazio. Oxford: The Voltaire Foundation, 1979, 253–62.

———. "Diderot et Buffon en 1749." *Diderot Studies* 4 (1963): 221–36.

———. "Histoire naturelle et biologie chez Buffon" in *Lazzaro Spallanzani e la biologia del Settecento: Teorie, esperimenti, istituzioni scientifiche*. Edited by Giuseppe Montalenti and Paolo Rossi. Florence: Leo S. Olschki, 1982, 353–62.

———. *The Life Sciences in Eighteenth-Century French Thought*. Stanford: Stanford University Press, 1997.

———. "La nature et l'histoire dans la pensée de Buffon." *Buffon 88*. Edited by Jean Gayon. Paris: Vrin, 1992, 193–205.

———. *Les sciences de la vie dans la pensée française au XVIIIe siècle*. Paris: Armand Colin, 1963.

Rosenfield, Leonora Cohen. *From Beast-Machine to Man-Machine: Animal Soul in French Letters from Descartes to La Mettrie*. Introduction by Paul Hazard. New York: Octagon Books, 1968.

Rostand, Jean. *L'Atomisme en biologie*. Paris: Gallimard, 1956.

———. "La conception de l'homme selon Helvétius et selon Diderot." *Revue d'histoire des sciences* 4 (1951): 213–22.

———. "XVIIIe siècle: Les Grands problèmes de la biologie" in Vol. 4 of *Histoire générale des sciences*, pp. 571–92. Edited by René Taton. Paris: Presses Universitaires de France, 1957–1964.

Rudwick, Martin J.S. *The Meaning of Fossils: Episodes in the History of Palaeontology*. 2nd ed. Chicago: The University of Chicago Press, 1972.

Russell, Edward Stuart. *Form and Function: A Contribution to the History of Animal Morphology*. Introduced by George V. Lauder. Chicago: University of Chicago Press, 1982.

Saigey, Emile. *Les sciences au 18e siècle; la physique de Voltaire*. Paris: Germer-Baillière, 1873.

Salaün, Franck. "L'identité personnelle selon Diderot." *Recherches sur Diderot et sur l'Encyclopédie* 26 (April 1999): 113–23.

Schérer, Jacques. *Le Cardinal et l'Orang-outang; Essai sur les inversions et les distances dans la pensée de Diderot*. Paris: Sedes, 1972.

Schiller, Joseph. "Queries, Answers, and Unsolved Problems in Eighteenth-Century Biology." *History of Science* 12 (1974): 184–99.

Schmidt, Johan Werner. "Diderot and Lucretius: the *De Rerum Natura* and Lucretius' Legacy in Diderot's Scientific, Aesthetic and Ethical Thought." *Studies on Voltaire and the Eighteenth Century* 208 (1982): 183–294.

Schneer, Cecil J. *Toward a History of Geology*. Cambridge, MA: MIT Press, 1969.

Shackleton, Robert. *Montesquieu: A Critical Biography*. Oxford: Oxford University Press, 1961.

Shklar, Judith N. *Montesquieu*. Oxford: Oxford University Press, 1987.

―――. "Virtue in a Bad Climate; Good Men and Good Citizens in Montesquieu's *L'Esprit des lois*" in *Enlightenment Studies in Honour of Lester G. Crocker*. Edited by Alfred J. Bingham and Virgil W. Topazio. Oxford: The Voltaire Foundation at the Taylor Institution, 1979, 315–28.

Singh, Christine M. "The *Lettre sur les aveugles*: Its Debt to Lucretius." *Studies in Eighteenth-Century French Literature Presented to Robert Niklaus*, edited by J.H. Fox, M.H. Waddicor and D.A. Watts (Exeter: University of Exeter, 1975), 233–42.

Skrzypek, Marian. "Les catégories centrales dans la philosophie de Diderot." *Recherches sur Diderot et sur l'Encyclopédie* 26 (1999): 27–36.

Sloan, Phillip R. "The Buffon-Linnaeus Controversy." *Isis* 67, no. 238 (September 1976): 356–75.

―――. "From Logical Universals to Historical Individuals: Buffon's Idea of Biological Species." *Histoire du concept d'espèce dans les sciences de la vie*. Edited by J. Roger and J.L. Fischer. Paris: Fondation Singer-Polignac, 1987, 101–40.

―――. "The Idea of Racial Degeneracy in Buffon's *Histoire naturelle*." *Racism in the Eighteenth Century*. Volume 3 of *Studies in Eighteenth-Century Culture*. Cleveland: Case Western Reserve University Press, 1973, 293–321.

―――. "Organic Molecules Revisited." *Buffon 88*. Edited by Jean Gayon. Paris: Vrin, 1992, 415–38.

Sorenson, Leonard. "Natural Inequality and Rousseau's Political Philosophy in his *Discourse on Inequality*." *Western Political Quarterly* 43, no. 4 (December 1990): 763–88.

Spangler, May. "Science, philosophie et littérature: le polype de Diderot." *Recherches sur Diderot et sur l'Encyclopédie* 23 (October 1997): 89–107.

Spary, Emma C. *Utopia's Garden: French Natural History from Old Regime to Revolution*. Chicago: University of Chicago Press, 2000.

Spink, John S. "L'échelle des êtres et des valeurs dans l'œuvre de Diderot." *Cahiers de l'Association Internationale des Etudes Françaises* (1961): 339–51.

Starobinski, Jean. *Montesquieu*. Paris: Seuil, 1989.

―――. "Le philosophe, le géomètre, l'hybride." *Poétique* (1975): 8–23.

―――. "Rousseau and Buffon" in *Jean-Jacques Rousseau: Transparency and Obstruction*. Translated by Arthur Goldhammer and introduced by Robert Morrisey. Chicago: University of Chicago Press, 1988, 323–32.

Stenger, Gerhardt. "L'ordre et les monstres dans la pensée philosophique, politique et morale de Diderot" in *Diderot et la question de la forme*. Paris: Presses Universitaires de France, 1999, 139–57.

―――. "La théorie de la connaissance dans la *Lettre sur les aveugles*." *Recherches sur Diderot et sur l'Encyclopédie* 26 (1999): 99–111.

Suratteau, Aurélie. "Les hermaphrodites de Diderot" in *Diderot et la question de la forme*. Paris: Presses Universitaires de France, 1999, 105–37.

Suratteau-Iberraken, Aurélie. "Diderot et la médecine, un matérialisme vitaliste?" *Recherches sur Diderot et sur l'Encyclopédie* 26 (1999): 173–95.

Suzuki, Mineko. "Chaîne des idées et chaîne des êtres dans *Le Rêve de d'Alembert*." *Dix-huitième siécle* 19 (1987): 327–38.

Szigeti József. *Denis Diderot: Une grande figure du matérialisme militant du XVIIIe siècle*. Budapest: Akadémiai Kiadó, 1962.

Tassy, Pierre. *L'ordre et la diversité du vivant*. Paris: Fayard, 1986.
Taylor, Kenneth. "Buffon's *Epoques de la nature* and Geology during Buffon's Latter Years." *Buffon 88*. Edited by Jean Gayon. Paris: Vrin, 1992, 371–85.
Thielemann, Leland J. "Diderot and Hobbes." *Diderot Studies* 2 (1952): 221–78.
Thomson, Ann. "Diderot, le matérialisme et la division de l'espèce humaine." *Recherches sur Diderot et sur l'Encyclopédie* 26 (1999): 197–211.
———. "From *L'Histoire naturelle de l'homme* to the Natural History of Mankind." *British Journal for Eighteenth-Century Studies* 9, no. 1 (Spring 1986): 73–80.
———. "Issues at Stake in Eighteenth-Century Racial Classification." http://www.cromohs. unifi.it/8_2003/thomson.html (Jan. 24, 2006).
———. "La Mettrie et Diderot." http://www.sigu7.jussieu.fr/ diderot/travaux/revseance2. htm (Jan. 24, 2006).
———. "L'unité matérielle de l'homme chez La Mettrie et Diderot." *Colloque International Diderot* (1985): 61–68.
Trousson, Raymond. "Diderot et l'antiquité grecque." *Diderot Studies* 6 (1964): 215–45.
Varloot, Jean. "Le projet 'antique' du *Rêve de d'Alembert* de Diderot: Légendes antiques et matérialisme au XVIIIe siècle." *Beiträge zur romanischen philologie* 2 (1963): 49–61.
Vartanian, Aram. "Buffon et Diderot." *Buffon 88*. Edited by Jean Gayon. Paris: Vrin, 1992, 119–33.
———. *Diderot and Descartes: A Study of Scientific Naturalism in the Enlightenment.* Princeton: Princeton University Press, 1953.
———. "Diderot and Maupertuis." *Revue internationale de philosophie* 148–49 (1984): 46–66.
———. "Diderot et Newton." *Nature, histoire, société, essais en hommage à Jacques Roger*. Paris: Klincksieck, 1995, 61–77.
———. "Diderot's Rhetoric of Paradox, or the Conscious Automaton Observed." *Eighteenth-Century Studies* 14 (1981): 379–405.
———. "The Enigma of Diderot's *Eléments de physiologie*." *Diderot Studies* 10 (1968): 285–301.
———. "From Deist to Atheist: Diderot's Philosophical Orientation, 1746–1749." *Diderot Studies* 1 (1949): 46–63.
———. "La Mettrie and Diderot Revisited: An Intertextual Encounter." *Diderot Studies* 21 (1983): 155–97.
———. *La Mettrie's L'Homme Machine: A Study in the Origins of an Idea; Critical Edition with an Introductory Monograph and Notes by Aram Vartanian* (Princeton: Princeton University Press, 1960).
———. "The Problem of Generation and the French Enlightenment." *Diderot Studies* 6 (1964): 339–52.
———. "The *Rêve de d'Alembert*: A Bio-Political View." *Diderot Studies* 17 (1973): 41–64.
———. "Trembley's Polyp, La Mettrie, and Eighteenth-Century French Materialism." *Journal of the History of Ideas* 2 (1950): 259–86.
Venturi, Franco. *La Jeunesse de Diderot (1713–1753)*. Paris: Skira, 1939.
Wartofsky, Marx W. "Diderot and the Development of Materialist Monism." *Diderot Studies* 2 (1952): 279–329.
Webster's Third New International Dictionary of the English Language Unabridged. Springfield: Merriam-Webster, Inc., 1993.
Winter, Ursula. "Quelques aspects de la méthode expérimentale chez Diderot." *Actes du VIIe Congrès International des Lumières*. Oxford: The Voltaire Foundation, 1989, 504–08.

Wokler, Robert. "The Ape Debates in Enlightenment Anthropology." *Studies on Voltaire and the Eighteenth Century* 192 (1980): 1164–75.

———. "Enlightenment Apes: Eighteenth-Century Speculation and Current Experiments on Linguistic Competence" in *Ape, Man, Apeman: Changing Views since 1600*. Edited by Raymond Corbey and Bert Theunissen. Leiden: Dept. of Prehistory, Leiden University, 1995, 87–100.

Wolfe, Charles T. "Machine et organisme chez Diderot." *Recherches sur Diderot et l'Encyclopédie* 26 (1999): 213–31.

———. "Qu'est-ce qu'un précurseur? ou La querelle du transformisme." http://www.sigu7.jussieu.fr/diderot/travaux/ revseance6.htm (April 21, 2006).

Index

A

Abbadie, Jacques 52
Abolition of slavery
    Condorcet and 240
    Declaration of the Rights of Man and the Citizen 13, 251
    Diderot and 215, 238–244
    Encyclopedia and 234–236
    Jaucourt and 234–236
    Montesquieu and 42–44, 230–232
    Rousseau and 11, 151–153, 165, 236–238
    Voltaire and 222, 232–234
Age of the earth 4, 6, 14, 19–22, 25, 29, 31, 167–168, 175, 247, 249
Albino Indian 96–99
Albino Negro 98–100, 110, 187–188, 227–228
Alembert, Jean Le Rond d' 4, 121, 157, 161
Animalcules 10, 12, 26, 79, 91, 113–114, 122, 128, 134–136, 169, 174, 194, 249
Anthropological metamorphosis, Rousseau and 6, 10–11, 16, 143, 157, 159, 166, 247, 249–250
Apes, similarity to man
    Battell and 4, 197, 200
    Beeckman and 198–199
    Buffon and 3–4, 78, 87, 154–158, 199, 203–209
    Diderot and 5, 12, 195–196, 212
    Encyclopedia and 4, 200
    Green and 4
    Jaucourt and 4, 200
    La Mettrie and 12, 46, 57–58, 61, 195, 199, 201–203
    Leguat and 198
    Lopez and 196
    Maillet and 24, 189, 199
    Perrault 201
    Petty 201
    Pigafetta and 196
    Prévost and 4
    Purchas and 4, 197–199
    Rousseau and 5, 11–12, 156–163, 166, 195, 199, 209–212, 250
    Smith and 199
    Tyson and 199, 201
Apes, possibility of speaking
    Buffon's opposition to 154–155, 203–209, 212
    Cartesians' opposition to 12–13, 196, 203–205, 251
    Diderot in favor of 5, 12, 195–196, 212–214
    La Mettrie in favor of 12, 49, 57, 195, 201–203
    materialists in favor of 12–13, 195–196, 209
    Perrault's opposition to 201
    Rousseau's opposition to 12, 209–212
    Tyson's opposition to 201
Aristotle 1–2, 7, 14, 34, 48, 75, 88, 90, 94, 111, 146–148, 152, 193, 201, 248
Arrangement of parental elements
    Diderot and 9–10, 131–132, 138, 141, 244, 249

Maupertuis and 17, 57, 94, 97, 100, 110–111, 113, 117, 120, 131, 138, 226, 228, 244, 249
Atheism
   Diderot and 3, 6, 52, 63, 65, 70, 121, 123–129, 139, 222, 247
   La Mettrie and 8, 47, 51–52, 54, 65, 67, 248
   Voltaire's opposition to 3, 106–107, 168–170, 172–174, 176–177, 184, 186, 194
Atomic motion 3–4, 8, 10, 12–13, 46, 53, 63–64, 67, 88, 112, 119–121, 124–125, 127–130, 132, 135, 141, 247–250
Attraction 105–106, 223

B

Battell, Andrew 4, 197, 200
Beeckman, Daniel 198–199
Beeson, David 109, 115–116
Belon, Pierre 76
Bergerac, Cyrano de 21
Bernasconi, Robert 219, 229
Bernier, François 13, 217–218
Biological engineering 212, 214, 226
Birth defects 11, 15, 56–57, 62–63, 70, 93, 95–96, 98–99, 108–110, 113, 117, 120–122, 126, 129, 131, 137, 140–141, 226, 249
Black Code (*Code Noir*) 229–231, 233, 245
Blacks 84, 93, 96, 100, 110, 187–188, 200, 218–224, 227–228, 232–235, 239–240
Boerhaave, Hermann 49
Bonnet, Charles 105, 184–185
Bordeu, Théophile de 15, 139
Bowler, Peter J. 29–30
Brain size 47
Brins, see Fibers
Buffon, Georges-Louis Leclerc, comte de
   "All that can be, is" 2, 9, 91, 248
   animalcules and 79, 91, 113, 134–135, 169
   apes and 3–4, 78, 87, 154–158, 199, 203–209
   The Ass 76, 85–90, 158, 218–219
   The Camel and the Dromedary 75, 81–82, 91
   Carnivorous Animals 207
   chain of beings and 2–3, 8–9, 15, 76–79, 83, 91, 120, 249
   classification and 13, 77
   climate and 70, 73–75, 79–84, 91, 120, 220–221
   cosmogony and 8, 29, 73–74
   creationism and 73, 145, 207–208, 247
   crossbreeding (hybridization) and 5, 82, 85–86, 91, 120, 134
   degeneration and 8, 15–16, 73–74, 80–83, 87–91 120, 133–136, 144, 146–147, 150–152, 163, 220–221, 247, 250
   Degeneration of Animals 220–221
   Diderot and 2, 9, 13–14, 16, 72–73, 78–79, 84, 88, 90, 113, 119–120, 122, 130, 133–135, 139–140, 174–175, 219, 248–249
   Discourse on the Nature of Animals 204–205, 207
   domestication and 15, 81–82, 120, 134, 151, 221
   epigenesis and 70–71, 90
   environment and 70, 73, 75, 80, 82, 91
   First Discourse 152
   flux and 88
   food and 70, 80–84, 91
   geographical determinism and 74, 82, 120
   The Goat 81, 85, 89
   God and 8, 73, 76–79, 82, 87
   genus, family, species and 77
   Harvey and 70
   The Hog, the Hog of Siam, and the Wild Boar 69
   homologies and 86–87, 91, 120, 122, 130
   The Horse 73–74, 81
   interior molding force and 15–17, 72–74, 79, 90–92, 133, 143

Index   337

land of origin and 73–75, 80, 82, 91, 221
legacy of 70, 74, 79
Linnaeus and 3, 13, 15, 77, 187
The Lion 74–75, 80
matrix theory and 77–80
Maupertuis and 2, 9, 13, 72–73, 78–79, 88, 90, 113–114, 248–249
mutability of species and 15, 81, 86–91
Natural History, General and Particular 2, 30, 69, 77, 122, 140, 162, 177–178, 219–220
Needham and 79, 90–91
Nomenclature of Apes 207–208
on blacks 84
On the Varieties of the Human Species 84
opposition to Voltaire 177–178, 181–183, 189
organic molecules and 71–72, 79, 90–91
overlapping physical characteristics and 3, 8–9, 14, 77–78, 120, 130
pangenesis and 71
perfection of species 80–82, 134
preformation, rejection of 70
prototype and 70, 73–74, 76, 78–81, 85–87, 91, 120
Purchas and 198
quadrupeds and 76–78
race and 13, 70–71, 81, 83–85, 87, 218–219
racial classification and 13
Recapitulation 71
Rousseau and 8, 11, 17, 143–163, 166, 209–210, 212, 247, 250
species, definition of 69, 85–86, 187, 218–219, 244
spontaneous generation and 72, 79, 90–91, 113, 134–135
Supplement to the Natural History 182
time and 69–70, 72–74, 78–79, 81–82, 84–86, 88–91

Theory of the Earth 14, 28, 177, 182, 189
transformism, denial of 16–17, 69–70, 85–92
unity of plan and 75–76
Voltaire's opposition to 25, 28–29, 169, 178, 184
way of life and 82–83, 91

C

Cartesianism 12–13, 51, 116, 153, 162, 193, 196, 203–205, 214, 236, 251
Cassirer, Ernst 48
Chabenat, P. 182–183
Chain of beings
    Aristotle and 1–2, 14
    Buffon and 2–3, 8–9, 15, 76–79, 83, 91, 120, 249
    continuity and the 1–2, 8
    definition of 1–2, 5
    Diderot and 3–4, 9, 14, 119, 122, 130, 138, 212, 214
    La Mettrie and 45–46, 49, 53, 56–57, 60–61, 66
    plenitude and the 1–2, 8
    Rousseau and 145, 147, 150–151, 160–162
    Voltaire's opposition to 3, 12, 168, 170–172
Chance, random 3–4, 8, 15, 46, 53–54, 59, 62–63, 99–100, 102–103, 107–109, 120–121, 124–125, 133, 139, 168–169, 178, 214, 222, 226–229, 248
Chaos (*chaos*) 8, 10, 37–38, 46, 62–63, 67, 88, 119, 124–128, 135, 142, 248
Civilized man, Rousseau and 152, 155, 163–164, 243, 250
Climate
    Buffon and 70, 73–75, 79–84, 91, 120, 220–221
    Diderot and 132–134, 214, 220
    La Mettrie and 8, 45, 55–56, 67, 201, 248
    Montesquieu and 7, 38–44, 231, 248

Code Noir, see Black Code
Condillac, Etienne Bonnot de 157, 161
Condorcet, Marie-Jean-Antoine-Nicolas de Caritat, marquis de 240
Consciousness (*sensibilité*)
  emergent
    Diderot and 9, 115, 119–120, 131–132, 249
    Maupertuis and 9, 111, 114–115, 117, 120, 249
    La Mettrie and 45–51, 53, 55–57, 63, 67
    memory and 9, 11, 51, 90, 107, 112–115, 117, 130–132, 137, 141, 153, 168, 206–207, 211–213, 249
  sentiment of existence 164–165, 210
Continuity 1–2, 8
Converts (*conversos*) 229
Coulet, Henri 43, 129, 139–140
Creationism
  Buffon and 73, 145, 207–208, 247
  Voltaire and 6, 12, 104, 167, 174, 184, 189, 247
Crocker, Lester 14–15, 53, 61, 65, 67, 121, 129, 139
Crossbreeding (hybridization)
  Buffon and 5, 82, 85–86, 91, 120, 134
  Diderot and 137–138, 212–213
  Maupertuis and 94–95, 99, 112–113, 222–223, 225, 227, 229
Cuna Indians 96–99

D

d'Alembert, see Alembert, Jean Le Rond d'
Dapper, Olfert 156, 158
Daubenton, Pierre 76
Da Vinci, Leonardo 96, 175–176
Declaration of the Rights of Man and the Citizen 13, 251
Degeneration 8, 15–16, 73–74, 80–83, 87–91 120, 133–136, 144, 146–147, 150–152, 163, 220–221, 247, 250
Deism

definition of 167
Diderot's period of 52, 65, 121–125, 127, 139
Rousseau and 143, 145, 149, 166
Voltaire and 3, 6, 8, 12, 17, 25, 27–28, 104, 106–107, 167–170, 173–174, 176, 180, 184, 186, 188, 194, 250
Democritus 111, 124, 135
Derham, William 52
Descartes, René 47–49, 51, 112, 153, 173, 193–194, 203–204, 213
Diderot, Denis 9–10, 14–17, 139–141
  animals and 5–6, 121, 127–128, 134–135, 138
  apes and 5, 12, 195–196, 212–214
  arrangement of parental elements and 9–10, 131–132, 138, 141, 244, 249
  atheism and 3, 6, 52, 63, 65, 70, 121, 123–129, 139, 222, 247
  atomic motion and 119–121, 124–125, 127–132, 135, 141
  biological engineering and 212, 214
  Bordeu and 15, 139
  Buffon and 2, 9, 13–14, 16, 72–73, 78–79, 84, 88, 90, 113, 119–120, 122, 130, 133–135, 139–140, 174–175, 219, 248–249
  chain of beings and 3–4, 9, 14, 119, 122, 130, 138, 212, 214
  chance, random and 4, 10, 17, 119–129, 131–134, 136, 139–141
  chaos and 119, 124–128, 135, 141
  climate and 132–134, 214, 220
  conscious property of atoms and 6, 9–10, 120, 129–132, 137–138, 141
  Conversation between d'Alembert and Diderot 122, 132, 136, 212
  crossbreeding (hybridization) and 137–138, 212–213
  D'Alembert's Dream 122, 128–129, 132–134, 136, 157, 174
  deistic period 52, 65, 121–125, 127, 139

Index

egg and 141
Elements of Physiology 129, 140, 212, 214, 246
emergent consciousness and 9, 115, 119–120, 131–132, 249
Encyclopedia and 4, 84, 119, 121, 125, 139, 214, 219–220
environment and 133, 214
epigenesis and 119–120, 136–137
fibers (*brins*) and 121, 129, 137, 140–141
final causes, opposition to 3, 129
flux and 4, 119, 121–122, 125–129, 133–136, 139–141
food and 120, 132–133
fractals and 128, 136
games of chance and 4, 46, 62, 121, 124, 133, 139
geographical determinism 132–133
Harvey and 120, 136
heredity and 9–10, 122, 129, 131, 133, 136–137
History of Two Indias, contributions to 216, 238
homologies and 120, 122, 130
inherited errors and 9–10, 120, 122, 131, 137–138, 140–141
La Mettrie and 16, 62–66, 119, 139–141
Leibniz and 15, 126, 139
Letter on the Blind 62–63, 65–66, 121, 125–126, 128–129, 136
Lucretius and 7, 119, 122–124, 127, 139, 141
Maillet and 139
materialism and 2–4, 6, 9, 12, 16, 52, 65, 70, 78, 90, 123, 139, 141
Maupertuis and 2, 6, 9, 13, 15, 17, 119–120, 122, 129–132, 136–140
mechanism and 126, 129
memory and 9–10, 130–132, 137, 141
monsters and 121–122, 126–129, 139–140
motive property of atoms and 119–121, 124–125, 127–132, 135, 141

mutability of species and 9–10, 13–15, 73–74, 119–121, 125, 129–137, 139, 141
natural man and 244
Needham and 119, 123, 134–135
overlapping physical characteristics and 14, 120, 122, 130
pantheism (Spinozism) and 121, 123, 139
permutations and 121, 124–128, 133, 141
Philosophic Thoughts
  condemnation by the Parlement de Paris 65
  Thought 18 52, 65, 121–122, 134
  Thought 19 123
  Thought 21 46, 62–63, 123–126, 128
preformation, rejection of 136
probability theory and 10, 14, 119, 121, 124–129, 139–141
racial classification and 13
Raynal and 238–239
Robinet and 15, 139
self-similarity and 128, 135
Sequel to the Conversation 5, 122, 133, 138, 196, 213–214
slavery, opposition to 13, 215, 238–244
spontaneous generation and 119, 123, 134–135
Supplement to the Voyage of Bougainville 215, 241–244
symmetry and 128, 135
Thoughts on the Interpretation of Nature 9, 46, 110, 122, 136, 249
  Thought 12 63, 79, 86, 114–115, 119, 129–130
  Thought 50 115, 130–131
time and 3–4, 10, 119, 121–128, 131, 133–136, 138–139, 141
transformism and 9–10, 13–15, 73–74, 119–121, 125, 129–137, 139, 141
way of life and 132–134
Diodorus of Sicily 224–225, 230

Domestication 15, 81, 120, 134, 151, 221
Duchet, Michèle 16, 140, 239
Duclos, Charles Pinot 157, 161

E

Eddy, John H. 16
Egg and Diderot 141
Ehrard, Jean 16, 39, 42, 140
Emboîtement (encasement), *see* Preformation
Encyclopedia 4, 84, 119, 121, 125, 139, 198, 214, 219–220
Environment
    Buffon 70, 73, 75, 80, 82, 91
    Diderot 133, 214
    La Mettrie 8, 45, 55–56, 64, 67, 201
    Maillet 7, 24, 29–31, 248
    Montesquieu 40–43
    Rousseau 144, 166
Epicurus 6, 8, 25, 27, 30, 46, 48, 61, 66, 112, 122, 124, 135
Epigenesis
    Buffon and 70–71, 90
    Diderot and 119–120, 136–137
    Maupertuis and 93–95, 97–100, 104–105, 112–113, 222–223
    Voltaire's opposition to 168, 184–186
Experimentation 11, 22, 79, 85, 87, 111, 113, 123, 134, 138, 157, 161, 166, 183, 217, 250
Extension 48, 51, 172

F

Farber, Paul Lawrence 16
Fellows, Otis E. 15–17, 28, 31, 37, 69–70, 72, 79, 155, 157–159, 165
Fénelon, François de Salignac de la Mothe- 37, 52
Fibers (*brins*) 121, 129, 137, 140–141
Final causes
    Diderot's opposition to 3, 129
    Maupertuis' opposition to 101, 104–105
    Voltaire's defense of 3, 169–170, 186–187, 190

Flux
    Buffon and 88
    Diderot and 4, 119, 121–122, 125–129, 133–136, 139–141
    La Mettrie and 46, 62–63, 66, 88
    Montesquieu 33, 38, 43–44
    Fontenelle, Bernard Le Bouyer de 6, 19, 25, 30–31, 43, 57, 133–134
Food
    Buffon and 70, 80–84, 91
    Diderot and 120, 132–133
    La Mettrie and 45, 52, 55–56, 58, 66
Fossils 4, 14
    articles in the *Encyclopedia* 4
    Diderot and 119, 121
    Maillet and 6, 19, 22, 29
    Voltaire's opposition to 12, 14, 168, 175–184
Fractals, Diderot and 128, 136
Free will, Rousseau and 143, 152, 158–159, 166

G

Games of chance 4, 46, 62, 121, 124, 133, 139
Gauthier, David 164–165, 210
Gellius, Aulus 34
Generation, spontaneous, *see* Spontaneous generation
Genus, family, and species 3, 77
Geographical determinism 7, 13, 19, 39, 42–43, 74, 82, 120, 132–133, 221, 244, 248
Glass, Bentley 14–15, 70, 104, 109–110, 140
God
    Buffon and 3, 73, 76–79, 82, 87, 91
    Voltaire and 3, 12, 17, 28, 167–174, 176, 184, 186–188, 191–192, 194
Gourevitch, Victor 160
Gregory, Mary Efrosini 14, 119–141

H

Haber, Francis 14–15

Index

Hartsoeker, Nicholas 52, 123
Harvey, William 10, 59, 70, 94, 97, 99, 120, 136, 184–185
Hereditary errors
    Diderot and 9–10, 120, 122, 131, 133, 137–138, 140–141
    Maupertuis and 9, 15, 17, 57, 93–100, 108, 110–111, 113, 116–117
Herodotus 7, 34–35, 146, 148, 248
Hill, Emita 139
Hobbes, Thomas 38
Homo duplex 196, 207
Homologies 86–87, 91, 120, 122, 130
Hooke, Robert 176
Huygens, Christian 59
Hybridization (crossbreeding)
    Buffon and 5, 82, 85–86, 91, 120, 134
    Diderot and 137–138, 212–213
    Maupertuis and 94–95, 99, 112–113, 222–223, 225, 227, 229

I

Imagination 51, 95, 153, 211
Inequality, Rousseau and 5, 10–12, 17, 144–147, 151–153, 160–161, 164–166
Inheritance
    Diderot and 9–10, 122, 129, 131, 133, 136–137
    La Mettrie and 8, 45, 55–56, 58, 67
    Maupertuis and 93, 97–8, 108–110
Instinct
    La Mettrie and 57–58, 60–61, 201–203
    Rousseau and 150–152, 162, 210–211
Interior molding force (*moule intérieure*) 15–17, 72–74, 79, 90–92, 133, 143 15–17, 72–74, 79, 90–92, 133, 143
Ivory fossils 4

J

Jaucourt, Louis de

article on pongos 4, 200
slavery, opposition to 234–236, 244

L

La Mettrie, Julien Offray de
    age and 45, 55–56, 66
    apes and language 12, 49, 57, 195, 201–203
    atheism and 8, 47, 51–52, 54, 65, 67, 248
    Boerhaave and 49
    brain size and 57, 201–203
    chain of beings and, 45–46, 49, 53, 56–57, 60–61, 66
    climate and 8, 45, 55–56, 67, 201, 248
    consciousness and 45–51, 53, 55–57, 63, 67
    continuity and 8, 53, 61, 66
    Descartes and 47–49, 51
    Diderot and 16, 62–66, 119, 139–141
    empiricism and 52
    environment and 8, 45, 55–56, 64, 67, 201
    Epicurus and 6, 25, 30, 46, 48, 53, 61, 63–66
    extension and 48, 51
    final causes, rejection of 64, 67
    five senses and 8, 50–52, 63, 66
    flux and 46, 62–63, 66, 88
    food and 45, 52, 55–56, 58, 66
    Harvey and 59
    imagination and 51
    immortal soul, rejection of 8, 45–52, 56, 60, 65
    inheritance and 8, 45, 55–56, 58, 67
    instinct and 57–58, 60–61, 201–203
    language and 46, 49, 57–58, 61, 64
    learning and 45, 55–56, 67
    Locke and 51
    Lucretius and 6, 25, 30, 46, 48, 53, 61, 63–66
    Maillet and 6, 25–27, 64–65
    Man a Machine 26, 45–46, 51, 195, 199, 201
    Man a Plant 46, 60

materialism and 12, 16, 47, 52, 65
Maupertuis and 45, 56–57, 65
memory and 51, 90
monism and 8, 47
monsters and 62
motive property of atoms and 46, 53, 63–64, 67
The Natural History of the Soul 45, 47, 49, 51–52, 65
Needham and 26
panspermia and 6, 8, 46, 59, 61, 64
Plato and 49
plenitude and 8
random chance and 8, 46, 53, 59, 62–64, 67
reliance on physiology 45–46, 48–49, 55–56, 58, 63, 67
self-organizing power of matter and 46–47, 49, 52–54, 60–61, 64
The System of Epicurus 8, 25, 27, 46, 61, 66
Trembley's polyp and 16, 26–27, 53–54, 61
Land of origin 73–75, 80, 82, 91, 221
Language
    La Mettrie and 45, 55–56, 67
    Rousseau and 11–12, 154–157
Learning, La Mettrie and 5, 45–47, 53
Least Action, Principle of 28, 105, 108, 205
Leeuwenhoek, Antonie van 185
Leguat, François 198
Leibniz, Gottfried Wilhelm 15, 30, 51, 53, 58, 76, 101, 106, 126, 139
Le Mascrier, Abbé Jean-Baptiste 21
Leroy's watch 59
Leucippus 124, 234–236, 238–239
Liberty 165
Linnaeus, Carolus 3, 13, 15, 61, 77, 153, 187, 219–220
Locke, John 30, 48, 51, 111, 162
Lopez, Odoardo 196
Lovejoy, Arthur O. 2, 14–16, 69, 71, 86–90, 140, 159, 171–172
Loy, J. Robert 37–38
Lucretius Carus, Titus 6–8, 24–27, 30, 35–36, 53, 61, 63–66, 102, 107, 119, 122–124, 127, 139, 141, 151, 248

M

Macé, Jean-Baptiste 137
Maillet, Benoît de
    age of the earth and 6, 19–22, 25, 29, 247
    apes and 23–24, 189
    Diderot and 139
    environment and 7, 24, 29–31, 248
    fossils and 6, 19, 22, 29
    La Mettrie and 6, 25–27, 64–65
    Lucretius and 24–27, 30
    mermaids, mermen and 16, 21–22, 24, 27–28, 31
    panspermia and 6, 8, 15–16, 24–27, 29–30, 64–65
    Telliamed 6–8, 15, 19–22, 24, 26–30, 64, 160, 179, 189–190, 247
    transformism and 7, 21–24, 25, 27, 30, 189–191
    Voltaire's ridicule of 179–180, 189–192, 194
Malebranche, Nicolas 48, 51
Malpighi, Marcello 52, 122, 185
Mammoths 4, 121, 175
Materialism
    definition of 47
    Diderot and 2–4, 6, 9, 12, 16, 52, 65, 70, 78, 90, 123, 139, 141
    La Mettrie and 12, 16, 47, 52, 65
Matrix theory 77–80
Maupertuis, Pierre-Louis de 9, 93, 116–117
    Albino Indian and 96–99
    Albino Negro and 98–100
    animalcules and 113–114
    attraction and 105–106, 223
    birth defects and 15, 95–96, 98, 110–111, 113, 117
    blacks and 93, 96, 100, 110
    Buffon and 2, 9, 13, 72–73, 78–79, 88, 90, 113–114, 248–249
    Cartesianism, against 112, 116

Index

crossbreeding (hybridization) and 94–95, 99, 112–113, 222–223, 225, 227, 229
Diderot and 2, 6, 9, 13, 15, 17, 119–120, 122, 129–132, 136–140
Dissertation on the Origin of Blacks 93, 95–101, 222–223
emergent consciousness and 9, 111, 114–115, 117, 120, 249
epigenesis, defense of 93–95, 97–100, 104–105, 112–113, 222–223
Essay on Cosmology 93, 101–105
Essay on the Formation of Organized Bodies 93, 107, 110–116
experimentation and 111, 113
final causes, against 101, 104–105
Harvey, influence of 94, 97, 99
hereditary errors and 9, 15, 17, 57, 93–100, 108, 110–111, 113, 116–117
homologies and 76
Inaugural Dissertation on Metaphysics 93, 100, 115, 129–130, 137
influence of 110
inherited errors and 93–100, 108, 110–111, 113, 116–117
La Mettrie and 45, 56–57, 65
least action and 28, 105, 108, 205
Letters 93, 106, 108–110
Lucretius and 93–101, 102, 107
mathematics and 105, 108–109, 117
monsters and 95
monogenesis and 13, 222
Needham and 113
null hypothesis and 109, 117
Physical Dissertation on the Origin of the White Negro 93, 95–101, 222–223
Physical Venus 105, 108, 112, 114, 185, 222–224
polydactyly and 9, 15, 93, 98, 108–112, 140, 249
preformation, rejection of 94–99, 102, 105, 111–112, 115, 117
probability theory and 108–109, 117

race and 94–96, 98–100, 104, 111, 115–116
random chance and 94, 98–100, 102–103, 107–108
Response to the Objections of Mr. Diderot 116
Ruhe family and 108–109, 111, 116
statistical probability of recurrence of inherited errors 9, 15, 108–109, 117, 140, 228, 249
System of Nature 93, 110
traits and 9, 94–95, 97, 100, 102, 106, 108–110, 113, 117, 129–132, 137, 223, 226–227
transformism and 85, 94, 96–99, 221
Voltaire's opposition to 28, 101, 104–108, 168–170, 184–185, 187
Mayer, Jean 65
Mechanism, Diderot and 126, 129
Mela, Pomponius 7, 34–35, 248
Memory
  Diderot and 9–10, 130–132, 137, 141
  La Mettrie and 51, 90
Milliken, Stephen F. 16, 69
Molding force, interior *moule intérieure*) 15–17, 72–74, 79, 90–92, 133, 143 15–17, 72–74, 79, 90–92, 133, 143
Monboddo, Lord 157–158
Monism 8, 16, 41, 47, 141, 196
Monogenesis 13, 221–222
Monsters 7–8, 15, 21, 35, 62, 95, 121–122, 126–129, 139–140, 147, 158, 179, 197, 220, 222–223, 225, 227, 236, 248
Montesquieu, Charles-Louis de Secondat de 138, 141
  climate and 7, 38–44, 231, 248
  environment and 40–43
  Fenelon and 37
  flux and 33, 38, 43–44
  Lucretius and 7, 35–36
  mutability of the physical form and 7, 33–35, 248
  Persian Letters 7, 33–34, 37–38, 230, 248

slavery, opposition to 42–44, 230–234, 243–244
Spirit of Laws 7, 38, 43, 231, 234, 248
Troglodytes 7, 33–38, 43, 145–146, 248
Moran, Francis 17, 159–163, 198–199
Morel, Jean 156
Motive property of atoms, random 3–4, 8, 10, 12–13, 46, 53, 63–64, 67, 88, 112, 119–121, 124–125, 127–130, 132, 135, 141, 247–250
Muller, G.-Fr. 177
Musschenbroek, Pieter van 52, 122
Mutability of the physical form
  Buffon's opposition to 16–17, 69–70, 85–92
  Diderot's defense of 9–10, 13–15, 73–74, 119–121, 125, 129–137, 139, 141
  Maillet and 7, 21–24, 25, 27, 30, 189–191
  Maupertuis and 85, 94, 96–99, 221
  Montesquieu 7, 33–35, 248
  Rousseau's opposition to 8, 10–12, 16–17, 143–145, 148–150, 157, 159, 166
  Voltaire's opposition to 8–9, 12, 167–168, 183, 186–187, 189–190, 192–194

N

Natural man, Rousseau and 11, 147, 151–155, 159–164, 195, 198, 209–210, 237–238
Needham, John Turberville 26
  Buffon and 79, 90–91
  Diderot and 119, 123, 134–135
  La Mettrie and 26
  Maupertuis and 113
  Voltaire's opposition to 25, 27–28, 169, 172–174, 184, 190, 194
Newton, Isaac 32, 52, 101–102, 105–106, 122–124, 126–127, 132, 167–168, 172, 174, 183, 186, 191, 193–194, 205

Nieuwentyt, Bernard 52, 123
Null hypothesis 109, 117

O

Oculocutaneous albinism 97–99
Organic molecules 28–29, 71–72, 79, 90–91, 189
Overlapping physical characteristics
  Buffon and 3, 8–9, 14, 77–78, 120, 130
  Diderot and 14, 120, 122, 130
Ovid 27, 190

P

Panspermia 4–5
  definition 24
  La Mettrie and 6, 8, 46, 59, 61, 64
  Maillet and 6, 8, 15–16, 24–27, 29–30, 64–65
  Voltaire's denial of 190
Pantheism, Diderot and 121, 123, 139
Pascal, Voltaire's ridicule of 191–192
Perfection of species
  Buffon and 80–82, 134
  Rousseau and 11, 145, 151–152, 156–157, 159, 199, 204, 206, 209, 211–212, 250
Perkins, Jean E. 16, 65, 140
Permutations, Diderot and 121, 124–128, 133, 141
Perrault, Claude 76, 201
Petty, William 201
Photophobia 97–99
Pigafetta, Philippo 196–197
Plato 14, 48–49, 171, 201
Plenitude 1–2, 8
Pliny the Elder 7, 21, 24, 31, 34–35, 47, 146, 148, 248
Plutarch 234, 243
Polydactyly, Maupertuis and 6, 10, 93–95
Polygenesis 13, 222, 228
Powell, James 20
Preformation
  Buffon's rejection of 70
  Diderot's rejection of 136

Index

Maupertuis' rejection of 85–86, 90–91, 95, 97, 99
    Voltaire's defense of 168, 184–186
Prévost, Antoine-François, abbé 4, 161, 198
Probability theory
    Diderot and 10, 14, 119, 121, 124–129, 139–141
    Maupertuis and 108–109, 117
    Rousseau and 10, 144, 146, 148–149, 154, 160
Prototype, Buffon and 70, 73–74, 76, 78–81, 85–87, 91, 120
Punctum saliens, see Rising point
Purchas, Samuel 4, 158–159, 197–200, 234–236
Pythagoras 111, 117

Q

Quadrupeds, Buffon and 76–78

R

Race
    as a foil for "civilized" Europe 13, 161, 215, 240–241
    Bernier and 13, 217–218
    Buffon and 13, 70–71, 81, 83–85, 87, 218–219
    Declaration of the Rights of Man and the Citizen 13, 251
    Diderot and 13, 215
    Linnaeus and 13, 187, 219–220
    Maupertuis and 94–96, 98–100, 104, 111, 115–116
    Montesquieu and 231–232
    monogenesis 13, 221–222
    polygenesis 13, 222, 228
    slavery 9, 11, 13, 42–44, 151–153, 165, 215, 222, 230–234, 236–240
    unity of the human 83–85, 161, 166, 168, 187, 189, 220, 230, 241, 244
    Voltaire and 13, 168, 170–171, 187–189
Random chance, *see* Chance, random

Ray, John 52
Raynal, Guillaume-Thomas, abbé de 238–239
Redi, Francesco 123, 134, 174
Richebourg, Marguerite 155
Rising point (*punctum saliens*) 59
Robinet, Jean-Baptiste-René 15, 139
Roe, Shirley A. 184
Roger, Jacques 14–16, 25, 27, 30, 45, 54, 61, 64–65, 69, 72, 74, 79, 82–83, 90, 101–102, 104, 106, 140, 153–154, 174, 189, 194, 207–208
Rosenfield, Leonora Cohen 203–204
Rousseau, Jean-Jacques
    anthropological metamorphosis and 6, 10–11, 16, 143, 157, 159, 166, 247, 249–250
    apes and 5, 11–12, 156–163, 166, 195, 199, 209–212, 250
    biological transformism, opposition to 8, 10–12, 16–17, 143–145, 148–150, 157, 159, 166
    Buffon, influence of 8, 11, 17, 143–163, 166, 209–210, 212, 247, 250
    Cartesianism and 12, 153
    chain of being and 145, 147, 150–151, 160–162
    childhood and 149, 151, 154–155
    civilized man and 11, 145–147, 152, 155, 160, 163–164
    Confessions 155
    convention and 147, 151, 164
    deism and 143, 145, 149, 166
    dependence on what others think of us 165
    Discourse on the Origin of Inequality 5, 10–11, 16, 143–144, 153, 156, 161–162, 164–165, 199, 209–210
    Domestication and 144, 151, 163
    Emile 143, 145, 150, 163
    Essay on the Origin of Languages 209
    environment 144, 166
    free will and 143, 152, 158–159, 166

inequality, moral and political 11–12, 17, 147, 151–152, 160–161, 164–166
inequality, natural or physical 17, 147, 152, 164
instinct and 150–152, 162, 210–211
Julie, or the New Heloïse 236–238
language and 11–12, 154–157
liberty and 152, 163, 165, 238
Monboddo and 157–158
natural man and 11, 147, 151–155, 159–164, 195, 198, 209–210, 237–238
perfectibility and 11, 145, 151–152, 156–157, 159, 199, 204, 206, 209, 211–212, 250
probability and 10, 144, 146, 148–149, 154, 160
Purchas and 158–159
self-love and 164–165
sentiment of existence 164–165
slavery, opposition to 151–153, 165, 236–238
Social Contract 238, 292–294
sociological change and 10, 166, 249
Rudwick, Martin J.S. 183–184
Ruhe family 108–109, 111, 116

S

San Blas Indians 96–99
Schmidt, Johan Werner *139*
Seguin, Jean-Pierre 65
Self-love 164–165
self-similarity, Diderot and 128, 135
Sensibilité, see Consciousness
Sentiment of existence 164–165, 210
Sesostris I 224
Severino, Marco Aurelio 76
Siberian mammoths 4, 121, 175, 250
Singh, Christine M. 139
Smith, William 199
Solinus, Gaius Julius 34
Slavery, abolition of
    Condorcet and 240
    Declaration of the Rights of Man and the Citizen 13, 251
    Diderot and 13, 215, 238–244

Encyclopedia and 234–236, 244
Jaucourt and 234–236, 244
Montesquieu and 42–44, 230–234, 243–244
Rousseau and 151–153, 165, 236–238
Voltaire and 222, 232–234
Sloan, Phillip R. 16
Sorenson, Leonard 17, 164
Spallanzani, Lazzaro 173–174
Species
    Buffon's definition of 69, 85–86, 187, 218–219, 244
    Voltaire on 167–168, 170–172, 176–177, 179, 181, 183, 186–189, 192, 195, 228
Spinoza, Benedict de 30, 116, 123, 169
Spitzer, Leo 65
Spontaneous generation
    Buffon and 72, 79, 90–91, 113, 134–135
    Diderot and 119, 123, 134–135
    Voltaire's opposition to 12, 28–29, 168–169, 172–174, 184
Starobinski, Jean 162–164
Stenger, Gerhardt 139
Steno (Niels Stensen) 176
Substantial forms 191–193
Suratteau, Aurélie 139
Swammerdam, Jan 76
Symmetry, Diderot and 128, 135

T

Talking apes, possibility of
    Buffon's opposition to 154–155, 203–209, 212194
    Diderot 195, 212–214
    La Mettrie 12, 49, 57, 195, 201–203
    Rousseau 156, 209–212
Teleology 101, 104
Thales 124
Thomson, Ann 16, 140
Thucydides 34
Time
    Buffon and 69–70, 72–74, 78–79, 81–82, 84–86, 88–91

Index

Diderot and 3–4, 10, 119, 121–128, 131, 133–136, 138–139, 141
Titus-Livy, 34
Traits, Maupertuis and 9, 94–95, 97, 100, 102, 106, 108–110, 113, 117, 129–132, 137, 223, 226–227
Transformism, biological
    Buffon's opposition to 16–17, 69–70, 85–92
    Diderot's defense of 9–10, 13–15, 73–74, 119–121, 125, 129–137, 139, 141
    Maillet and 7, 21–24, 25, 27, 30, 189–191
    Maupertuis and 85, 94, 96–99, 221
    Montesquieu 7, 33–35, 248
    Rousseau's opposition to 8, 10–12, 16–17, 143–145, 148–150, 157, 159, 166
    Voltaire's opposition to 8–9, 12, 167–168, 183, 186–187, 189–190, 192–194
Trembley's polyp 16, 26–27, 53–54, 61, 140
Tripartite soul 49–50, 193
Troglodytes 7, 33–38, 43, 145–146, 220, 248
Tyson, Edward 198–199, 201

U

Unity of the human race 83–85, 161, 166, 168, 187, 189, 220, 230, 241, 244
Unity of plan 75–76
Ussher, James 20–21

V

Vallisneri, Antonio 176, 185
Vartanian, Aram 16, 51–52, 54, 65–66, 140
Vernière, Paul 34, 37, 65, 123
Versini, Laurent 239
Virgil 63
Vital forces 223
Vital laws 223
Voltaire, François-Marie Arouet de

ABC 174
age of the earth and 167–168, 175–184
Age of Louis XIV 174
albino Negro and 187–188
animalcules, denial of 169, 174–175, 194
atheism, opposition to 3, 106–107, 168–170, 172–174, 176–177, 184, 186, 194
Bonnet, defense of 184–185
Buffon, opposition to 25, 28–29, 169, 178, 184
chain of beings, opposition to 3, 12, 168, 170–172
creationism and 6, 12, 104, 167, 174, 184, 189, 247
Defense *of My Uncle* 167, 173, 178, 180, 189, 194
deism and 3, 6, 8, 12, 17, 25, 27–28, 104, 106–107, 167–170, 173–174, 176, 180, 184, 186, 188, 194, 250
Descartes, ridicule of 173–174, 193–194
Dialogues of Euhemerus 175
Diderot, opposition to 3–4, 168–169, 172, 174–175
epigenesis, opposition to 168, 184–186
Essay on Morals 233–234
final causes and 3, 169–170, 186–187, 190
fossils, opposition to 12, 14, 168, 175–184
God and 3, 12, 17, 28, 167–174, 176, 184, 186–188, 191–192, 194
Harvey, opposition to 184–185
La Henriade 63, 125
History of Doctor Akakia and the Native of Saint-Malo 106
Maillet, ridicule of 179–180, 189–192, 194
The Man of Forty Ecus 27, 189–190
Maupertuis, opposition to 28, 101, 104–108, 168–170, 184–185, 187

Needham, opposition to 25, 27–28, 169, 172–174, 184, 190, 194
On the Changes that Have Taken Place on Earth and on Petrifications that are Pretended to Still BearWitness 177–178
on blacks 187–188
on species 167–168, 170–172, 176–177, 179, 181, 183, 186–189, 192, 195, 228
panspermia, denial of 190
Pascal, ridicule of 191–192
Philosophical Dictionary articles on
  atheism 106–108, 168–170, 186
  chain of beings 170–172
  soul 193
  theism 188
Poem on the Lisbon Disaster 172
polygenesis and 13, 222, 228
preformation, defense of 168, 184–186
Questions on the Encyclopedia 174–175
Questions on Miracles 172–174
race and 13, 168, 170–171, 187–189
random chance, opposition to 3, 12, 167–170, 176, 186, 194
Report concerning a White Moor brought from Africa to Paris in 1744 187–188, 228
slavery, opposition to 222, 232–234
science, opposition to 12, 17, 168–169, 172–186, 189, 193–194, 250
Singularities of Nature 180–181, 186, 190–194
The Snails of Reverend Father Dung-beetle-monger 174
Summary of the Age of Louis XV 174
Spallanzani and 173–174
spontaneous generation, opposition to 12, 28, 168–169, 172–175, 184, 194
Supplement to the Age of Louis XIV 174
transformism, opposition to 8–9, 12, 167–168, 183, 186–187, 189–190, 192–194

Travels of Scarmentado 232–233
Treatise on Metaphysics 186–187, 222
We Must Take Sides, or the Principle of Action 172
Von Wolf, Johann Christian 101

W

Wafer, Lionel 96–99
Wartofsky, Marx W. 16, 140–141
Way of life
  Buffon and 82–83, 91
  Diderot 132–134
Wokler, Robert 200–201

Currents in Comparative Romance Languages and Literatures

This series was founded in 1987, and actively solicits book-length manuscripts (approximately 200–400 pages) that treat aspects of Romance languages and literatures. Originally established for works dealing with two or more Romance literatures, the series has broadened its horizons and now includes studies on themes within a single literature or between different literatures, civilizations, art, music, film and social movements, as well as comparative linguistics. Studies on individual writers with an influence on other literatures/civilizations are also welcome. We entertain a variety of approaches and formats, provided the scholarship and methodology are appropriate.

For additional information about the series or for the submission of manuscripts, please contact:

> Tamara Alvarez-Detrell and Michael G. Paulson
> c/o Dr. Heidi Burns
> Peter Lang Publishing, Inc.
> P.O. Box 1246
> Bel Air, MD 21014-1246

To order other books in this series, please contact our Customer Service Department:

> 800-770-LANG (within the U.S.)
> 212-647-7706 (outside the U.S.)
> 212-647-7707 FAX

or browse online by series at:

> www.peterlang.com